Proteostasis and Proteolysis

Oxidative Stress and Disease
Series Editors:
Enrique Cadenas, MD, PhD,
University of Southern California School of Pharmacy,
Los Angeles, California
Helmut Sies, PhD,
Heinrich Heine Universität Düsseldorf,
Düsseldorf, Germany

For more information about this series, please visit: https://www.crcpress.com/Oxidative-Stress-and-Disease/book-series/CRCOXISTRDIS

Proteostasis and Proteolysis

Edited by

Niki Chondrogianni, Elah Pick, and Anna Gioran

CRC Press is an imprint of the
Taylor & Francis Group, an **informa** business

First edition published 2022
by CRC Press
6000 Broken Sound Parkway NW, Suite 300, Boca Raton, FL 33487-2742
and by CRC Press

2 Park Square, Milton Park, Abingdon, Oxon, OX14 4RN

Library of Congress Cataloging in Publication Data

Chondrogianni, Niki, editor. | Pick, Elah, editor. | Gioran, Anna, editor.
... osis and proteolysis / edited by Niki Chondrogianni, Elah Pick, Anna Gioran.
... st edition. | Boca Raton : Taylor & Francis, 2021. | Includes bibliographical references and index.
... 2021026112 (print) | LCCN 2021026113 (ebook) | ISBN 9780367499327 (hardback) |
... 34 (paperback) | ISBN 9781003048138 (ebook)
... eomics. | Proteins. | Proteolytic enzymes.
... P551 .P7572 2021 (print) | LCC QP551 (ebook) | DDC 572/.6—dc23
... tps://lccn.loc.gov/2021026112 LC ebook record available at https://lccn.loc.gov/2021026113

CONTENTS

CONTENTS

PREFACE

Proteins are the key decision-making workforce for practically all cellular signals. The precise equilibrium between protein translation, folding, function and timely degradation, also known as "proteostasis", determines cellular health as well as organismal maintenance, healthspan and survival. Loss of cellular proteostasis is linked to physiological processes of deterioration such as aging, to conditions characterized by supra-physiological oxidative stress as well as to illness, including age-related diseases, neurodegeneration, inflammation, dystrophies and cancer.

This book includes 18 chapters logically organized to allow comprehensive understanding of how maintenance of proteostasis protects cellular and organismal health and how environmental and metabolic pressure can impair proteostasis and lead to disequilibrium and disease. Each chapter contains up-to-date information on its respective topic while some of them review the interplay of certain proteostatic mechanisms, a newly arising topic. Importantly, most chapters include tangible examples of how failure of proteostasis can underlie aging and disease.

We hope that the compilation of the above topics will assist both novice and experienced researchers and students to become more familiar with the subject of proteostasis. In the long run, we hope that this book will inspire its readers and eventually promote new ideas and new research studies.

The first part of the book (**Chapters 1–5**) presents proteostatic mechanisms acting at the beginning of a protein's life (synthesis and folding) as well as those acting at a transcriptional and post-transcriptional level. In **Chapter 1**, *Ferrando et al.* describe recent discoveries on protein quality control (PQC) during protein translation and specifically a key regulatory process that removes defective polypeptides when ribosomes slowdown or arrest due to stress. Following translation or in parallel with that, proteins need to assume their final conformation. In **Chapter 2**, *Riemer et al.* present the executors of protein folding. They also describe how failure to adhere to a strict final protein conformation may result in detrimental proteostatic imbalance that characterizes diseases such as the age-related proteinopathies. Accumulation of unfolded or misfolded proteins due to various causes is a widely studied trigger of several proteostatic mechanisms. In **Chapter 3**, *Kapetanou et al.* describe the transcriptional regulation of the heat shock response (HSR), the oxidative stress response (OXR) and the unfolded protein response (UPR). These sensors of protein dyshomeostasis regulate in turn the transcription of components of the cellular proteolytic machineries, namely the ubiquitin-proteasome system (UPS) and autophagy-lysosomal pathway (ALP); this regulation is described as well.

In recent years, it has become clear that gene expression is additionally regulated at the epigenetic level. Emerging data along these lines are reported in **Chapter 4** by *Kanakis et al.* who review the paradigm of microRNA-mediated regulation of proteostasis in the regenerating adult muscle. Moreover, they recapitulate the association of microRNAs with stressful conditions occurring in the muscle leading to proteostasis loss and sarcopenia. Other RNA molecules are also involved in proteostatic responses through various mechanisms, among which the formation of mRNA granules during stress is included. In **Chapter 5**, *Borbolis and Syntichaki* focus on recent discoveries that reveal how the assembly and disassembly of these granules may be interconnected with the proteostasis network and thus with proteostasis maintenance.

Notably, accumulation of unfolded or misfolded proteins is not the only event triggering proteostatic responses. Phospholipid bilayer stress that has been associated with numerous pathologies, such as cancer and diabetes, can trigger the UPR. This is a rather unexplored and exciting emerging field, and it could not be missing from a contemporary book on proteostasis. Recent findings on phospholipid synthesis and the UPR signaling are described by Gkikas and Tavernarakis in **Chapter 6**.

An intriguing mechanism that regulates proteostasis and assures the specificity of the involved mechanisms is the reversible tagging of proteins by members of the ubiquitin-like (Ubl) family of modifiers (ubiquitin, SUMO and NEDD8), as summarized in the second part of the book (**Chapters 7–9**). Ubiquitin, which regulates the lifespan of numerous proteins and is thus involved in a plethora of cell functions, is a key player in various aspects of proteostasis. Therefore, it is not surprising that ubiquitin ligases play pivotal roles in proteostasis. In fact, in **Chapter 7**, *Fechtner and Pfirrmann* discuss how specific ubiquitin ligases are involved in proteostasis imbalance associated with normal aging, age-related diseases and premature aging syndromes. SUMO has also emerged as a regulator of proteostasis. In **Chapter 8**, *Parra-Peralbo et al.* demonstrate that SUMO is essential for proper neuronal function and that it exerts a pivotal role in the progression of neurodegenerative diseases such as Alzheimer's, Parkinson's and polyglutamine diseases either through the modification of key for the diseases proteins or through the regulation of neuroinflammation. Completing the second part, in **Chapter 9**, *Pick and Serino* describe NEDD8 function, the way that it is affected by the cellular redox state and how NEDD8 has an effect on key cell cycle and oxidative stress regulators.

Innovative therapeutic strategies based on NEDD8 regulation are also presented.

Once the proteotoxic stress is sensed and the corresponding proteostatic mechanisms are activated, misfolded, abnormal or damaged (in any way) proteins need to be removed. This occurs via the cellular proteolytic machineries. The third part of the book (**Chapters 10–16**) focuses on the proteostatic control exerted by the UPS that is one of the two main proteolytic mechanisms of the cell. Nevertheless, references to the ALP also appear in many of these chapters since it is rather difficult to isolate one mechanism from the other. In **Chapter 10**, *Jung and Höhn* present a comprehensive description of the structure and function of the proteasome, the main core particle of the UPS and of its regulators. In **Chapter 11**, *Studencka-Turski and Krüger* review the effects of proteasome impairment on cellular signaling pathways. Moreover, they present the potential crosstalk between impaired proteasome activity and inflammation. On top of inflammation, insufficient capacity of proteasomal degradation may lead to accumulation of harmful aggregate-prone proteins leading to protein misfolding diseases or else termed as proteinopathies. Activation of the proteasome as a preventive strategy against proteinopathies is presented in **Chapter 12** by *Vasilopoulou et al.*, along with the UPS status in aging and selected proteinopathies. Moreover, a brief description of the ALP implication in these conditions is presented. The opposite strategy, namely proteasome inhibition, is a widely used therapeutic avenue for cancer as discussed in **Chapter 13** by *Arslan-Eseryel et al.* along with the central role of the proteasome system in carcinogenesis.

The implication of the proteasome has been studied extensively in many pathologies as well as in the aging of various tissues. In **Chapter 14**, *Bulteau and Friguet* review how protein degradation and its failure may underlie skin aging. They discuss the role of mitochondrial proteases, UPS and ALP in skin aging, and they reveal the current knowledge with regard to their interplay. Similarly, in **Chapter 15**, *Lourenço dos Santos et al.* focus on the redox regulation and proteostasis (in terms of both proteasome and autophagy) of satellite cells in muscle regeneration and aging. **Chapter 16** authored by *Wang* closes the third part of the book presenting yet another paradigm of how UPS dysfunction, mainly in terms of proteasome failure, is responsible for cardiac pathogenesis. The phosphoregulation of proteasome function is suggested as a potential therapeutic strategy for various heart disorders.

While decline in the UPS is observed with aging and it has been associated with many age-related diseases, it is not the only proteolytic mechanism that exhibits these traits as described in the fourth and final part of the book (**Chapters 17–18**). As described in **Chapter 17** by *Ranti et al.*, autophagy declines with aging and its activation is sufficient to promote longevity in various model organisms. The protective effect of autophagy against oxidative stress is also discussed along with the basic molecular mechanisms comprising autophagy. Finally, **Chapter 18** by *Häseli and Ott* depicts an emerging role of autophagy in the maintenance of cardiac homeostasis during aging. This chapter further focuses on mitophagy that has a decisive role in cardiac aging.

As editors of this book, we are extremely grateful to the leading researchers in the field of proteostasis who contributed to this collaborative effort. We would also like to express our gratitude to our Editorial Assistant, Mrs. Ana Lucia Eberhart, and to our Editor, Dr. Chuck Crumly, for their professional support in the preparation of the book. Moreover, we would like to cordially thank Dr. Margarita Theodoropoulou for using her impeccable organizational skills to help us handle various administrative issues that tagalong with the endeavor of book editing. Last but not least, we would like to thank our families and beloved ones. Some of the hours we spent poring over this book were taken away from time we would otherwise have spent with them.

This book was written while the whole world was in a joint confrontation with the COVID-19 pandemic and lockdowns disrupted the daily lives of the authors as well as, in some cases, their research activities. In light of all this, we consider ourselves lucky for having a team of contributors who were willing to deliver high-quality content in a timely fashion. Like everyone else, we are relieved that the new vaccines are here to shine some light at the finish line of this bizarre period. Although humanity still has a considerable amount of time ahead dealing with the pandemic and the ramifications it will leave on its path through our lives, we are hoping that this will be put behind us once and for all.

DR. NIKI CHONDROGIANNI
PROF. ELAH PICK
DR. ANNA GIORAN

CONTRIBUTORS

ARSLAN-ESERYEL, SEMA
Department of Biochemistry
School of Medicine
Genetic and Metabolic Diseases Research and
    Investigation Center
Marmara University
Istanbul, Turkey

ATHANASOPOULOU, SOPHIA
Institute of Chemical Biology
National Hellenic Research Foundation
Athens, Greece

BARRIO, ROSA
Center for Cooperative Research in Biosciences
    (CIC bioGUNE)
Basque Research and Technology Alliance (BRTA)
Derio, Spain

BARROSO-GOMILA, ORHI
Center for Cooperative Research in Biosciences
    (CIC bioGUNE)
Basque Research and Technology Alliance (BRTA)
Derio, Spain

BORBOLIS, FIVOS
Biomedical Research Foundation of the Academy
    of Athens
Center of Basic Research
Athens, Greece

BULTEAU, ANNE-LAURE
LVMH Recherche
Life Science Department
Saint Jean de Braye, France

CHONDROGIANNI, NIKI
Institute of Chemical Biology
National Hellenic Research Foundation
Athens, Greece

FARRÀS, ROSA
Oncogenic Signalling Laboratory
Centro de Investigación Príncipe Felipe (CIPF)
Valencia, Spain

FATHINAJAFABADI, ALIHAMZE
Oncogenic Signalling Laboratory
Centro de Investigación Príncipe Felipe (CIPF)
Valencia, Spain

FECHTNER, LISA
Institute of Physiological Chemistry
Martin-Luther University Halle-Wittenberg
Halle, Germany

FERRANDO, ALEJANDRO
Instituto de Biología Molecular y Celular de
    Plantas CSIC
Universidad Politécnica de Valencia
Valencia, Spain

FRIGUET, BERTRAND
Sorbonne Université, CNRS, INSERM
Institut de Biologie Paris-Seine
Biological Adaptation and Ageing
Paris, France

GIORAN, ANNA
Institute of Chemical Biology
National Hellenic Research Foundation
Athens, Greece

GKIKAS, ILIAS
Institute of Molecular Biology and
    Biotechnology
Foundation for Research and
    Technology-Hellas
Department of Biology
University of Crete
Heraklion, Crete, Greece

GOLJANEK-WHYSALL, KATARZYNA
Institute of Life Course and Medical Sciences
University of Liverpool
Liverpool, UK
Department of Physiology
School of Medicine
NUI Galway, Galway, Ireland

GONOS, EFSTATHIOS S.
Institute of Chemical Biology
National Hellenic Research Foundation
Athens, Greece

GUVEN, ULKUGUL
Department of Biochemistry
School of Medicine
Genetic and Metabolic Diseases Research and
    Investigation Center
Marmara University
Istanbul, Turkey

HÄSELI, STEFFEN
Department of Molecular Toxicology
German Institute of Human Nutrition
    Potsdam-Rehbruecke (DIfE)
Nuthetal, Germany
German Center for Cardiovascular Research
    DZHK-partner site
Berlin, Germany

HÖHN, ANNIKA
Department of Molecular Toxicology
German Institute of Human Nutrition Potsdam-
    Rehbruecke (DIfE)
Nuthetal, Germany
German Center for Diabetes Research (DZD)
Muenchen-Neuherberg, Germany

JUNG, TOBIAS
Department of Molecular Toxicology
German Institute of Human Nutrition Potsdam-
    Rehbruecke (DIfE)
Nuthetal, Germany

KANAKIS, IOANNIS
Chester Medical School
University of Chester
Chester, UK
Institute of Life Course and Medical Sciences
    University of Liverpool
Liverpool, UK

KAPETANOU, MARIANNA
Institute of Chemical Biology
National Hellenic Research Foundation
Athens, Greece

KARADEMIR-YILMAZ, BETUL
Department of Biochemistry
School of Medicine
Genetic and Metabolic Diseases Research and
    Investigation Center
Marmara University
Istanbul, Turkey

KRÜGER, ELKE
Institute of Medical Biochemistry and Molecular
    Biology
Universitätsmedizin Greifswald
Greifswald, Germany

L'HONORÉ, AURORE
Sorbonne Université, CNRS, INSERM
Institut de Biologie Paris-Seine
Biological Adaptation and Ageing
Paris, France

LOURENÇO DOS SANTOS, SOFIA
Sorbonne Université, CNRS, INSERM
Institut de Biologie Paris-Seine
Biological Adaptation and Ageing
Paris, France

MARTÍNEZ-FÉRRIZ, ARANTXA
Oncogenic Signalling Laboratory
Centro de Investigación Príncipe Felipe (CIPF)
Valencia, Spain

MURATORE, VERONICA
Center for Cooperative Research in Biosciences
    (CIC bioGUNE)
Basque Research and Technology Alliance (BRTA)
Derio, Spain

MYRTZIOU, IOANNA
Chester Medical School
University of Chester
Chester, UK

OLZSCHA, HEIDI
Martin Luther University Halle-Wittenberg
Faculty of Medicine
Institute of Physiological Chemistry
Halle, Germany

OTT, CHRISTIANE
Department of Molecular Toxicology
German Institute of Human Nutrition Potsdam-
    Rehbruecke (DIfE)
Nuthetal, Germany
German Center for Cardiovascular Research
DZHK-Partner Site
Berlin, Germany

PANFILOVA, DIANA
Martin Luther University Halle-Wittenberg
Faculty of Medicine
Institute of Physiological Chemistry
Halle, Germany

PAPAEVGENIOU, NIKOLETTA
Institute of Chemical Biology
National Hellenic Research Foundation
Athens, Greece

PARRA-PERALBO, ESMERALDA
Department of Nutrition and Nursery
Faculty of Biomedical Sciences
Universidad Europea
Madrid, Spain

PETROPOULOS, ISABELLE
Sorbonne Université, CNRS, INSERM
Institut de Biologie Paris-Seine
Biological Adaptation and Ageing
Paris, France

PFIRRMANN, THORSTEN
Institute of Physiological Chemistry
Martin-Luther University Halle-Wittenberg
Halle, Germany
HMU Health and Medical University Potsdam
Potsdam, Germany

PICK, ELAH
Department of Biology and Environment at Oranim
Faculty of Natural Sciences
University of Haifa
Haifa, Israel

RANTI, DIMITRA
Institute of Chemical Biology
National Hellenic Research Foundation
Athens, Greece

RIEMER, JUDITH-ELISABETH
Martin Luther University Halle-Wittenberg
Faculty of Medicine
Institute of Physiological Chemistry
Halle, Germany

SERINO, GIOVANNA
Department of Biology and Biotechnology
Sapienza University of Rome
Rome, Italy

STUDENCKA-TURSKI, MAJA
Institute of Medical Biochemistry and Molecular
    Biology
Universitätsmedizin Greifswald
Greifswald, Germany

SUTHERLAND, JAMES D.
Center for Cooperative Research in Biosciences
    (CIC bioGUNE)
Basque Research and Technology Alliance (BRTA)
Derio, Spain

SYNTICHAKI, POPI
Biomedical Research Foundation of the Academy
    of Athens
Center of Basic Research
Athens, Greece

TALAMILLO, ANA
Center for Cooperative Research in Biosciences
    (CIC bioGUNE)
Basque Research and Technology Alliance (BRTA)
Derio, Spain

TAVERNARAKIS, NEKTARIOS
Department of Biology
Department of Basic Sciences
Faculty of Medicine
University of Crete, Heraklion
Crete, Greece

VASILAKI, APHRODITE
Department of Physiology
School of Medicine
NUI Galway
Galway, Ireland

VASILOPOULOU, MARY A.
Institute of Chemical Biology
National Hellenic Research Foundation
Athens, Greece

WANG, XUEJUN
Division of Basic Biomedical Sciences University of
    South Dakota
Sanford School of Medicine
Vermillion, SD, USA

THE EDITORS

Niki Chondrogianni

Research areas: Aging, longevity, age-related diseases, proteolysis, proteasome, redox regulation, cellular senescence, Caenorhabditis elegans

Dr. Niki Chondrogianni graduated and obtained her Ph.D. from the Department of Biology of the National and Kapodistrian University of Athens in Greece, while she was a visiting fellow at the Universities Paris 7 in France and Bristol in the United Kingdom. She conducted her postdoctoral studies at the National Hellenic Research Foundation (NHRF) in Athens focusing on the role of proteasome activation on cellular and organismal lifespan extension. She was trained in the use of *C. elegans* at the Foundation of Research and Technology-Hellas, Institute of Molecular Biology and Biotechnology in Greece. She established her lab at NHRF in 2009. She focuses on the genetic and environmental factors that govern aging, longevity and age-related diseases with emphasis on proteasome regulation. She is seeking for natural or chemically synthesized compounds that may act as proteasome activators and thus may serve as anti-aging agents and/or as anti-aggregation compounds that can decelerate the progression of various proteinopathies with emphasis on Alzheimer's disease.

Elah Pick

Research areas: Cell biology, protein biochemistry, proteolysis, redox, the ubiquitin-proteasome system

Prof. Elah Pick performed her graduate and postdoctoral training in cell biochemistry, membrane traffic and protein degradation at the Technion – Israel Institute of Technology. She performed a second postdoctoral training at Yale University on a family of genes that regulate the molecular and biochemical mechanisms responsible for cellular responses to external stimuli and stresses. She established her own lab in Israel, at the department for Biology and Environment of the University of Haifa, located at Oranim. Her lab investigates the effect of metabolic and environmental stresses such as oxidation, on the regulation of ubiquitin-like proteins and the protein degradation machineries.

Anna Gioran

Research areas: Mitochondria, neurobiology, metabolism, proteostatic mechanisms, Caenorhabditis elegans

Dr. Anna Gioran carried out her graduate training at the German Center of Neurodegenerative Diseases

(DZNE) in Bonn, Germany. During this time, she studied the effects of mitochondrial deficiency on the morphology of the nematode's neurons. In her first postdoctoral fellowship, she continued at the DZNE and she focused more on mitochondrial deficiencies and specifically on their metabolic implications and manners to rescue their detrimental effects at organismal level. At the time this book was written, she was conducting her second postdoctoral research under the supervision of Dr. Niki Chondrogianni at the National Hellenic Research Foundation in Athens, Greece, focusing on the interplay between proteostatic mechanisms and mitochondrial function.

THE EDITORS

CHAPTER ONE

Ribosomal Pauses during Translation and Proteostasis

Alejandro Ferrando, Alihamze Fathinajafabadi, Arantxa Martínez-Férriz and Rosa Farràs

CONTENTS

1.1 INTRODUCTION

To ensure a healthy regulation of protein homeostasis (proteostasis), the cell needs to fulfill several steps such as the optimum production of proteins, their correct folding and targeting to the appropriate subcellular localization. In the case of a faulty event, surveillance mechanisms should act immediately to remove incorrect polypeptides that otherwise could lead to pathogenesis (Chiti and Dobson, 2017).

The process of translation or protein biosynthesis includes four phases, namely initiation, elongation, termination and ribosome recycling. The initiation stage is considered the rate-limiting step in protein biogenesis as it is highly regulated at various levels like the phosphorylation of 4E-BP by mTORC1 and the eIF2a phosphorylation by stress-specific kinases (Sonenberg and Hinnebusch, 2009). Nevertheless, recent data suggest that proper functioning of the four phases is important for a healthy proteostasis. In particular, the translation elongation phase is getting increased attention, thanks to breakthrough technologies like the ribosome profiling (Ribo-seq or RiboProfiling), that allows the assessment of the ribosome movement on the mRNA molecule to study ribosome dynamics at genome-wide scale with single codon resolution (Ingolia et al., 2009). The Ribo-seq technique is based on the collection of the footprints generated after in vitro RNase digestion of ribosome-protected mRNA fragments (RPFs) and their subsequent identification via sequencing (Brar and Weissman, 2015). The bioinformatic determination of the ribosomal P-site for every footprint allows the positioning of the ribosomes along the mRNA to demonstrate the periodicity of the ribosome-protected fragments. Finally, the quantification of the ribosomal occupancy facilitates the calculation of the translational efficiency for every mRNA (Ingolia et al., 2009). The use of this powerful technology allows the visualization of ribosome queuing under certain stressful conditions (Darnell et al., 2018; Wu et al., 2019), and under alterations of the translational machinery (Choe et al., 2016; Schuller et al., 2017; Woolstenhulme et al., 2015).

It is not surprising that translation elongation, as a readout of protein biogenesis, can provide molecular information for pathways involved in the control of proteostasis. During the translation elongation step, successive cycles of aminoacyl-tRNA accommodation, followed by peptide bond formation

DOI: 10.1201/9781003048138-1

1

with the nascent peptidyl-tRNA chain and translocation from the A-site to the P-site of the ribosome, take place at a rate of approximately 6 amino acids/second (Ingolia et al., 2011). However, translation elongation does not proceed always at the same rate since different factors can have strong influence on the process leading, occasionally, to ribosomal pauses. The presence of several ribosomes on a single mRNA molecule may lead to ribosomal "traffic jams" when the leading ribosome stops reading the mRNA, producing di-ribosomes (disomes), tri-ribosomes (trisomes) or even larger ribosome collisions. The translational machinery and accessory elements must decide whether the ribosome arrest is irreversible, thus leading to the dismantling of the translational apparatus and the removal of inaccurate or toxic polypeptides, or it is just a transient pause that may be promptly solved to resume translation. Inappropriate removal of faulty polypeptides may lead to fatal consequences for the cell and the organism (Chu et al., 2009; Martin et al., 2020). In other occasions, halting the ribosome may be beneficial for cellular processes or used as a regulatory mechanism (Panasenko et al., 2019; Pechmann et al., 2014). To support the required precision of this process, the cell utilizes highly sensitive molecular mechanisms to monitor the translational status and to make quick decisions on how to proceed when the ribosome is transiently paused or irreversibly arrested. These mechanisms are intimately linked to the mRNA surveillance pathways that lead to mRNA degradation upon translational disabilities like no-go-decay (NGD) and non-stop-decay (NSD) pathways (Chen et al., 2010; Doma and Parker, 2006; Passos et al., 2009; Simms et al., 2017).

In this chapter, we focus on the recent discoveries of the eukaryotic pathways involved in the quality control of the protein biosynthesis process during translation elongation and termination. The ribosome-associated quality-control pathway (RQC) has emerged as a key regulatory process that removes defective polypeptides when ribosomes arrest. Priming signaling events such as ribosome collisions, captured by the ribosome profiling technology and its variants, and the recently identified downstream-activated stress pathways are also the focus of this chapter. Furthermore, transient ribosomal pauses resolved by supporting mechanisms that allow resuming translation without drastic measures are considered. As an example, how the translation factor eIF5A assists stalled ribosomes

when decoding polyproline sequences is described in detail. Finally, other cases where ribosome slowdowns are required for specific functions such as protein folding, organellar targeting and multicomplex assembly are also presented.

1.2 IRREVERSIBLE RIBOSOMAL STALLING AND THE RIBOSOME-ASSOCIATED QUALITY-CONTROL PATHWAY

The translation of proteins is affected by different factors that can interrupt the movement of the ribosome irreversibly. Some of these factors are excessive mRNA secondary structure (Doma and Parker, 2006), truncated or damaged mRNA (Shao et al., 2013; Tsuboi et al., 2012), the presence of non-optimal or rare codons (Letzring et al., 2013), the presence of ribosomes on 3′ Untranslated Region (3′UTR) polyadenylated sequence (Arthur et al., 2015; Juszkiewicz and Hegde, 2017; Sundaramoorthy et al., 2017) or the exposure of cells to adverse conditions such as amino acid starvation, tRNA deficiency, oxidative stress and genetic mutations (Guydosh and Green, 2014; Ishimura et al., 2014; Simms et al., 2014).

These challenging conditions for the ribosomal movement can lead to ribosome collisions caused by the arrest of the leading ribosome or by stalling a subset of ribosomes (Chih-Chien Wu et al. 2020). Detection of the collided ribosomes initiates the RQC pathway in eukaryotes. With this process, collision complexes formed by disomes or trisomes due to prolonged ribosome stalling are specifically recognized, leading to dissociation of the ribosome into 40S and 60S subunits, targeting of the defective mRNA and nascent peptide for degradation and recycling of the ribosome (Ikeuchi et al., 2019). The ubiquitin pathway targets the arrested peptide for ubiquitin-proteasome-mediated degradation and recruits factors for ribosome dissociation, thus facilitating its recycling (Figure 1.1).

The initiation of the RQC process has been studied extensively in vitro by inducing severe ribosome stalling and collision at the poly(A) tail (Brandman and Hegde, 2016; Ito-Harashima et al., 2007). This results in the formation of a disome in which the leading ribosome is stalled and the next ribosome collides with the former (Figure 1.1, step 1).

The collided ribosomes are recognized by the RING-type E3 ligase ZNF598 in mammals and Hel2 in yeast (Juszkiewicz et al., 2018; Matsuo et al., 2017). The ribosome-associated protein RACK1

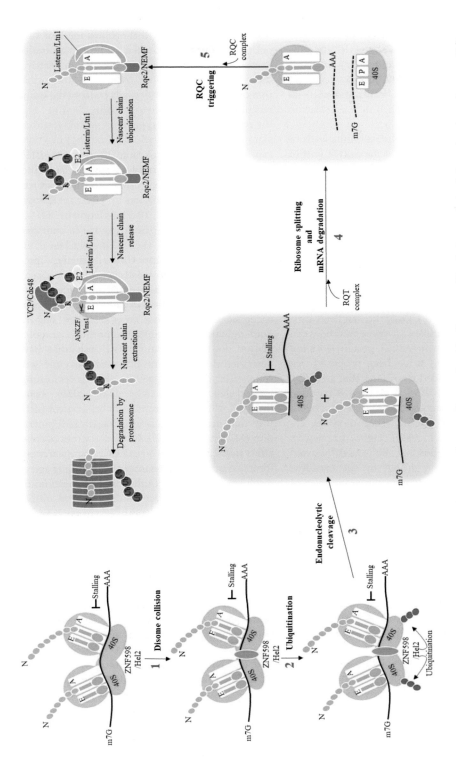

Figure 1.1 **Irreversible ribosomal stalling in eukaryotic organisms.** Upon disome collision induced by ribosome stalling during translational elongation phase, the sensor E3 ligase protein ZNF598/Hel2 detects the collision site between two ribosomes (**1**), which is an essential step to resolve this situation by ubiquitination of ribosomal 40S subunits of the two colliding ribosomes (**2**). This in turn induces endonucleolytic cleavage of mRNA at the collision site (**3**). The two ribonucleoprotein portions generated by the mRNA cleavage will be recognized by specific protein factors (**4**) in order to separate the two ribosomal subunits and induce mRNA degradation. The ribosomal 60S subunit bound to the nascent peptide is recognized by the ribosome quality control (RQC) system (**5**), which polyubiquitinates the nascent peptide to trigger its subsequent degradation by the proteasomal machinery. *Abbreviation:* RQT: RQC-trigger complex.

(Asc1 in yeast), located on the 40S, contributes to the E3 ligase recognition (Sundaramoorthy et al., 2017). The E3 ligase ZNF598/Hel2 ubiquitinates specific ribosomal proteins of the small 40S subunit, such as eS10 (at K138 and K139) in mammalian cells (Garzia et al., 2017; Juszkiewicz et al., 2018; Sundaramoorthy et al., 2017) and uS10 (at K4,K8 and K63) and uS3 (at K63) in yeast (Matsuo et al., 2017) (Figure 1.1, step 2).

The ubiquitination of the 40S subunits leads to the endonucleolytic cleavage of the mRNA within the disome (Figure 1.1, step 3) separating the collided trailing and leading ribosomes.

The next commitment step for RQC activation is the ribosomal subunit dissociation mediated by the protein complex named RQC-trigger complex, which in humans is the ASC-1 complex. This protein complex is formed by the ATP-dependent helicase ASCC3, the ubiquitin-binding protein ASCC2, the zinc-finger-type protein TRIP4 or ASC-1, and the ASCC1 protein (Hashimoto et al., 2020; Matsuo et al. 2017). Only the first three proteins of the ASC-1 complex are conserved between humans and yeast, as the ASCC1 protein does not have a yeast homolog. It has been proposed that ASCC2 binds to collided ribosomes that are polyubiquitinated by ZNF598 and then ribosomal subunits are dissociated by the ATPase-dependent helicase activity of ASCC3 (Hashimoto et al., 2020; Juszkiewicz et al., 2020) (Figure 1.1, step 4). Disassembly of the ribosome subunits is followed by degradation of the mRNA by the exosome complex, a multi-protein intracellular complex of 10 or more $3'-5'$ exonucleases capable of degrading various types of RNA (Mitchell et al., 1997).

The released 60S subunit contains the nascent peptide in the form of peptidyl-tRNA producing a surface recognized by the RQC complex (Figure 1.1, step 5). The NEMF (Rqc2) protein, a subunit of the RQC, prevents re-association of the 40S subunit and facilitates binding of the RING-type E3 ubiquitin ligase Listerin (LTN1 in yeast) to the 60S (Lyumkis et al., 2014; Shao et al., 2015). Listerin/LTN1 polyubiquitinates the nascent peptide to target it for degradation by the proteasome (Brandman et al., 2012). The TCF25 (Rqc1 in yeast) protein accelerates this polyubiquitination (Kuroha et al., 2018). Ubiquitinated peptides are released from the 60S by the tRNA endonuclease ANKZF1 (Vms1 in yeast) (Kuroha et al., 2018; Su et al., 2019), and extracted by VCP/p97 (Cdc48 in yeast), together with its cofactors, for degradation

by the proteasome (Verma et al., 2013). Listerin can efficiently ubiquitinate only nascent chain lysines immediately proximal to the ribosome exit tunnel. Some substrates do not have lysines accessible to Listerin because they have been sequestered in the ribosome exit tunnel. When the ubiquitination of the nascent peptides fails because the lysine residues are inaccessible to Listerin, the Rqc2 tags the partially synthesized peptides with C-terminal alanine and threonine tails (CAT-tailing) in a non-canonical elongation reaction (Shen et al., 2015). The addition of these residues pushes the lysines out of the ribosome exit tunnel and makes them accessible for ubiquitination by Listerin (Kostova et al., 2017; Osuna et al., 2017). Deletion of LTN1 in yeast results in proteotoxic stress due to increased protein aggregation by the CAT tails (Choe et al., 2016), indicating the importance of the RQC process in maintaining proteostasis.

Defects in the RQC also contribute to mouse and human disease. Mutation of Listerin leads to neurodegeneration in mice (Chu et al., 2009). NEMF/Rqc2 mutations in mice lead to progressive motor neuron degeneration and NEMF mutations have been identified in patients presenting juvenile neuromuscular disease suggesting that its loss causes neurodegeneration (Martin et al., 2020). Thus, the RQC is a critical protein quality-control pathway that may protect neurons against degeneration (Martin et al., 2020). Oxidative stress can also affect translation and trigger RQC. Chemical modifications caused by alkylating and oxidizing agents result in damaged mRNA that can lead to ribosome collision and degradation of the oxidized mRNA by NGD, while the resulting aberrant peptides are degraded by the RQC (Yan et al., 2019). Therefore, the RQC belongs to the cellular mechanisms that prevent accumulation of toxic products generated by oxidative stress thus preserving cell viability.

Recently, ribosome collisions have been linked to the activation of ribotoxic stress response (RSR) that stimulates MAP kinase signaling. Cellular response to stress caused, for example, by amino acid starvation and UV irradiation, can lead to RNA damage and induce ribosome collisions in human cells (Chih-Chien Wu et al., 2020). Collision-mediated stress responses can also activate an RSR that responds to this stress, thus maintaining cellular homeostasis. The sensor of a collision-mediated stress response is the ZAK protein, a member of the mitogen-activated protein kinase kinase kinase (MAPKKK) family of signal transduction molecules. ZAK phosphorylates and activates the stress-activated protein kinases (SAPKs) that are the p38 mitogen-activated protein kinases and Jun N-terminal kinase (JNK). SAPKs are sensors of multiple stress stimuli including environmental stresses such as ionizing radiation, heat shock, oxidative stress, UV irradiation, protein synthesis inhibitors, DNA damage, inflammatory cytokines and growth factors. Activation of SAPKs is involved in the control of cell proliferation, differentiation and death (Morrison, 2012). It has been shown that translation elongation inhibitors, such as anisomycin and emetine (Iordanov et al., 1997), or UV irradiation (Iordanov et al., 1998) can lead to RSR mediated by p38/JNK. These stresses can cause severe ribosome collisions that lead to activation of ZAK by autophosphorylation that in turn activates p38 and JNK and induces the GCN2-mediated stress response pathways, which can trigger cell cycle arrest and apoptosis (Chih-Chien Wu et al., 2020).

How GCN2-mediated stress response pathway and RQC activation due to ribosome stalling are coordinated is still unclear, but it has been proposed that since ribosome collisions are more severe under stress conditions, it is likely that ZNF598/Hel2 becomes oversaturated and ZAK mediates Gcn2 activation. Future experiments aimed at dissecting the thermodynamic and kinetic contributions of both processes will yield important information on how they are coordinated (Yan and Zaher, 2020).

1.3 REVERSIBLE RIBOSOMAL PAUSES

1.3.1 eIF5A Rescues Poly-Pro Stalling Motifs

The standard ribosome profiling technology was originally designed to identify mRNA fragments protected from degradation by monosomes (mono-ribosome-protected fragments or mRPFs). However, the discovery of ribosome collisions (Wolin and Walter, 1988) allowed the study of disome-protected fragments (dRPFs) or even trisome-protected fragments (tRPFs) by RiboProfiling, as a means to visualize and characterize the ribosome stalling. These approaches (disome or trisome profiling) have shown widespread distribution of ribosomal collisions in the yeast translatome (Diament et al., 2018; Guydosh and Green, 2014; Meydan and Guydosh, 2020). Noticeably, similar studies in humans and

zebrafish have shown specific amino acid motifs and stop codons as the main causes of ribosome stalling as in the case of Pro-Pro/Gly/Asp and Arg-X-Lys motifs, stop codons and polyA sequences present in 3′ UTR regions (Han et al., 2020). Of special interest is the presence of poly-Pro motifs, as ribosome stalling signals that require the activity of highly specialized non-canonical translation factors.

The chemical features of the amino acid proline explain its problematic behavior for peptide bond formation, in particular, when two or more proline residues appear consecutively. As a secondary amine with a cyclic lateral chain, proline is a poor peptidyl donor and acceptor (Melnikov et al., 2016; Pavlov et al., 2009). The association of poly-Pro stretches with problematic translation elongation contrasts with their abundance in protein interaction networks ranging from yeast and mouse to higher plants (Belda-Palazón et al., 2016; Mandal et al., 2014; Pällmann et al., 2015). The evolutionary solution to this paradox is the presence of a non-canonical translation factor named eIF5A, originally described as an initiation factor based on its ability to stimulate the formation of methionyl-puromycin in vitro (Cooper et al., 1983; Kemper et al., 1976; Thomas et al., 1979), and later shown to act as an elongation factor also involved in translation termination (Saini et al., 2009; Schuller et al., 2017). The key feature of this translation factor is the unique post-translational modification of a highly conserved lysine into the unusual residue hypusine or Nε-(4-amino-2-hydroxybutyl)lysine. This post-translational modification occurs in two steps, first by the transfer of the aminobutyl moiety of spermidine to the conserved eIF5A lysine by the enzyme deoxyhypusine synthase and second by its irreversible hydroxylation through the activity of the deoxyhypusine hydroxylase (Park et al., 1981). Spermidine is an essential polyamine metabolite for eukaryote cell viability as it is also the modification of eIF5A by hypusination (Nishimura et al., 2012; Pagnussat et al., 2005; Pällmann et al., 2015). The localization of hypusinated eIF5A to the ribosomes highlights its role in the stabilization of the acceptor arm of the P-site peptidyl-tRNA, thus facilitating peptide bond formation at the peptidyl transferase center (PTC) in particular in the presence of poly-Pro stretches (Schmidt et al., 2016) (Figure 1.2A).

A consequence of this particular reversible ribosome stalling during elongation is that the

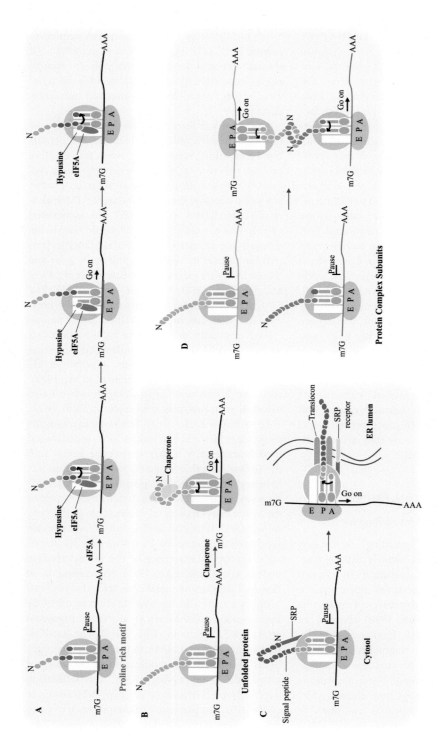

Figure 1.2 **Reversible ribosomal pauses in eukaryotic organisms.** There are several cases of identified reversible ribosomal pauses in eukaryotic organisms where the elongating ribosome undergoes a transitory pause. **(A)** In the case of coding sequences for polyproline motifs, the ribosome undergoes a pause until the recruitment of the hypusinated eIF5A protein to the E site of the paused ribosome, which facilitates the translation of these motifs and resumption of translation elongation. **(B)** Another example of useful ribosomal pause is when the nascent polypeptide chain needs chaperone recruitment to facilitate protein folding. The folded nascent chain serves as a signal to resume translation elongation of the paused ribosome. **(C)** Co-translational protein targeting to subcellular organelles is another case where the elongating ribosome undergoes a useful pause. For instance, the arrested translational elongation of the protein targeted to the endoplasmic reticulum (ER) where the binding of signal recognition particle (SRP) to the signal peptide of nascent chain disrupts translational elongation. Protein synthesis is restarted again when the translation machinery has been directed to the ER membrane. **(D)** In some cases, transitory arrest of elongating ribosomes allows synchronized translation of the different subunits of the same protein complex. The co-translational folding and interaction of these subunits resume translation elongation eventually leading to the correctly folded and functional protein complex.

PROTEOSTASIS AND PROTEOLYSIS

vacant E-site position activates completely different mechanisms to help resume translation. This is different from irreversible arrests where ribosomes present vacant A-sites that may accelerate the selective elimination of the mRNA and the truncated polypeptide, as discussed before. In the following section, we summarize other examples of reversible ribosomal pauses that occur with functional purposes.

1.3.2 Translational Pauses during Protein Folding, Organellar Targeting and Protein Complex Assembly

Translation rules drive the organized departure of the polypeptide chain out of the ribosome from the N to C terminus that facilitates the folding of the nascent amino acid chain particularly for eukaryotic proteins that frequently contain multiple domains. Adequate proteostasis demands effective folding pathways to prevent protein misfolding, and this is facilitated by the directional nature of the translation process from the N-terminal to the C-terminal sites (Frydman et al., 1994, 1999; Kim et al., 2013). The collateral risk of this strategy is that the nascent polypeptide chain is susceptible to misfolding and aggregation until the protein is completely ejected out of the ribosome, as folding is much faster than translation (milliseconds vs. seconds, respectively) (Holtkamp et al., 2015; Kaiser and Liu, 2018).

The elongation phase of translation plays a role in directly monitoring the status of the nascent polypeptide chain in collaboration with modification enzymes and targeting factors that, altogether, dictate the appropriate folding, assembly and targeting of the synthesized polypeptide (Kramer et al., 2009). Translation elongation kinetics may be modulated by several factors like the codon usage and its surrounding sequence, the tRNA abundance, and the secondary structure of the mRNA and protein sequence as explained above (Chaney and Clark, 2015; Choi et al., 2018). Codon usage, or the frequency of codons in mRNA sequence, co-participates with the tRNA availability in the regulation of translation elongation kinetics (Fluitt et al., 2007; Ikemura, 1985), while alterations of the balance between the pools of mRNA and tRNA may disturb proteostasis (Yona et al., 2013). A correct balance of these two elements does not necessarily need to be optimized for massive expression. Instead, it has been shown that optimal codons are enriched in structural protein domains, whereas rare codons that slow down translation elongation often encode residues located in the boundaries of structural domains as in the case of linker regions, probably facilitating sub-domain protein folding (Chaney et al., 2017; Zhou et al., 2009).

Ribosome-associated chaperones are the first ones in contact with the nascent polypeptide chain, identified in yeast as the Nascent polypeptide Associated Complex (NAC) and Ribosome Associated Complex (RAC) protein complexes (Rospert et al., 2002; Zhang et al., 2017). Ribo-seq studies with the RAC chaperone showed specific target selection and timing binding to ribosome-nascent chain complexes, and a potential impact of this chaperone on translation elongation (Döring et al., 2017) (Figure 1.2B).

In addition to the chaperones involved in protein folding, the ribosome dynamics also facilitate the recruitment of other factors that recognize the nascent polypeptide chains and drive their correct subcellular targeting. One example is the SRP that allows co-translational targeting to the ER. In some cases, once the targeting signals recognized by the SRP emerge from the ribosome, a cluster of rare codons at the PTC slows translation, thus facilitating recognition by the SRP and the subsequent ER targeting (Pechmann et al., 2014) (Figure 1.2C).

Proteins can achieve high structural order by the assembly of quaternary structures through homo- or hetero-oligomeric complexes. In some cases, the protein complexes form post-translationally, but in others co-translational folding should occur in a timely manner to fulfill the required protein stoichiometry and this may require functional ribosomal pausing. One example is the translational pauses that take place during assembly of the proteasome as revealed by the analysis of ribosome footprinting data of the proteasomal subunits Rpt1 and Rpt2 (Panasenko et al., 2019). The derived model suggests that translation elongation kinetics may facilitate the interaction of the nascent polypeptide chains before their translation is complete (Figure 1.2D).

1.4 CONCLUDING REMARKS

This chapter summarized the recent outstanding findings on the importance of translation elongation for a healthy proteostasis. A large amount of evidence indicates that during the elongation phase of translation, the ribosomes are subjected to multiple intracellular or extracellular signals that impact their performance. One of the most dramatic situations for the translational machinery is when ribosomes stop and a decision is taken to either resolve the situation and resume translation or dismantle the machinery with the clearance of defective and potentially toxic molecules. This is a crucial point considering the particularly high energetic cost of translation (Kafri et al., 2016), and the fact that in some instances a slowdown of the ribosome may be functionally required.

The majority of the recent discoveries on translation elongation pauses have been found thanks to the development of the Ribo-seq technology, widely used nowadays in any model organism with an annotated genome. Analysis of immunoprecipitated proteins associated to the ribosome with the Ribo-seq has provided much insight on the role of co-translational processes like protein folding and protein complex assembly. On the other hand, the discovery of ribosome collisions has opened new avenues for Ribo-seq approaches aimed to identify dRPFs and tRPFs to elucidate with more precision the mRNA sequences involved in ribosome stalling.

The knowledge on molecular events associated to irreversible ribosomal arrests has also advanced rapidly and has helped to understand how ribosome collisions lead to activation of the RQC pathway to eliminate defective polypeptides and mRNAs. Although the number of human pathologies associated with defects in these pathways is still scarce, we can envision that a more profound knowledge of these processes will provide a better understanding of proteostasis at the cellular and organismal level in the near future.

ACKNOWLEDGMENTS

We thank the support given by the Spanish grants PI15/206, PI20/194 and SAF2017-90900-REDT. A.M.F. acknowledges Asociación Española Contra el Cáncer for her predoctoral contract.

REFERENCES

Arthur, L.L., Pavlovic-Djuranovic, S., Koutmou, K.S., et al., 2015. Translational control by lysine-encoding A-rich sequences. Sci. Adv. 1, e1500154. doi:10.1126/sciadv.1500154

Belda-Palazón, B., Almendáriz, C., Martí, E., et al., 2016. Relevance of the axis spermidine/eIF5A for plant growth and development. Front. Plant Sci. 7, 245. doi:10.3389/fpls.2016.00245

Brandman, O., Hegde, R.S., 2016. Ribosome-associated protein quality control. Nat. Struct. Mol. Biol. doi:10.1038/nsmb.3147

Brandman, O., Stewart-Ornstein, J., Wong, D., et al., 2012. A ribosome-bound quality control complex triggers degradation of nascent peptides and signals translation stress. Cell 151, 1042–1054. doi:10.1016/j.cell.2012.10.044

Brar, G.A., Weissman, J.S., 2015. Ribosome profiling reveals the what, when, where and how of protein synthesis. Nat. Rev. Mol. Cell Biol. doi:10.1038/nrm4069

Chaney, J.L., Clark, P.L., 2015. Roles for synonymous codon usage in protein biogenesis. Annu. Rev. Biophys. 44, 143–166. doi:10.1146/annurev-biophys-060414-034333

Chaney, J.L., Steele, A., Carmichael, R., et al., 2017. Widespread position-specific conservation of synonymous rare codons within coding sequences. PLoS Comput. Biol. 13, e1005531. doi:10.1371/journal.pcbi.1005531

Chen, L., Muhlrad, D., Hauryliuk, V., et al., 2010. Structure of the Dom34-Hbs1 complex and implications for no-go decay. Nat. Struct. Mol. Biol. 17, 1233–1240. doi:10.1038/nsmb.1922

Chih-Chien Wu, C., Peterson, A., Zinshteyn, B., et al., 2020. Ribosome collisions trigger general stress responses to regulate cell fate. Cell 182, 404–416. doi:10.1016/j.cell.2020.06.006

Chiti, F., Dobson, C.M., 2017. Protein misfolding, amyloid formation, and human disease: a summary of progress over the last decade. Annu. Rev. Biochem. doi:10.1146/annurev-biochem-061516-045115

Choe, Y.J., Park, S.H., Hassemer, T., et al., 2016. Failure of RQC machinery causes protein aggregation and proteotoxic stress. Nature 531, 191–195. doi:10.1038/nature16973

Choi, J., Grosely, R., Prabhakar, A., et al., 2018. How messenger RNA and nascent chain sequences regulate translation elongation. Annu. Rev. Biochem. 87, 421–449. doi:10.1146/annurev-biochem-060815-014818

Chu, J., Hong, N.A., Masuda, C.A., et al., 2009. A mouse forward genetics screen identifies Listerin as an E3 ubiquitin ligase involved in neurodegeneration. Proc. Natl. Acad. Sci. U. S. A. 106, 2097–2103. doi:10.1073/pnas.0812819106

Cooper, H.L., Park, M.H., Folk, J.E., et al., 1983. Identification of the hypusine-containing protein Hy+ as translation initiation factor eIF-4D. Proc.

Natl. Acad. Sci. U. S. A. 80, 1854–1857. doi:10.1073/pnas.80.7.1854

Darnell, A.M., Subramaniam, A.R., O'shea, E.K., 2018. Translational control through differential ribosome pausing during amino acid limitation in mammalian cells. Mol. Cell 71, 229–243. doi:10.1016/j.molcel.2018.06.041

Diament, A., Feldman, A., Schochet, E., et al., 2018. The extent of ribosome queuing in budding yeast. PLoS Comput. Biol. 14, e1005951. doi:10.1371/journal.pcbi.1005951

Doma, M.K., Parker, R., 2006. Endonucleolytic cleavage of eukaryotic mRNAs with stalls in translation elongation. Nature 440, 561–564. doi:10.1038/nature04530

Döring, K., Ahmed, N., Riemer, T., et al., 2017. Profiling Ssb-nascent chain interactions reveals principles of Hsp70-assisted folding. Cell 170, 298–311.e20. doi:10.1016/j.cell.2017.06.038

Fluitt, A., Pienaar, E., Viljoen, H., 2007. Ribosome kinetics and aa-tRNA competition determine rate and fidelity of peptide synthesis. Comput. Biol. Chem. 31, 335–346. doi:10.1016/j.compbiolchem.2007.07.003

Frydman, J., Nimmesgern, E., Ohtsuka, K., et al., 1994. Folding of nascent polypeptide chains in a high molecular mass assembly with molecular chaperones. Nature 370, 111–117. doi:10.1038/370111a0

Frydman, J., Erdjument-Bromage, H., Tempst, P., et al., 1999. Co-translational domain folding as the structural basis for the rapid de novo folding of firefly luciferase. Nat. Struct. Biol. 6, 697–705. doi:10.1038/10754

Garzia, A., Jafarnejad, S.M., Meyer, C., et al., 2017. The E3 ubiquitin ligase and RNA-binding protein ZNF598 orchestrates ribosome quality control of premature polyadenylated mRNAs. Nat. Commun. 8, 1–10. doi:10.1038/ncomms16056

Guydosh, N.R., Green, R., 2014. Dom34 rescues ribosomes in 3' untranslated regions. Cell 156, 950–962. doi:10.1016/j.cell.2014.02.006

Han, P., Shichino, Y., Schneider-, T., et al., 2020. Genome-wide survey of ribosome collision. Cell Rep. doi:10.1016/j.celrep.2020.107610

Hashimoto, S., Sugiyama, T., Yamazaki, R., et al., 2020. Identification of a novel trigger complex that facilitates ribosome-associated quality control in mammalian cells. Sci. Rep. 10, 1–12. doi:10.1038/s41598-020-60241-w

Holtkamp, W., Kokic, G., Jäger, M., et al., 2015. Cotranslational protein folding on the ribosome monitored in real time. Science (80-.). 350, 1104–1107. doi:10.1126/science.aad0344

Ikemura, T., 1985. Codon usage and tRNA content in unicellular and multicellular organisms. Mol. Biol. Evol. doi:10.1093/oxfordjournals.molbev.a040335

Ikeuchi, K., Tesina, P., Matsuo, Y., et al., 2019. Collided ribosomes form a unique structural interface to induce Hel2-driven quality control pathways. EMBO J. 38. doi:10.15252/embj.2018100276

Ingolia, N.T., Ghaemmaghami, S., Newman, J.R.S., et al., 2009. Genome-wide analysis in vivo of translation with nucleotide resolution using ribosome profiling. Science (80-) 324, 218–223. doi:10.1126/science.1168978

Ingolia, N.T., Lareau, L.F., Weissman, J.S., 2011. Ribosome profiling of mouse embryonic stem cells reveals the complexity and dynamics of mammalian proteomes. Cell 147, 789–802. doi:10.1016/j.cell.2011.10.002

Iordanov, M.S., Pribnow, D., Magun, J.L., et al., 1997. Ribotoxic stress response: activation of the stress-activated protein kinase JNK1 by inhibitors of the peptidyl transferase reaction and by sequence-specific RNA damage to the alpha-sarcin/ricin loop in the 28S rRNA. Mol. Cell. Biol. 17, 3373–3381. doi:10.1128/mcb.17.6.3373

Iordanov, M.S., Pribnow, D., Magun, J.L., et al., 1998. Ultraviolet radiation triggers the ribotoxic stress response in mammalian cells. J. Biol. Chem. 273, 15794–15803. doi:10.1074/jbc.273.25.15794

Ishimura, R., Nagy, G., Dotu, I., et al., 2014. Ribosome stalling induced by mutation of a CNS-specific tRNA causes neurodegeneration. Science (80-) 345, 455–459. doi:10.1126/science.1249749

Ito-Harashima, S., Kuroha, K., Tatematsu, T., et al., 2007. Translation of the poly(A) tail plays crucial roles in nonstop mRNA surveillance via translation repression and protein destabilization by proteasome in yeast. Genes Dev. 21, 519–524. doi:10.1101/gad.1490207

Juszkiewicz, S., Hegde, R.S., 2017. Initiation of quality control during poly(A) translation requires site-specific ribosome ubiquitination correspondence. Mol. Cell 65, 743–750.e4. doi:10.1016/j.molcel.2016.11.039

Juszkiewicz, S., Chandrasekaran, V., Lin, Z., et al., 2018. ZNF598 is a quality control sensor of collided ribosomes correspondence. Mol. Cell 72, 469–481. doi:10.1016/j.molcel.2018.08.037

Juszkiewicz, S., Speldewinde, S.H., Wan, L., et al., 2020. The ASC-1 complex disassembles collided ribosomes. Mol. Cell. doi:10.1016/j.molcel.2020.06.006

Kafri, M., Metzl-Raz, E., Jona, G., et al., 2016. The cost of protein production. Cell Rep. 14, 22–31. doi:10.1016/j.celrep.2015.12.015

Kaiser, C.M., Liu, K., 2018. Folding up and moving on—nascent protein folding on the ribosome. J. Mol. Biol. doi:10.1016/j.jmb.2018.06.050

Kemper, W.M., Berry, K.W., Merrick, W.C., 1976. Purification and properties of rabbit reticulocyte protein synthesis initiation factors M2Bα and M2Bβ. J. Biol. Chem. 251, 5551–5557.

Kim, Y.E., Hipp, M.S., Bracher, A., et al., 2013. Molecular chaperone functions in protein folding and proteostasis. Annu. Rev. Biochem. doi:10.1146/annurev-biochem-060208-092442

Kostova, K.K., Hickey, K.L., Osuna, B.A., et al., 2017. CAT-tailing as a fail-safe mechanism for efficient degradation of stalled nascent polypeptides. Science (80–) 357, 414–417. doi:10.1126/science.aam7787

Kramer, G., Boehringer, D., Ban, N., et al., 2009. The ribosome as a platform for co-translational processing, folding and targeting of newly synthesized proteins. Nat. Struct. Mol. Biol. doi:10.1038/nsmb.1614

Kuroha, K., Zinoviev, A., Hellen, C.U.T., et al., 2018. Release of ubiquitinated and non-ubiquitinated nascent chains from stalled mammalian ribosomal complexes by ANKZF1 and PTRH1. Mol. Cell 72, 286–302.e8. doi:10.1016/j.molcel.2018.08.022

Letzring, D.P., Wolf, A.S., Brule, C.E., et al., 2013. Translation of CGA codon repeats in yeast involves quality control components and ribosomal protein L1. RNA 19, 1208–1217. doi:10.1261/rna.039446.113

Lyumkis, D., Dos Passosb, D.O., Tahara, E.B., et al., 2014. Structural basis for translational surveillance by the large ribosomal subunit-associated protein quality control complex. Proc. Natl. Acad. Sci. U. S. A. 111, 15981–15986. doi:10.1073/pnas.1413882111

Mandal, A., Mandal, S., Park, M.H., 2014. Genome-wide analyses and functional classification of proline repeat-rich proteins: potential role of eIF5A in eukaryotic evolution. PLoS One 9, e111800. doi:10.1371/journal.pone.0111800

Martin, P.B., Kigoshi-Tansho, Y., Sher, R.B., et al., 2020. NEMF mutations that impair ribosome-associated quality control are associated with neuromuscular disease. Nat. Commun. 11, 1–12. doi:10.1038/s41467-020-18327-6

Matsuo, Y., Ikeuchi, K., Saeki, Y., et al., 2017. Ubiquitination of stalled ribosome triggers ribosome-associated quality control. Nat. Commun. 8, 1–14. doi:10.1038/s41467-017-00188-1

Melnikov, S., Mailliot, J., Rigger, L., et al., 2016. Molecular insights into protein synthesis with proline residues. EMBO Rep. 17, 1776–1784. doi:10.15252/embr.201642943

Meydan, S., Guydosh, N.R., 2020. Disome and trisome profiling reveal genome-wide targets of ribosome quality control. Mol. Cell 79, 588–602.e6. doi:10.1016/j.molcel.2020.06.010

Mitchell, P., Petfalski, E., Shevchenko, A., et al., 1997. The exosome: a conserved eukaryotic RNA processing complex containing multiple 3′→5′ exoribonucleases. Cell 91, 457–466. doi:10.1016/S0092-8674(00)80432-8

Morrison, D.K., 2012. MAP kinase pathways. Cold Spring Harb. Perspect. Biol. doi:10.1101/cshperspect.a011254

Nishimura, K., Lee, S.B., Park, J.H., et al., 2012. Essential role of eIF5A-1 and deoxyhypusine synthase in mouse embryonic development. Amino Acids 42, 703–710. doi:10.1007/s00726-011-0986-z

Osuna, B.A., Howard, C.J., Subheksha, K.C., et al., 2017. In vitro analysis of RQC activities provides insights into the mechanism and function of CAT tailing. Elife 6. doi:10.7554/eLife.27949.001

Pagnussat, G.C., Yu, H.J., Ngo, Q.A., et al., 2005. Genetic and molecular identification of genes required for female gametophyte development and function in Arabidopsis. Development 132, 603–614. doi:10.1242/dev.01595

Pällmann, N., Braig, M., Sievert, H., et al., 2015. Biological relevance and therapeutic potential of the hypusine modification system. J. Biol. Chem. doi:10.1074/jbc.M115.664490

Panasenko, O.O., Somasekharan, S.P., Villanyi, Z., et al., 2019. Co-translational assembly of proteasome subunits in NOT1-containing assemblysomes. Nat. Struct. Mol. Biol. 26, 110–120. doi:10.1038/s41594-018-0179-5

Park, M.H., Cooper, H.L., Folk, J.E., 1981. Identification of hypusine, an unusual amino acid, in a protein from human lymphocytes and of spermidine as its biosynthetic precursor. Proc. Natl. Acad. Sci. U. S. A. 78, 2869–2873. doi:10.1073/pnas.78.5.2869

Passos, D.O., Doma, M.K., Shoemaker, C.J., et al., 2009. Analysis of Dom34 and its function in No-Go decay. Mol. Biol. Cell 20, 3025–3032. doi:10.1091/mbc.E09-01-0028

Pavlov, M.Y., Watts, R.E., Tan, Z., et al., 2009. Slow peptide bond formation by proline and other N-alkyl amino acids in translation. Proc. Natl. Acad. Sci. U. S. A. 106, 50–54. doi:10.1073/pnas.0809211106

Pechmann, S., Chartron, J.W., Frydman, J., 2014. Local slowdown of translation by nonoptimal codons promotes nascent-chain recognition by SRP in vivo. Nat. Struct. Mol. Biol. 21, 1100–1105. doi:10.1038/nsmb.2919

Rospert, S., Dubaquié, Y., Gautschi, M., 2002. Nascent-polypeptide-associated complex. Cell. Mol. Life Sci. 59, 1632–1639. doi:10.1007/PL00012490

Saini, P., Eyler, D.E., Green, R., et al., 2009. Hypusine-containing protein eIF5A promotes translation elongation. Nature 459, 118–121. doi:10.1038/nature08034

Schmidt, C., Becker, T., Heuer, A., et al., 2016. Structure of the hypusinylated eukaryotic translation factor eIF-5A bound to the ribosome. Nucleic Acids Res. 44, 1944–1951. doi:10.1093/nar/gkv1517

Schuller, A.P., Wu, C.C.C., Dever, T.E., et al., 2017. eIF5A functions globally in translation elongation and termination. Mol. Cell 66, 194–205.e5. doi:10.1016/j.molcel.2017.03.003

Shao, S., Von der Malsburg, K., Hegde, R.S., 2013. Listerin-dependent nascent protein ubiquitination relies on ribosome subunit dissociation. Mol. Cell 50, 637–648. doi:10.1016/j.molcel.2013.04.015

Shao, S., Brown, A., Santhanam, B., et al., 2015. Structure and assembly pathway of the ribosome quality control complex. Mol. Cell 57, 433–444. doi:10.1016/j.molcel.2014.12.015

Shen, P.S., Park, J., Qin, Y., et al., 2015. Rqc2p and 60S ribosomal subunits mediate mRNA-independent elongation of nascent chains. Science (80–) 347, 75–78. doi:10.1126/science.1259724

Simms, C.L., Hudson, B.H., Mosior, J.W., et al., 2014. An active role for the ribosome in determining the fate of oxidized mRNA. Cell Rep. 9, 1256–1264. doi:10.1016/j.celrep.2014.10.042

Simms, C.L., Thomas, E.N., Zaher, H.S., 2017. Ribosome-based quality control of mRNA and nascent peptides. Wiley Interdiscip. Rev. RNA. doi:10.1002/wrna.1366

Sonenberg, N., Hinnebusch, A.G., 2009. Regulation of translation initiation in eukaryotes: mechanisms and biological targets. Cell. doi:10.1016/j.cell.2009.01.042

Su, T., Izawa, T., Thoms, M., et al., 2019. Structure and function of Vms1 and Arb1 in RQC and mitochondrial proteome homeostasis. Nature 570, 538–542. doi:10.1038/s41586-019-1307-z

Sundaramoorthy, E., Leonard, M., Mak, R., et al., 2017. ZNF598 and RACK1 regulate mammalian ribosome-associated quality control function by mediating regulatory 40S ribosomal ubiquitylation. Mol. Cell 65, 751–760.e4. doi:10.1016/j.molcel.2016.12.026

Thomas, A., Goumans, H., Amesz, H., et al., 1979. A comparison of the initiation factors of eukaryotic protein synthesis from ribosomes and from the postribosomal supernatant. Eur. J. Biochem. 98, 329–337. doi:10.1111/j.1432-1033.1979.tb13192.x

Tsuboi, T., Kuroha, K., Kudo, K., et al., 2012. Dom34: Hbs1 plays a general role in quality-control systems by dissociation of a stalled ribosome at the 3′ end of aberrant mRNA. Mol. Cell 46, 518–529. doi:10.1016/j.molcel.2012.03.013

Verma, R., Oania, R.S., Kolawa, N.J., et al., 2013. Cdc48/p97 promotes degradation of aberrant nascent polypeptides bound to the ribosome. Elife 2013. doi:10.7554/eLife.00308

Wolin, S.L., Walter, P., 1988. Ribosome pausing and stacking during translation of a eukaryotic mRNA. EMBO J. 7, 3559–3569. doi:10.1002/j.1460-2075.1988.tb03233.x

Woolstenhulme, C.J., Guydosh, N.R., Green, R., et al., 2015. High-precision analysis of translational pausing by ribosome profiling in bacteria lacking EFP. Cell Rep. 11, 13–21. doi:10.1016/j.celrep.2015.03.014

Wu, C.C.C., Zinshteyn, B., Wehner, K.A., et al., 2019. High-resolution ribosome profiling defines discrete ribosome elongation states and translational regulation during cellular stress. Mol. Cell 73, 959–970. e5. doi:10.1016/j.molcel.2018.12.009

Yan, L.L., Zaher, H.S., 2020. Ribosome quality control antagonizes the activation of the integrated stress response on colliding ribosomes. Mol. Cell. doi:10.1016/j.molcel.2020.11.033

Yan, L.L., Simms, C.L., McLoughlin, F., et al., 2019. Oxidation and alkylation stresses activate ribosome-quality control. Nat. Commun. 10, 1–15. doi:10.1038/s41467-019-13579-3

Yona, A.H., Bloom-Ackermann, Z., Frumkin, I., et al., 2013. tRNA genes rapidly change in evolution to meet novel translational demands. Elife 2013. doi:10.7554/eLife.01339

Zhang, Y., Sinning, I., Rospert, S., 2017. Two chaperones locked in an embrace: structure and function of the ribosome-associated complex RAC. Nat. Struct. Mol. Biol. doi:10.1038/nsmb.3435

Zhou, T., Weems, M., Wilke, C.O., 2009. Translationally optimal codons associate with structurally sensitive sites in proteins. Mol. Biol. Evol. 26, 1571–1580. doi:10.1093/molbev/msp070

CHAPTER TWO

Protein Folding and Misfolding: Deciphering Mechanisms of Age-Related Diseases

Judith-Elisabeth Riemer, Diana Panfilova, Heidi Olzscha

CONTENTS

2.1 INTRODUCTION

2.1.1 Kinetic and Thermodynamic Aspects of Protein Folding

Protein folding refers to the physical process by which a protein obtains its native state and becomes functionally active. As correct protein folding is a prerequisite for proteostasis in biological systems, determining the mechanisms of this process is crucial for understanding age-related diseases, especially proteinopathies, which are also named proteopathies or protein-folding diseases. A breakthrough in determining three-dimensional structures of proteins occurred in

DOI: 10.1201/9781003048138-2

13

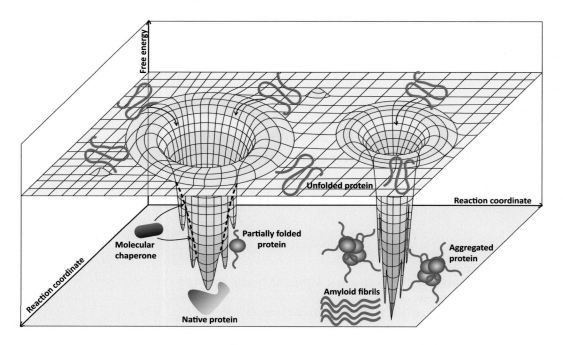

Figure 2.1 Protein-folding and misfolding energy landscape. During protein folding, unfolded proteins may have to surpass high free energy stages in order to "fall" into the folding funnel and reach their native conformation, thereby minimizing their free energy level. Molecular chaperones help to prevent trapping of partially folded proteins in low energy states. In case of protein misfolding, aggregated proteins can reach low free energy states and even form highly stable amyloid fibrils with minimal free energy levels.

the 1940s and 1950s, when Kendrew et al. (1958) and Pauling et al. (1951) discovered elements of secondary, tertiary and quaternary protein structures. Then, Anfinsen et al. (1961) demonstrated in 1961 for the small protein ribonuclease A that the information for its folding is inherent in its primary sequence. Another important theorem was given by Levinthal with the "Levinthal's paradox", which states that a protein has an astronomical number of possible conformations due to the high number of degrees of freedom in its chain (Levinthal, 1969). However, according to biophysical measurements, some small simple proteins can be folded in their active form in less than 50 μs, although the number of their possible conformations is extremely high (Mayor et al., 2003). This contradiction was explained by Anfinsen's thermodynamic hypothesis (Anfinsen's dogma) in 1973. It states that the unique native structure that is formed in the environment at which protein folding takes place is the most thermodynamically stable conformation (Anfinsen, 1973). Thus, according to Levinthal's and Anfinsen's hypothesis, many proteins are believed to self-assemble under thermodynamic control (Varela et al., 2019).

In general, the process of protein folding has two stages. Firstly, a molten globule is formed from the random coil of an unfolded protein, losing entropy with little change in energy. In the second, slower stage, the molten globule evolves into the native conformation with reduced free energy and only a small change in entropy (Shirdel and Khalifeh, 2019). These two stages combined form a "folding funnel" (Figure 2.1).

Before a protein reaches its native state, the protein folding goes through a stochastic search of conformations. In this process, the number of possible conformations is irrelevant. Of importance are the kinetically controlled conformations with distinct transition states. In other words, the kinetic control of folding events allows the existence of thermodynamically unstable molecular structures, which is called kinetic trapping (Dill, 1999). In order to prevent kinetic trapping of partially folded proteins, molecular chaperones assist in protein folding (Figure 2.1).

Although most proteins require a defined three-dimensional structure to fulfill their functions, there are some functional proteins in cells that can be either fully unstructured or have only

TABLE 2.1

Frequently used methods to analyze protein folding.

Methods	Physical principles	Output
X-ray crystallography	Scattering of X-ray waves through the atoms' electrons of a protein crystal	Three-dimensional electron density map of electrons within a crystal. Gives information about protein structure
Nuclear magnetic resonance spectroscopy	Atomic nuclei respond to oscillating magnetic fields depending on the local molecular environment	A map of chemical bonds between the protein's atoms and their relative position in three-dimensional protein structure
Cryo-electron microscopy	A beam of electrons is transmitted through a sample cooled to cryogenic temperature	Thousands of pictures taken from various perspectives are combined to form a 3D model
Small-angle X-ray scattering	Scattering of a monochromatic X-ray beam through the structures with 1–100 nm dimensions	Function of electron density distribution within a protein. Gives information about protein shape and size
Circular dichroism	Different absorption of left and right circular polarized light through the chiral systems like proteins	Circular dichroism spectrum that is unique for each conformation of a single protein
Dynamic light scattering	Light scattering through particles like protein molecules	Information about protein size. Changes in molecular size of proteins during folding/unfolding can be observed
Fluorescence spectroscopy	Aromatic amino acids like Trp and Tyr emit light that is absorbed, while emission peaks depend on the polarity of the local environment	A fluorescence spectrum gives information about protein-folding state and conformation

unstructured regions; they are termed intrinsically disordered proteins (IDP) and intrinsically disordered regions (IDR), respectively (Dyson and Wright, 2005). Notably, many proteins involved in proteinopathies contain IDRs, including Aβ, α-synuclein and the pathologic form of prion proteins (PrP^sc) (Knowles et al., 2014).

A variety of methods exists that allows to study protein folding and misfolding as summarized in Table 2.1.

2.2 MOLECULAR CHAPERONES

2.2.1 Definition of the Term Molecular Chaperone

Molecular chaperones can be defined as proteins occurring in all taxonomic domains of life that assist other cellular proteins (clients) to adopt their final physiologically active form without being part of their functional structure (Hartl, 1996). Molecular chaperones should be distinguished from chemical chaperones, which are small physiological or non-natural molecules including osmolytes and

hydrophobic compounds that can enhance protein stability and folding (Ignatova and Gierasch, 2006). Molecular chaperones do not have a built-in-concept of their clients resembling a catalytic center, which discriminates them from folding catalysts. They are enzymes such as peptidyl-prolyl-isomerases or protein disulfide isomerases, which can break and build covalent bonds (Wang et al., 2015).

2.2.2 Significance of Molecular Chaperones in the Cell

Most molecular chaperones can be upregulated and activated by cells in response to stressful conditions including abnormal temperature and redox states, whereas some molecular chaperones are constitutively expressed. Many, but not all heat shock proteins (Hsp), are molecular chaperones, and *vice versa* many, but not all molecular chaperones, are Hsps (Nollen and Morimoto, 2002). Constitutively expressed molecular chaperones are often abbreviated as Hsc (heat shock cognate proteins). Compared to a simple folding experiment with a single protein in a test tube (Anfinsen

TABLE 2.2
Major molecular chaperone families, their structural characteristics and selected functions.

Molecular chaperone family	Examples prokaryotes	Examples humans	Nomenclature for human genes	Oligomerization degree	Selected functions
Hsp110	–	Hsp110	HSPH	Monomer (interacting with Hsp70)	Recognition of unfolded proteins; nucleotide exchange factors (NEF) for Hsp70
Hsp100	Clp family	Hsp104	–	6-/7-mer	Folding of newly synthesized proteins, disaggregation of aggregated proteins; stress tolerance
Hsp90	HtpG	Hsp90	HSPC	Dimer	Regulation of HSR; signal transduction, e.g. transport of hormone receptors; stabilizing proteins
Hsp70	DnaK	Hsp70, Hsc70	HSPA	Monomer (interacting with Hsp40 and Hsp110)	Binding and folding of nascent chains and unfolded proteins; protein transport; refolding of proteins; heat shock regulation
Chaperonins (Hsp60)	GroEL/ES	Hsp60, TRiC/CCT	HSPE	14-/16-mer	Folding of non-nascent proteins; specialized in their substrates, e.g. TRiC/CCT with actin and tubulin
Hsp40	DnaJ	Hsp40, Hdj	DNAJ	Monomer (interacting with Hsp70)	Substrate holding (holdase); essential co-factor for the ATP-dependent Hsp70 chaperones
Small Hsps	Ipb, Hsp20	Hsp25, Hsp27	HSPB	8- to 24-mer	Protein folding; preventing misfolding and aggregation; microfilament stabilization

et al., 1961), folding of proteins in cells is much more complex making the presence of molecular chaperones inevitable. Firstly, protein folding occurs in a crowded environment, leading to an excluded volume effect. This stems from the fact that macromolecules occupy a large proportion of the available volume in the cell, which reduces the volume of solvent that is available for protein folding. As a result, proteins have fewer degrees of freedom for folding (Ellis, 2001). Secondly, the protein emerges slowly from the ribosome tunnel during translation, and exposed hydrophobic patches may interact with each other intra- or intermolecularly (Etchells and Hartl, 2004). Molecular chaperones protect these exposed regions and prevent misfolding and tagging for degradation. Thirdly, the proteome, especially in eukaryotic cells, is diverse and needs different folding microenvironments, to assist different classes of proteins in folding (Kerner et al., 2005). Especially chaperonins, that are large cylindrical

molecular machines, can discharge the clients into their cavity and facilitate folding.

2.2.3 A Network of Molecular Chaperones

Molecular chaperones are grouped into families and named according to their molecular weight (Table 2.2). To ensure that the folding of the proteome in its entirety is covered and the flux through the network of molecular chaperones is maintained, protein folding is often organized in folding pathways. It is therefore widely accepted that one molecular chaperone can hand over its clients to the next molecular chaperone, sometimes belonging to a different family (Figure 2.2). Some molecular chaperone systems are ATP-dependent, including Hsp60, Hsp70 and Hsp90. Chaperonins belong to the Hsp60 family and they are special in that they provide a protected environment as a nanocage for ATP-assisted folding cycles (Hayer-Hartl et al., 2016). Members of

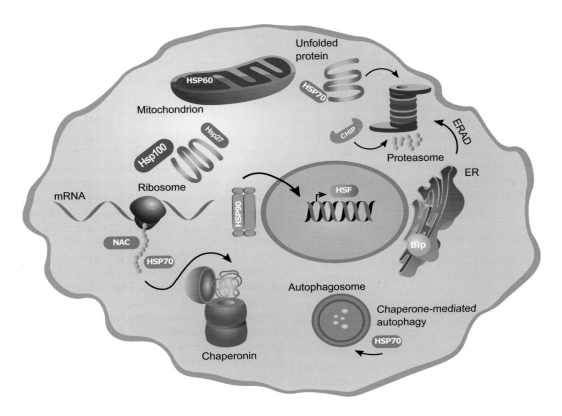

Figure 2.2 **The network of molecular chaperones.** The figure displays a network of the major molecular chaperone families in a eukaryotic cell and interactions with other protein quality control (PQC) systems. Molecular chaperones are depicted in blue, nucleic acids in magenta, protein chains in yellow, transcription factors in green and biomolecular degradation systems in purple.

other Hsp families can simply bind and stabilize the client in an ATP-independent manner such as Hsp40 and some of the small Hsps. The respective chaperones are sometimes termed foldases or holdases, respectively. The concerted action of the molecular chaperones needs to be tightly regulated with the help of heat shock factors (HSFs). HSFs are transcription factors that are bound by molecular chaperones, which can be released upon cellular stress, bind onto special regions of the genome called heat shock response elements (HSE) and initiate the transcription of stress response genes (Sarge et al., 1993).

2.2.4 Functions of Molecular Chaperones and Interplay with Other Protein Quality Control Systems

Aside from assisting in co-translational and post-translational *de novo* protein folding, molecular chaperones can refold misfolded proteins and destroy protein aggregates (Nillegoda et al., 2015). Molecular chaperones do not act only in the cytoplasm; they are important proteostatic factors in the nucleus, endoplasmic reticulum (ER), mitochondria and chloroplasts, where they are also involved in protein transport.

A stress response similar to the heat shock response in the cytoplasm can occur in the ER. The cell handles misfolded proteins in the ER by initiating an unfolded protein response (UPR), where ER-residing chaperones such as Bip (binding immunoglobulin [Ig] protein) help with protein folding but also retrograde transport through the ER membrane (Walter and Ron, 2011). This is tightly coupled with the ER-associated protein degradation (ERAD) (Christianson et al., 2011). During ERAD, misfolded proteins are recognized and transported back into the cytoplasm, where they are handed over to the ubiquitin-proteasome system (UPS). Chaperones binding misfolded proteins in the cytoplasm can be recognized by adaptor proteins which label the protein with ubiquitin for degradation, like the E3 ligase CHIP (C-terminus of HSC70 interacting protein) (Smith et al., 2013). Some proteasomal shuttling factors belonging to the UbL/UbA (ubiquitin-like/ubiquitin-associated protein) family, such as HR23B, contain chaperone-binding domains and may interact with molecular chaperones (New et al., 2013). Molecular chaperones also play a role in targeting misfolded proteins for autophagy, especially for chaperone-mediated autophagy (CMA) (Kaushik and Cuervo, 2012).

2.3 PROTEIN MISFOLDING

As mentioned above, the free energy of a protein is determined by intramolecular interactions between amino acid residues. Thus, even small changes in the protein chain, for example due to mutations or aging, can cause reshaping of the folding funnel landscape, which may result in the formation of a new global free-energy minimum. This new stable state can initiate protein misfolding and therefore lead to different adverse effects including loss of protein function and aggregation (Clark, 2004).

2.3.1 Process of Protein Misfolding

Most proteins can fold from their native state toward misfolded species, crossing different intermediate states. However, misfolded species and aggregates can originate from both intermediate and native states of a protein (Clark, 2004) (Figure 2.3).

Due to the high-energy barrier that needs to be surpassed to obtain the native conformation, molecules can get trapped in partially folded states. These proteins might interact with other partially folded proteins and form aggregates, as they often expose hydrophobic amino acid residues and domains on their surface (Liu and Eisenberg, 2002). The process of aggregation is concentration-dependent, driven by hydrophobic forces and can result in amorphous or highly organized structures (Chiti and Dobson, 2006). Interestingly, some misfolded proteins are able to convert their native protein variants into misfolded proteins; these types of proteins are called prions (proteinaceous infectious particle derived from the words **pr**otein and infect**ion**) (Prusiner, 1982). Protein aggregation can also be caused by

aberrant posttranslational modifications (PTM) including proteolytic cleavage, glycosylation, phosphorylation, ubiquitination or acetylation (Olzscha, 2019; Olzscha et al., 2017).

2.3.1.1 Amyloid Proteins

Protein aggregates can be more stable than native proteins due to the depth of the kinetic trap (Gregersen et al., 2006). Through nucleation processes, aggregates can initiate the formation of pre-amyloid oligomers, which can grow into protofilaments and eventually mature fibrils with cross-β-structures. These structures are characterized by a perpendicular arrangement of β-strands along the fibril axis. Once a certain size of a protein-consisting nucleus is reached ("critical nucleus"), aggregates grow exponentially, due to partial dissociation, providing an increasing number of seeds that repeatedly form new aggregates. In case of amyloid fibrils that are highly organized and stable structures, spontaneous dissociation into monomers is unlikely (Lashuel et al., 2002). Abnormal deposition of such amyloid structures in extracellular compartments can be observed in a wide group of diseases called amyloidoses (Benson et al., 2018).

2.3.2 Factors Leading to Protein Misfolding

2.3.2.1 Mutations

Genetic alterations can cause changes in protein structure, function or localization within the cell. Even minor modifications can make a protein prone to misfold or aggregate which leads to a number of inheritable diseases (Gámez et al., 2018) (Figure 2.4).

For example, a single nucleotide polymorphism can lead to the manifestation of cystic fibrosis (CF), as demonstrated by the phenotypes associated with the ΔF508-CFTR (CF transmembrane conductance regulator protein)-mutated allele. This allele carries a deletion of a phenylalanine, leading to the misfolding of CFTR in the ER and its impaired trafficking toward the plasma membrane (Goor et al., 2006).

2.3.2.2 Nongenetic Causes

During aging, PQC systems begin to malfunction, which can lead to the accumulation of unfolded, misfolded and aggregated proteins.

PROTEOSTASIS AND PROTEOLYSIS

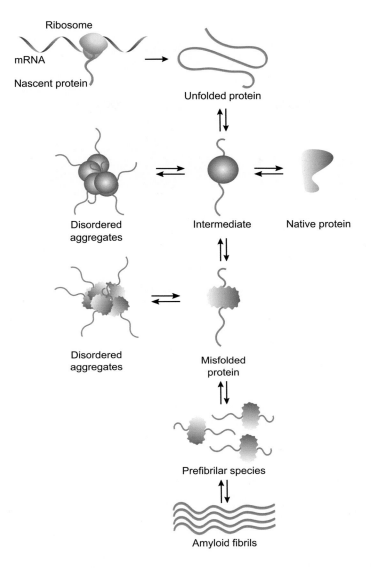

Figure 2.3 **Protein folding and misfolding.** Different steps in protein folding, misfolding and aggregation starting with the nascent polypeptide chain are displayed.

For instance, the heat shock response (HSR) becomes less sensitive with age, leading to slower and reduced activation of molecular chaperones (Calderwood et al., 2009). The increased amount of toxic protein species can then lead to cellular dysfunction, further impairment of the PQC systems and disease acceleration (Gidalevitz et al., 2006) (Figure 2.3).

Like aging and mutations, high levels of stress, including oxidative stress, can cause protein misfolding. High amounts of reactive oxygen species (ROS) in a cell can affect biomolecules and lead to lipid peroxidation as well as a range of different PTMs, which may make proteins prone to misfolding or aggregation (Levy et al., 2019). Aldehyde products resulting from lipid peroxidation such as 4-hydroxy-2-nonenal (HNE) can then further trigger formation of aberrant PTMs and this has been shown to give rise to toxic oligomeric species formed by the affected proteins (Xiang et al., 2015). Neurons in particular tend to be more vulnerable to oxidative stress than other cell types due to several reasons. Firstly, they are highly dependent on glucose oxidation as a major

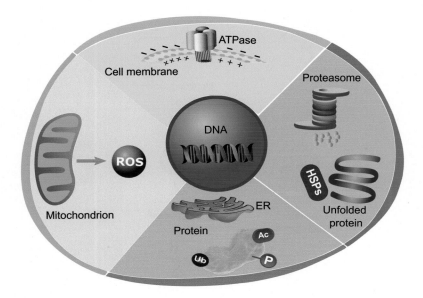

Figure 2.4 Factors leading to protein misfolding. There is a variety of factors affecting the protein-folding process and thus leading to misfolded protein species including mutations (dark blue section), chemiosmotic stress (green section), declining protein quality control (light blue section), aberrant posttranslational modifications (red section) and redox reactions (yellow section).

energy source. An imbalance in energy-producing processes, for example through aging mechanisms, can significantly affect redox homeostasis. Secondly, due to their post-mitotic state, neurons become more prone to the accumulation of ROS (Cobley et al., 2018). In Parkinson's disease (PD) for example, mitochondrial dysfunction and increase of intracellular ROS play a key role in pathogenesis (Schapira and Gegg, 2011) (Figure 2.4).

2.4 AGE-RELATED PROTEINOPATHIES

Misfolding and aggregation of proteins can cause cytotoxicity through many different pathways. Considering that these proteins do not reach their native conformation, they cannot execute their normal functions. This can lead to disturbances of cellular processes, cell death and can trigger disease onset.

Moreover, protein misfolding may have harmful effects based on the gain-of-function hypothesis of misfolded proteins. Research also suggests that oligomeric species that precede fibril formation in aggregation processes induce greater cytotoxic effects than mature fibrils and play a key role in driving neurodegeneration (Rockenstein et al., 2014). They characteristically expose hydrophobic domains, singular β-strands and IDRs on their surface, interacting with proteins of various functions, often IDRs themselves and involved in key cellular processes (Olzscha et al., 2011) (see Table 2.3).

2.4.1 Systemic Proteinopathies

Amyloidoses are diseases with various clinical manifestations in which amyloid structures accumulate in tissues. Proteinopathies and amyloidoses can be discriminated, based on their localization, in systemic and localized diseases. One example of a systemic form of amyloidosis is β$_2$-microglobulin amyloidosis (Aβ$_2$M). High concentrations but also mutations of the protein (hereditary Aβ$_2$M-amyloidosis) can make it prone to misfolding and lead to consequent deposition in different connective tissues (Stoppini and Bellotti, 2015). Clinical signs are carpal-tunnel syndrome and other joint pains as well as bone cysts and consequent pathological fractures.

Multiple myeloma (MM) is a hematologic malignancy in which monoclonal plasma cells in the bone marrow produce high amounts of Igs uncontrollably. In the case of Bence-Jones protein (BJP)-MM, mostly Ig light chains accumulate, which can cause kidney damage (Umberto et al.,

TABLE 2.3

TABLE 2.3

Selected array of age-related conformational diseases.

Proteinopathy	Pathophysiology	Aggregated/ Misfolded protein	Age of onset	Clinical features
I. Systemic				
Systemic amyloidosis (Aβ₂M)	Impaired renal clearance/ dialysis therapy/ misfolding mutations cause accumulation of Aβ₂M	β₂-Microglobulin	Usually following chronic kidney disease	Arthralgia, bone cysts and pathological fractures
Amyloid transthyretin amyloidosis (ATTR)	TTR conversion into amyloid fibrils Familial amyloid polyneuropathy (FAP): Val30Met mutation Familial amyloid cardiomyopathy (FAC): Val122Ile mutation Wild-type ATTR amyloidosis (ATTRwt): acquired during aging	Transthyretin	50–60 years ATTRwt: >60 years	FAP: polyneuropathy, neurogenic bladder, autonomic dysfunction FAC: arrhythmia, heart failure ATTRwt: cardiological manifestations, arrhythmia, carpal tunnel syndrome
Multiple myeloma (MM)	Malignant degeneration of plasma cells, producing high amounts of Ig/Ig light chains	Igs/Ig light chains Bence Jones protein (BJP)	50–70 years	B-symptomatology, anemia, bone pains, osteolysis
IIa. Localized (Non–nervous system related)				
Primary localized cutaneous amyloidosis (PLCA)	Mutations in OSMR or IL31RA, amyloid deposits in skin, mechanisms unknown	IL-31 receptor	Adulthood	Pruritus, skin scratching, discolored skin
Cataract	Loss of transparency of eye lens, aggregates form through accumulation of oxidative stress	γ-Crystalline	>65 years	Blurred/Impaired vision
IIb. Localized (nervous system related)				
Alzheimer's disease (AD)	Inadequate clearance/ increased production of Aβ, hyperphosphorylation of tau	Aβ Tau	>65 years	Dementia, loss of cognitive functions
Frontotemporal dementia (FTD)/ Frontotemporal lobar degeneration (FTLD)	FTLD-tau: astrocytic plaques, corticobasal degeneration, Pick bodies FTLD-TDP: cytoplasmic inclusions, dystrophic neurites	Tau TDP-43	45–64 years "Early-onset dementia"	Dementia, prominent behavioral features, language deficit
Parkinson's disease (PD)	Degeneration of dopaminergic neurons, neuronal inclusions of aggregated α-synuclein	α-Synuclein	50–60 years	Bradykinesia, muscular rigidity, tremor

(*Continued*)

TABLE 2.3 (CONTINUED)

Proteinopathy	Pathophysiology	Aggregated/ Misfolded protein	Age of onset	Clinical features
Lewy body dementia (LBD)	Majority sporadic, rare mutations: SNCA, LRRK2 gene. Neuronal inclusions of α-synuclein, with or without Aβ Cerebrovascular pathology	α-Synuclein Aβ	65 years	Dementia, deficit in attention span, executive function Visual hallucinations, spontaneous parkinsonism
Multiple system atrophy (MSA)	Glial and neuronal inclusions of α-synuclein, demyelination, gliosis	α-Synuclein	55–65 years	MSA-P (parkinsonian): parkinsonism MSA-C(cerebellar): ataxia Autonomic failure, sleep disorders
Spinocerebellar ataxia (SCA)	PolyQ expansions in different proteins, aggregation in cerebellar neurons	Ataxin-1/2/3/7 TBP (TATA-binding protein) CACNA1A (calcium channel)	Adulthood Depending on SCA type	Ataxia, neuronal atrophy, specific symptoms in different SCA forms
Huntington's disease (HD)	Mutation in IT-15 gene PolyQ expansions and aggregation of Htt	Huntingtin (Htt)	30–50 years	Abnormal motor movements, personality change
Prion's disease	Infectious proteinaceous particles (prions) lead to misfolding/aggregation of other proteins	Prion protein	Around 60 years Latent onset (20-30 years after infection)	Psychopathological abnormalities, severe dementia, myoclonus
Amyotrophic Lateral Sclerosis (ALS)	Aberrant modification TDP-43 protein Mutation in SOD1 gene	TDP-43 Cu/Zn superoxide dismutase (SOD1)	Familial: 50 years Other: 60–80 years	Progressive loss of muscular strength, muscle cramps
Familial neurohypophyseal diabetes insipidus (FNDI)	Mutation in AVP gene	Neurophysin II	N/A	Polydipsia, polyuria

2016). BJPs can form aggregates like amyloid fibrils and precipitate in tissues. The underlying mechanisms are not fully elucidated, although some mutations have been found to change the aggregation propensity of the proteins (Timchenko and Timchenko, 2018). Clinical features of MM can encompass fatigue, anemia, skeletal pain as well as osteolysis that results in hypercalcemia. One therapeutic strategy includes administration of proteasome inhibitors, which lead to an abundance of misfolded Ig within the mutated plasma cells due to their high production rate, resulting in ER-stress and subsequent apoptosis (Gandolfi et al., 2017).

2.4.2 Localized Proteinopathies

Proteinopathies can also manifest in specific cell types and therefore affect only distinct organ systems.

2.4.2.1 Non-nervous System-Related Diseases

A prominent example of a localized non-nervous system-related and age-dependent protein-misfolding disease is cataract. The crystalline proteins within the eye lens serve to transmit and focus light into a bundle, in order to be projected onto the retina at the back of the eye. During aging, different toxic agents accumulate, in particular through UV radiation, oxidation and deamination which make crystalline proteins prone to misfolding and aggregation (Moreau and King, 2012). The protein solution loses its transparency, the lens becomes cloudy and vision is impaired.

2.4.2.2 Nervous System-Related Diseases

An abundance of proteinopathies manifests in the nervous system with increasing age. With little or no ability to regenerate or replace affected cells, the capacity of the nervous system to cope with increasing amounts of misfolded or aggregated species is reached with time leading to disease onset (Cenini et al., 2019).

Alzheimer's disease (AD) is one of the most common age-related neurodegenerative diseases, characterized by progressive cognitive impairment and dementia. Major histological manifestations include senile plaques consisting of pathological forms of the Aβ peptide in the extracellular matrix and misfolded tau species arranged in intracellular neurofibrillary tangles (NFT) (Lane et al., 2018). Aβ is produced through cleavage of membrane-associated amyloid precursor protein (APP) by β- and γ-secretases. An inadequate clearance of the peptide or mutations in either APP or APP cleaving secretase complexes can trigger accumulation and aggregation of Aβ and amyloid plaque formation (Gouras et al., 2015). Oligomeric Aβ peptides have been found to induce neuroinflammation, mitochondrial dysfunction and oxidative damage, as well as influence synaptic efficacy and kinase/phosphatase activities (Lane et al., 2018). In addition, hyperphosphorylation of microtubule-associated tau protein occurs and results in misfolding and the development of NFT (Iqbal et al., 2009).

Another highly prevalent neurodegenerative disorder is PD, which is predominantly characterized by motor deficits including bradykinesia, muscular rigidity and resting tremor. Pathological hallmarks include the degeneration of dopaminergic neurons in the *substantia nigra* and intracellular inclusions called Lewy bodies (LB) and Lewy neurites (LN). These formations consist primarily of misfolded or aggregated α-synuclein, which is a presynaptic protein considered an IDP (Benskey et al., 2016). Mutations in the SNCA

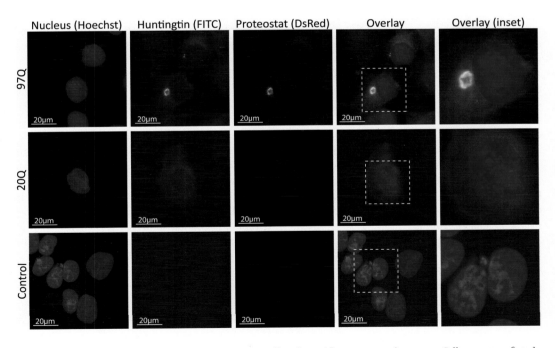

Figure 2.5 **Detection of aggregated Htt in human U2OS cells using epifluorescence microscopy.** Cells were transfected with HA-tagged exon 1 Htt constructs expressing different polyQ stretch lengths. The cells were stained using an HA antibody (green), Hoechst (blue) and Proteostat (red), a dye that binds aggregates. Cells transfected with the 97Q construct displayed Htt aggregates while the 20Q-transfected cells had an even distribution of HA-tagged Htt without aggregates.

gene, which encodes for α-synuclein, lead to a family of diseases called synucleinopathies (Stefanis, 2012). Synucleinopathy-related toxicity may be caused by impairment of the UPS (Kanaan and Manfredsson, 2012), mechanical damage of cellular compartments due to the involvement of α-synuclein in many cellular pathways including synaptic vesicle trafficking and mitochondrial dysfunction (Breydo et al., 2012).

Another late-onset proteinopathy, causing abnormal motor movement, depression, personality change and early death is Huntington's disease (HD). It is the result of a mutation of the HTT gene, leading to the extension of CAG repeats and a consequential expanded polyglutamine (polyQ) stretch in the encoded protein huntingtin (Htt). PolyQ expansions are considered pathological at >35 repeats (Shacham et al., 2019) while the CAG-repeat length determines the age of onset that is between 30 and 50 years. The mechanism underlying neurotoxicity in HD is the misfolding and aggregation of mutated huntingtin due to its polyQ expansion (Figure 2.5).

Amyotrophic lateral sclerosis (ALS) is one of the most common motor neuron diseases in adults. A determining clinical feature is the progressive muscle weakness leading to loss of motor movement and inevitably to respiratory failure (Scotter et al., 2015). The TART-DNA binding protein 43 (TDP-43) shows aberrant modifications in 97% of ALS cases and can be found aggregated in ubiquitinated inclusions in the brain and spinal cord of ALS, as well as frontotemporal lobar degeneration (FLD) patients (Prasad et al., 2019). In some cases, the superoxide dismutase-1 (SOD1) gene is mutated and leads to SOD1 accumulation in inclusions. Although there is no definite data on how TDP-43 or SOD1 might affect neurotoxicity in ALS, it is suggested that a combination of toxic gain and loss of function might be crucial (Mackenzie et al., 2007).

Prion diseases represent a group of neurodegenerative diseases where protein-only infectious agents cause fatal brain damage. PrP^{Sc} are misfolded species that form through conformational changes of the normally folded protein PrP^{C} (Collinge, 2016). PrP^{Sc} forms perineural plaques, causing a typical pathological pattern, also known as spongiform encephalopathy. This process is amplified, because PrP^{Sc} can cause misfolding and conformational changes in PrP^{C} and eventually in other proteins. This so-called templating plays a

key role in the prion hypothesis, which suggests that the same processes could underlie the pathological mechanisms of other neurodegenerative diseases (Soto and Pritzkow, 2018). Specifically, it has been hypothesized that misfolded proteins can propagate from cell to cell and initiate templating in neighboring cells (Guo and Lee, 2011).

2.5 CURRENT AND FUTURE TREATMENTS

Although there have been many discoveries in the field of age-related proteinopathies, development of efficient therapy strategies has not been entirely successful. Nonetheless, there have been several approaches particularly focusing on modulating PQC systems because of their major role in disease development during aging (Kulka et al., 2020). It is important to note that not only reduction of protein aggregation but also management of toxic oligomeric species should be the target of a therapy (see Table 2.4).

Proteostasis regulators (PRs) are molecules that can interfere with PQC functions and signaling pathways. They can modulate cellular response mechanisms like the cytoplasmic HSR and ER-associated UPR through a number of different mechanisms (Muntau et al., 2014).

Modulation of HSFs has been shown to reduce levels of aggregated proteins and restore proteostasis in several disease models. Non-steroidal anti-inflammatory drugs (NSAIDs), for example, can induce HSF-1 or ensure complete activation of HSR under stress conditions (Jurivich et al., 1992).

Another class of PRs is Hsp90 inhibitors like geldanamycin and its derivatives, which are presently used in cancer therapy. Shifting the conformation of Hsp90 from an ATP-binding to an ADP-binding complex, its chaperone function is inhibited and its association with the UPS is enabled. This leads to disassociation and degradation of Hsp90 client proteins, including steroid receptors, tyrosine kinases and HSF-1, which can initiate the HSR (Sittler et al., 2001).

Proteins that act as sensors of misfolded/unfolded species and conduct the UPR^{ER} are the inositol-requiring protein 1 (IRE1), the translation factor 6 (ATF6) and the protein kinase RNA (PKR)-like ER kinase (PERK). ATF6 activates genes involved in protein folding (i.e. ER-associated chaperones), PERK inhibits mRNA translation and IRE-1 upregulates the ERAD pathway and ER-associated chaperones via the X-box-protein 1

TABLE 2.4

Novel therapeutics in clinical trials affecting protein-misfolding pathologies.

Drug category	Drug/Substance	Phase	Mode of action	Application	References
HSP inhibitors	1) XL888 2) Vemurafenib	I	1) Hsp90 inhibitor 2) BRAF kinase inhibitor	BRAF-mutated stage III/ IV melanoma	NCT01657591
	Onalespib	I	Hsp90 inhibitor	Advanced triple negative breast cancer	NCT02474173
	PU-AD	II	Small molecule epichaperome inhibitor	ALS AD	NCT04505358 NCT04311515
Proteasome inhibitors	Ixazomib	I/II	Proteasome inhibitor	Light chain amyloidosis	NCT03236792
Proteostasis regulators	AMX0035	II	Combination: tauroursodeoxycholic acid (TUDCA): inhibits ER/ mitochondrial stress-mediated apoptosis, reduces formation of ROS Sodium phenylbutyrate (PB): chemical chaperone, histone deacetylase inhibitor	ALS	NCT03488524
NSAID	ALZT-OP1	III	Combination: ibuprofen (NSAID) Cromolyn (mast cell stabilizer)	AD	NCT02547818
ASO	RG6042	I	Antisense oligonucleotide (ASO), reduces concentration of Htt mRNA	HD	NCT04000594
Small kinetic stabilizer	AG10	III	Transthyretin stabilizer	ATTR – cardiomyopathy Transthyretin amyloid polyneuropathy (ATTR-PN)	NCT03860935 NCT04418024
Antibodies	Lecanemab (BAN2401)	III	Monoclonal anti-Aβ protofibril antibody	AD	NCT04468659
	PRX004	I	Monoclonal anti-amyloid transthyretin antibody	ATTR	NCT03336580

(Ron and Walter, 2007). Small molecule IRE-1 inhibitors have been discovered to have cytotoxic effects in hematological pathologies. STF-083010, for example, inhibits IRE-1 endonuclease activity and increases the inherent ER stress in MM cells, which translates into cytotoxicity (Papandreou et al., 2011).

Complementing the aforementioned approaches, chemical chaperones can aid to sustain protein stability by broadening the free energy gap between the partially folded and native states of a protein, therefore decreasing the number of unfolded species of aggregation-prone proteins (Hekmatimoghaddam et al., 2017). Additionally, pharmacological chaperones are specific small

molecules that can bind to proteins via van der Waals or electrostatic forces and through hydrogen bonds and can facilitate protein folding and trafficking (Beerepoot et al., 2017).

2.6 CONCLUSIONS

Protein folding is an essential process in virtually all living systems. It has become clear over the past decade that aging has an adverse effect on protein folding toward the native structure and the capacity of PQC systems, including the system of molecular chaperones. One hypothesis suggests that the decline of crucial PQC components leads to the occurrence of misfolded cellular proteins. Therefore, protein-misfolding diseases are likely to be increasing in aging societies. There are many attempts to cure these age-related diseases; however, most of them have failed so far. It is conceivable that the critical time point for disease intervention has already passed when protein misfolding and aggregation occur, and an upregulation of the capacity of PQC systems has to take place far earlier. That means genetic testing has to be increased and the likelihood of a disease onset needs to be determined. Besides ethical considerations, stratification of patients and assessing the individual risk score for a disease needs the establishment of reliable biomarkers or even molecular signatures to initiate personalized treatment.

AUTHOR CONTRIBUTIONS

Conceptualization, H.O.; writing – original draft preparation, J.-E. R., D.P. and H.O.; writing – review and editing, J.-E. R., D.P. and H.O; visualization J.-E. R. and D.P.; supervision, H.O.; project administration, H.O.; funding acquisition, H.O. All authors have read and agreed to the manuscript.

FUNDING

This work was funded by the Deutsche Forschungsgemeinschaft (DFG, Germany, RTG 2155, ProMoAge).

ACKNOWLEDGMENT

The authors acknowledge Thorsten Pfirrmann for critical reading of the manuscript.

CONFLICTS OF INTEREST

The authors declare no conflict of interest.

REFERENCES

Anfinsen, C. B. 1973. Principles that govern the folding of protein chains. *Science*, 181(4096), 223–230.

Anfinsen, C. B., Haber, E., Sela, M., et al. 1961. The kinetics of formation of native ribonuclease during oxidation of the reduced polypeptide chain. *Proc Natl Acad Sci U S A*, 47, 1309–1314. doi:10.1073/pnas.47.9.1309

Beerepoot, P., Nazari, R., & Salahpour, A. 2017. Pharmacological chaperone approaches for rescuing GPCR mutants: current state, challenges, and screening strategies. *Pharmacol Res*, 117, 242–251. doi:10.1016/j.phrs.2016.12.036

Bensky, M. J., Perez, R. G., & Manfredsson, F. P. 2016. The contribution of alpha synuclein to neuronal survival and function – implications for Parkinson's disease. *J Neurochem*, 137(3), 331–359. doi:10.1111/jnc.13570

Benson, M. D., Buxbaum, J. N., Eisenberg, D. S., et al. 2018. Amyloid nomenclature 2018: recommendations by the International Society of Amyloidosis (ISA) nomenclature committee. *Amyloid*, 25(4), 215–219. doi:10.1080/13506129.2018.1549825

Breydo, L., Wu, J. W., & Uversky, V. N. 2012. α-Synuclein misfolding and Parkinson's disease. *Biochim Biophys Acta – Mol Basis Dis*, 1822(2), 261–285. doi:10.1016/j.bbadis.2011.10.002

Calderwood, S. K., Murshid, A., & Prince, T. 2009. The shock of aging: molecular chaperones and the heat shock response in longevity and aging – a mini-review. *Gerontology*, 55(5), 550–558. doi:10.1159/000225957

Cenini, G., Lloret, A., & Cascella, R. 2019. Oxidative stress in neurodegenerative diseases: from a mitochondrial point of view. *Oxidat Med Cell Longev*, 2019, 2105607. doi:10.1155/2019/2105607

Chiti, F., & Dobson, C. M. 2006. Protein misfolding, functional amyloid, and human disease. *Annu Rev Biochem*, 75, 333–366. doi:10.1146/annurev.biochem.75.101304.123901

Christianson, J. C., Olzmann, J. A., Shaler, T. A., et al. 2011. Defining human ERAD networks through an integrative mapping strategy. *Nat Cell Biol*, 14(1), 93–105. doi:10.1038/ncb2383

Clark, P. L. 2004. Protein folding in the cell: reshaping the folding funnel. *Trends Biochem Sci*, 29(10), 527–534. doi:10.1016/j.tibs.2004.08.008

Cobley, J. N., Fiorello, M. L., & Bailey, D. M. 2018. 13 Reasons why the brain is susceptible to oxidative stress. *Redox Biol*, 15, 490–503. doi:10.1016/j.redox.2018.01.008

Collinge, J. 2016. Mammalian prions and their wider relevance in neurodegenerative diseases. *Nature*, 539(7628), 217–226. doi:10.1038/nature20415

Dill, K. A. 1999. Polymer principles and protein folding. *Protein Sci*, 8(6), 1166–1180. doi:10.1110/ps.8.6.1166

Dyson, H. J., & Wright, P. E. 2005. Intrinsically unstructured proteins and their functions. *Nat Rev: Mol Cell Biol*, 6(3), 197–208. doi:10.1038/nrm1589

Ellis, R. J. 2001. Macromolecular crowding: an important but neglected aspect of the intracellular environment. *Curr Opin Struct Biol*, 11(1), 114–119. doi:10.1016/s0959-440x(00)00172-x

Etchells, S. A., & Hartl, F. U. 2004. The dynamic tunnel. *Nat Struct Mol Biol*, 11(5), 391–392. doi:10.1038/nsmb0504-391

Gámez, A., Yuste-Checa, P., Brasil, S., et al. 2018. Protein misfolding diseases: prospects of pharmacological treatment. *Clin Genet*, 93(3), 450–458. doi:10.1111/cge.13088

Gandolfi, S., Laubach, J. P., Hideshima, T., et al. 2017. The proteasome and proteasome inhibitors in multiple myeloma. *Cancer Metastasis Rev*, 36(4), 561–584. doi:10.1007/s10555-017-9707-8

Gidalevitz, T., Ben-Zvi, A., Ho, K. H., et al. 2006. Progressive disruption of cellular protein folding in models of polyglutamine diseases. *Science*, 311(5766), 1471–1474. doi:10.1126/science.1124514

Goor, F. V., Straley, K. S., Cao, D., et al. 2006. Rescue of ΔF508-CFTR trafficking and gating in human cystic fibrosis airway primary cultures by small molecules. *Am J Physiol: Lung Cell Mol Physiol*, 290(6), L1117–L1130. doi:10.1152/ajplung.00169.2005

Gouras, G. K., Olsson, T. T., & Hansson, O. 2015. β-Amyloid peptides and amyloid plaques in Alzheimer's disease. *Neurotherapeutics*, 12(1), 3–11. doi:10.1007/s13311-014-0313-y

Gregersen, N., Bross, P., Vang, S., et al. 2006. Protein misfolding and human disease. *Annu Rev Genomics Hum Genet*, 7, 103–124. doi:10.1146/annurev.genom.7.080505.115737

Guo, J. L., & Lee, V. M. 2011. Seeding of normal Tau by pathological Tau conformers drives pathogenesis of Alzheimer-like tangles. *J Biol Chem*, 286(17), 15317–15331. doi:10.1074/jbc.M110.209296

Hartl, F. U. 1996. Molecular chaperones in cellular protein folding. *Nature*, 381(6583), 571–579. doi:10.1038/381571a0

Hayer-Hartl, M., Bracher, A., & Hartl, F. U. 2016. The GroEL-GroES chaperonin machine: a nano-cage for protein folding. *Trends Biochem Sci*, 41(1), 62–76. doi:10.1016/j.tibs.2015.07.009

Hekmatimoghaddam, S., Zare-Khormizi, M. R., & Pourrajab, F. 2017. Underlying mechanisms and chemical/biochemical therapeutic approaches to ameliorate protein misfolding neurodegenerative diseases. *Biofactors*, 43(6), 737–759. doi:10.1002/biof.1264

Ignatova, Z., & Gierasch, L. M. 2006. Inhibition of protein aggregation in vitro and in vivo by a natural osmoprotectant. *Proc Natl Acad Sci U S A*, 103(36), 13357–13361. doi:10.1073/pnas.0603772103

Iqbal, K., Liu, F., Gong, C.-X., et al. 2009. Mechanisms of tau-induced neurodegeneration. *Acta Neuropathol*, 118(1), 53–69. doi:10.1007/s00401-009-0486-3

Jurivich, D. A., Sistonen, L., Kroes, R. A., et al. 1992. Effect of sodium salicylate on the human heat shock response. *Science*, 255(5049), 1243–1245. doi:10.1126/science.1546322

Kanaan, N. M., & Manfredsson, F. P. 2012. Loss of functional alpha-synuclein: a toxic event in Parkinson's disease? *J Parkinson's Dis*, 2(4), 249–267. doi:10.3233/JPD-012138

Kaushik, S., & Cuervo, A. M. 2012. Chaperone-mediated autophagy: a unique way to enter the lysosome world. *Trends Cell Biol*, 22(8), 407–417. doi:10.1016/j.tcb.2012.05.006

Kendrew, J. C., Bodo, G., Dintzis, H. M., et al. 1958. A three-dimensional model of the myoglobin molecule obtained by x-ray analysis. *Nature*, 181(4610), 662–666. doi:10.1038/181662a0

Kerner, M. J., Naylor, D. J., Ishihama, Y., et al. 2005. Proteome-wide analysis of chaperonin-dependent protein folding in Escherichia coli. *Cell*, 122(2), 209–220. doi:10.1016/j.cell.2005.05.028

Knowles, T. P., Vendruscolo, M., & Dobson, C. M. 2014. The amyloid state and its association with protein misfolding diseases. *Nat Rev: Mol Cell Biol*, 15(6), 384–396. doi:10.1038/nrm3810

Kulka, L. A. M., Fangmann, P. V., Panfilova, D., et al. 2020. Impact of HDAC inhibitors on protein quality control systems: consequences for precision medicine in malignant disease. *Front Cell Dev Biol*, 8, 425. doi:10.3389/fcell.2020.00425

Lane, C. A., Hardy, J., & Schott, J. M. 2018. Alzheimer's disease. *Eur J Neurol*, 25(1), 59–70. doi:10.1111/ene.13439

Lashuel, H. A., Petre, B. M., Wall, J., et al. 2002. Alpha-synuclein, especially the Parkinson's disease-associated mutants, forms pore-like annular and tubular protofibrils. *J Mol Biol*, 322(5), 1089–1102. doi:10.1016/s0022-2836(02)00735-0

Levinthal, C. 1969. How to fold graciously. In P. Debrunner, J. C. M. Tsibris, & E. Münck (Eds.), *Proceedings of a Meeting held at Allerton House* (pp. 22–24). University of Illinois, Urbana. Curr Opin Struct Biol.

Levy, E., El Banna, N., Baille, D., et al. 2019. Causative links between protein aggregation and oxidative stress: a review. *Int J Mol Sci*, 20(16). doi:10.3390/ijms20163896

Liu, Y., & Eisenberg, D. 2002. 3D domain swapping: as domains continue to swap. *Protein Sci*, 11(6), 1285–1299. doi:10.1110/ps.0201402

Mackenzie, I. R., Bigio, E. H., Ince, P. G., et al. 2007. Pathological TDP-43 distinguishes sporadic amyotrophic lateral sclerosis from amyotrophic lateral sclerosis with SOD1 mutations. *Ann Neurol*, 61(5), 427–434. doi:10.1002/ana.21147

Mayor, U., Guydosh, N. R., Johnson, C. M., et al. 2003. The complete folding pathway of a protein from nanoseconds to microseconds. *Nature*, 421(6925), 863–867. doi:10.1038/nature01428

Moreau, K. L., & King, J. A. 2012. Protein misfolding and aggregation in cataract disease and prospects for prevention. *Trends Mol Med*, 18(5), 273–282. doi:10.1016/j.molmed.2012.03.005

Muntau, A. C., Leandro, J., Staudigl, M., et al. 2014. Innovative strategies to treat protein misfolding in inborn errors of metabolism: pharmacological chaperones and proteostasis regulators. *J Inherit Metab Dis*, 37(4), 505–523. doi:10.1007/s10545-014-9701-z

New, M., Olzscha, H., Liu, G., et al. 2013. A regulatory circuit that involves HR23B and HDAC6 governs the biological response to HDAC inhibitors. *Cell Death Differ*, 20(10), 1306–1316. doi:10.1038/cdd.2013.47

Nillegoda, N. B., Kirstein, J., Szlachcic, A., et al. 2015. Crucial HSP70 co-chaperone complex unlocks metazoan protein disaggregation. *Nature*, 524(7564), 247–251. doi:10.1038/nature14884

Nollen, E. A., & Morimoto, R. I. 2002. Chaperoning signaling pathways: molecular chaperones as stress-sensing 'heat shock' proteins. *J Cell Sci*, 115(Pt 14), 2809–2816.

Olzscha, H. 2019. Posttranslational modifications and proteinopathies: how guardians of the proteome are defeated. *Biol Chem*, 400(7), 895–915. doi:10.1515/hsz-2018-0458

Olzscha, H., Fedorov, O., Kessler, B. M., et al. 2017. CBP/p300 bromodomains regulate amyloid-like protein aggregation upon aberrant lysine acetylation. *Cell Chem Biol*, 24(1), 9–23. doi:10.1016/j.chembiol.2016.11.009

Olzscha, H., Schermann, S. M., Woerner, A. C., et al. 2011. Amyloid-like aggregates sequester numerous metastable proteins with essential cellular functions. *Cell*, 144(1), 67–78. doi:10.1016/j.cell.2010.11.050

Papandreou, I., Denko, N. C., Olson, M., et al. 2011. Identification of an Ire1alpha endonuclease specific inhibitor with cytotoxic activity against human multiple myeloma. *Blood*, 117(4), 1311–1314. doi:10.1182/blood-2010-08-303099

Pauling, L., Corey, R. B., & Branson, H. R. 1951. The structure of proteins: two hydrogen-bonded helical configurations of the polypeptide chain. *Proc Natl Acad Sci U S A*, 37(4), 205–211. doi:10.1073/pnas.37.4.205

Prasad, A., Bharathi, V., Sivalingam, V., et al. 2019. Molecular mechanisms of TDP-43 misfolding and pathology in amyotrophic lateral sclerosis. *Front Mol Neurosci*, 12, 25–25. doi:10.3389/fnmol.2019.00025

Prusiner, S. B. 1982. Novel proteinaceous infectious particles cause scrapie. *Science*, 216(4542), 136–144. doi:10.1126/science.6801762

Rockenstein, E., Nuber, S., Overk, C. R., et al. 2014. Accumulation of oligomer-prone alpha-synuclein exacerbates synaptic and neuronal degeneration in vivo. *Brain*, 137(Pt 5), 1496–1513. doi:10.1093/brain/awu057

Ron, D., & Walter, P. 2007. Signal integration in the endoplasmic reticulum unfolded protein response. *Nat Rev Mol Cell Biol*, 8(7), 519–529. doi:10.1038/nrm2199

Sarge, K. D., Murphy, S. P., & Morimoto, R. I. 1993. Activation of heat shock gene transcription by heat shock factor 1 involves oligomerization, acquisition of DNA-binding activity, and nuclear localization and can occur in the absence of stress. *Mol Cell Biol*, 13(3), 1392–1407. doi:10.1128/mcb.13.3.1392

Schapira, A. H., & Gegg, M. 2011. Mitochondrial contribution to Parkinson's disease pathogenesis. *Parkinson's Dis*, 2011, 159160. doi:10.4061/2011/159160

Scotter, E. L., Chen, H.-J., & Shaw, C. E. 2015. TDP-43 proteinopathy and ALS: insights into disease mechanisms and therapeutic targets. *Neurotherapeutics*, 12(2), 352–363. doi:10.1007/s13311-015-0338-x

Shacham, T., Sharma, N., & Lederkremer, G. Z. 2019. Protein misfolding and ER stress in Huntington's disease. *Front Mol Biosci*, 6, 20. doi:10.3389/fmolb.2019.00020

Shirdel, S. A., & Khalifeh, K. 2019. Thermodynamics of protein folding: methodology, data analysis and interpretation of data. *Eur Biophys J*, 48(4), 305–316. doi:10.1007/s00249-019-01362-7

Sittler, A., Lurz, R., Lueder, G., et al. 2001. Geldanamycin activates a heat shock response and inhibits huntingtin aggregation in a cell culture model of Huntington's disease. *Hum Mol Genet*, 10(12), 1307–1315. doi:10.1093/hmg/10.12.1307

Smith, M. C., Scaglione, K. M., Assimon, V. A., et al. 2013. The E3 ubiquitin ligase CHIP and the molecular chaperone Hsc70 form a dynamic, tethered complex. *Biochemistry*, 52(32), 5354–5364. doi:10.1021/bi4009209

Soto, C., & Pritzkow, S. 2018. Protein misfolding, aggregation, and conformational strains in neurodegenerative diseases. *Nat Neurosci*, 21(10), 1332–1340. doi:10.1038/s41593-018-0235-9

Stefanis, L. 2012. Alpha-synuclein in Parkinson's disease. *Cold Spring Harb Perspect Med*, 2(2), a009399. doi:10.1101/cshperspect.a009399

Stoppini, M., & Bellotti, V. 2015. Systemic amyloidosis: lessons from beta2-microglobulin. *The Journal of biological chemistry*, 290(16), 9951–9958. doi:10.1074/jbc.R115.639799

Timchenko, M. A., & Timchenko, A. A. 2018. Influence of a single point mutation in the constant domain of the Bence-Jones protein bif on its aggregation properties. *Biochemistry (Mosc)*, 83(2), 107–118. doi:10.1134/S0006297918020037

Umberto, B., Francesca, G., Eleonora, T., et al. 2016. Evaluation of screening method for Bence Jones protein analysis. *Clin Chem Lab Med*, 54(11), e331–e333. doi:10.1515/cclm-2015-1239

Varela, A. E., England, K. A., & Cavagnero, S. 2019. Kinetic trapping in protein folding. *Protein Eng Des Sel*, 32(2), 103–108. doi:10.1093/protein/gzz018

Walter, P., & Ron, D. 2011. The unfolded protein response: from stress pathway to homeostatic regulation. *Science*, 334(6059), 1081–1086. doi:10.1126/science.1209038

Wang, L., Wang, X., & Wang, C. C. 2015. Protein disulfide-isomerase, a folding catalyst and a redox-regulated chaperone. *Free Radic Biol Med*, 83, 305–313. doi:10.1016/j.freeradbiomed.2015.02.007

Xiang, W., Menges, S., Schlachetzki, J. C., et al. 2015. Posttranslational modification and mutation of histidine 50 trigger alpha synuclein aggregation and toxicity. *Mol Neurodegener*, 10, 8. doi:10.1186/s13024-015-0004-0

CHAPTER THREE

Transcriptional Regulation of Proteostatic Mechanisms

Marianna Kapetanou, Sophia Athanasopoulou, Efstathios S. Gonos

CONTENTS

3.1 INTRODUCTION

A functional proteome is fundamental for all living systems. Eukaryotic cells have evolved highly sophisticated protein quality-control pathways to secure proteome integrity, which are referred as the proteostasis network (PN). This network dynamically coordinates protein synthesis, folding, modification and degradation, and responds to proteotoxic stress to either rescue or degrade misfolded or non-native polypeptides. The PN dynamically adapts to meet cellular requirements and to prevent proteostasis imbalance through integrated processes involving the coordinated transcriptional regulation of protein folding and degradation in response to diverse signals. This chapter focuses on the transcriptional regulation of the PN under physiological or proteostasis-challenging conditions.

3.2 TRANSCRIPTIONAL REGULATION OF PROTEOME INTEGRITY

In response to proteotoxic stress, three conserved stress response pathways, the heat shock response (HSR), the oxidative stress response (OxR) and the unfolded protein response (UPR), induce protective transcriptional responses to curate the basal function and composition of the proteome and to assist to overcome the challenges of protein folding and solubility (Brehme et al., 2019; Miles et al., 2019) (Figure 3.1).

3.2.1 The Heat Shock Response

The HSR is a well-characterized, rapid and transient gene expression program that is activated when misfolded proteins accumulate in the cytosol or nucleus (Miles et al., 2019). The HSR prevents or reverses protein misfolding through the transcriptional induction of molecular chaperones. The expression of chaperone genes is mainly regulated by the highly conserved transcription factor (TF) heat shock factor 1 (HSF1). In total nine human HSF isoforms are known; however, we will focus here on HSF1. Under basal conditions, HSF1 is maintained in an inert monomeric state in the cytosol through binding to various heat shock proteins (HSPs)/chaperones including HSP90, HSP70 and HSP40. Upon proteotoxic stress

DOI: 10.1201/9781003048138-3

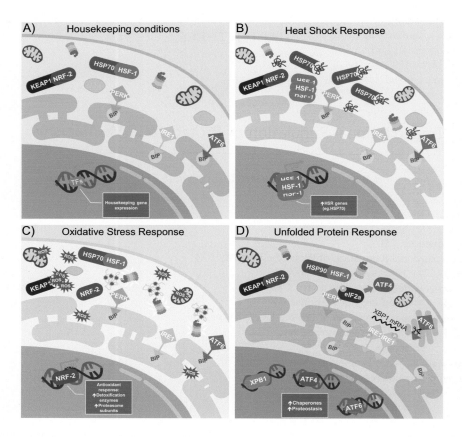

Figure 3.1 Schematic diagram of (A) housekeeping conditions, in comparison to the transcriptional activation of (B) heat shock response, (C) oxidative stress response and (D) unfolded protein response.

conditions, the chaperones within the repressive HSF1-containing multi-chaperone complexes preferentially bind to the unfolded proteins and dissociate from HSF1, enabling it to undergo phosphorylation, homotrimerization and translocation into the nucleus. HSF1 trimeric forms bind to a consensus heat shock element (HSE) located at the promoter regions of target genes promoting rapid upregulation of HSR genes, increased protein folding and proteostasis restoration (Labbadia and Morimoto, 2015; Miles et al., 2019; Sims et al., 2011). Chaperones, including HSP70, are among the most significantly upregulated heat shock genes, as they are required in stoichiometric ratios relative to the unfolded client proteins (Akerfelt et al., 2010). The activity of the HSF-1 trimer is further regulated by post-transcriptional modifications, like sumoylation (Hietakangas et al., 2003) and acetylation (Westerheide et al., 2009). Once the stress is resolved, excessive HSP70 exerts a negative feedback on HSF1 by binding to a site

in its transactivation domain to attenuate its DNA-binding affinity and to restore the system to its original resting state (Kmiecik et al., 2020).

3.2.2 Oxidative Stress Response

Under normal conditions, reactive oxygen species (ROS) or reactive nitrogen species (RNS) are generated as natural by-products of many essential biological processes. Besides the physiological participation of ROS to several signaling pathways, an excessive elevation of their levels may have deleterious effects on cellular components (Gonos et al., 2018). Consequently, the exposure to high ROS levels triggers the expression of a wide array of genes encoding antioxidant proteins and detoxifying enzymes to offer protection to proteins and other macromolecules against oxidative damage (Niforou et al., 2014).

A key TF coordinating OxR is the nuclear factor erythroid 2-related factor 2 (Nrf2 in mammals,

SKN-1 in *Caenorhabditis elegans*). Under basal conditions, Nrf2 is retained in the cytoplasm, bound to the endogenous inhibitor Kelch-like ECH-associated protein1 (Keap1), which targets Nrf2 for ubiquitination and degradation by the proteasome (Kansanen et al., 2013). The modification of cysteine residues of Keap1 by oxidants or electrophiles acts as a redox sensor and alters the interaction of Keap1 with Nrf2 (Holland and Fishbein, 2010). As such, oxidative stress interrupts the Keap1-Nrf2 complex, leading to the translocation of Nrf2 to the nucleus. There it forms heterodimers with other TFs such as c-Jun and small musculoaponeurotic fibrosarcoma (Maf) proteins and binds to the antioxidant response elements (ARE), on the promoters of genes with a crucial role in redox homeostasis and cytoprotection during oxidative stress.

Nrf2 acts on AREs to stimulate the expression of more than 200 genes involved in the cellular antioxidant defense such as the detoxification enzymes NAD(P)H quinone oxidoreductase 1 (NQO1), heme-oxygenase 1 (HMOX1), glutamate-cysteine ligase (GCL), glutathione S transferases (GSTs), extracellular superoxide dismutase, glutamate-6-phosphate-dehydrogenase cyclooxygenase-2 (COX-2) and inducible nitric oxide synthase (iNOS), ultimately protecting the cells from damage (Niforou et al., 2014; Raghunath et al., 2018). The different interacting partners confer functional complexity in the Keap1/Nrf2/ARE pathway either homodimerize or heterodimerize to bind the AREs. c-Jun mainly acts as transcriptional activator while the small Mafs and c-Myc suppress gene transcription after Nrf2 binding (Kansanen et al., 2013). Nrf2 activity is also regulated at the transcriptional level by phosphorylation (Li et al., 2019).

A similar activation pattern is observed also in another key regulatory molecule of OxR, the nuclear factor-kappa B (NF-κB). NF-κB is normally sequestered in the cytoplasm of non-stimulated cells and it translocates into the nucleus to regulate effector gene expression. A family of inhibitory proteins, the IκBs, binds to NF-κB and masks its nuclear localization signal domain, thus controlling the translocation of NF-κB. Extracellular stimuli perturbing redox balance result in rapid phosphorylation, ubiquitination and proteolytic degradation of IκBs. This process releases NF-κB and enables it to translocate to the nucleus where it regulates the transcription of target genes. Antioxidant targets induced by NF-κB include manganese superoxide dismutase (MnSOD),, thioredoxin, glutathione S-transferase, metallothionein-3, NAD(P)H dehydrogenase [quinone]-1, HMOX1 and glutathione peroxidase-1. Alternatively, the activation of NF-κB signaling can exert pro-oxidant effects by induction of genes such as NADPH oxidase NOX2 subunit gp91phox (Lingappan, 2018) or by repression of the peroxisome proliferator-activated receptor-gamma coactivator 1-alpha (PGC-1α), which is well known for its ability to reduce ROS production and promote detoxification during inflammation (Rius-Pérez et al., 2020).

3.2.3 Unfolded Protein Response

The endoplasmic reticulum (ER) is a main site for protein folding and maturation in eukaryotes. The accumulation of unfolded proteins in the ER may result in a proteostasis imbalance, known as ER stress. The UPR is a signaling pathway that regulates ER folding capacity to meet the cellular folding requirements and to restore protein homeostasis (Adams et al., 2019). In metazoans, the three UPR^{ER} effectors, namely the activating TF 6 (ATF6), protein kinase R-like endoplasmic reticulum kinase (PERK) and inositol-requiring kinase 1 (IRE1), act concomitantly to increase ER folding capacity and to counteract ER stress by triggering the expression of several chaperones and eventually by reducing protein load (Gardner et al., 2013). ATF6 is a transmembrane TF that in response to ER stress is transported from the ER to the Golgi and processed by site-1 and site-2 proteases, resulting in the 50-kDa p50ATF-6 fragment. This cytoplasmic fragment of ATF6 is then released from the membrane, translocates to the nucleus and activates the expression of target chaperone genes (Conn and Qian, 2011). PERK is a protein kinase that is autophosphorylated and activated when BiP dissociates from ER lumen in response to ER stress. PERK phosphorylates the α-subunit of eukaryotic initiation factor 2 (eIF2α), thereby switching off general protein synthesis and starting selected and stress-related protein production, such as the one of the activating transcription factor 4 (ATF4). ATF4 activates downstream target genes that exert either pro-survival or pro-apoptotic activities (Conn and Qian, 2011). Similarly, once released from BiP, IRE1 homodimerizes and trans-autophosphorylates, thus activating its endoribonuclease activity. Activated IRE1 processes the X-box binding protein 1 (XBP1) mRNA to produce a potent TF that promotes the expression of target genes that reduce the ER folding load and restore proteostasis (Hetz and Glimcher, 2009;). Under physiological conditions, ATF6, PERK and IRE1 are repressed though binding of their domains that are projected into the ER lumen by an HSP70 chaperone, known as BiP/GRP78. Upon conditions of increased misfolded protein levels in the ER lumen, BiP is titrated away from these stress sensors leading to their activation (Amin-Wetzel et al., 2017).

Similarly, to the ER lumen, the accumulation of damaged or misfolded proteins in the mitochondrial matrix initiates an unfolded protein response (UPR^{mt}) that alters the expression of mitochondrial genes encoded by nuclear DNA. Factors triggering the UPR^{mt} include mtDNA depletion, impaired mitochondrial protein quality control or oxidative phosphorylation dysregulation. Mitochondrial proteostasis can be disturbed by increased levels of ROS, which are generated from the electron transport chain and directly damage proteins. Mitochondrial stress activates the proteolytic activity of ClpXP, resulting in the transport of misfolded proteins to the cytoplasm by the peptide transporter HAF1. Upon such signal, activating transcription factor-associated with stress 1 (ATFS-1) is not able to enter mitochondria anymore and is guided back to the nucleus. There it triggers a transcriptional response to induce the expression of mitochondrial chaperones and to promote a functional protein-folding environment in the organelle, thus preventing further damage (Jensen and Jasper, 2014; Nargund et al., 2015).

3.3 TRANSCRIPTIONAL REGULATION OF PROTEIN DEGRADATION

3.3.1 Ubiquitin-Proteasome System

Clearance of short-lived proteins or soluble misfolded proteins is governed by the ubiquitin-proteasome system (UPS). Substrate recognition and degradation by the 26S/30S proteasome involves its conjugation with ubiquitin via an ATP-dependent formation of isopeptide bonds that engage three enzyme families: the E1 ubiquitin-activating enzymes, the E2 ubiquitin-conjugating enzymes and the E3 ubiquitin-protein ligases. The

selective substrate polyubiquitination will eventually lead to its recognition by the ubiquitin-binding proteasome subunits or related shuttle factors (Hjerpe et al., 2016; Marshall and Vierstra, 2019; Samant et al., 2018). Ubiquitin chains, precursors and substrate proteins also interact with another important UPS enzyme family, the deubiquitylating enzymes (DUBs), responsible for the removal of ubiquitin moieties from substrates targeted to the proteasome and their recycling. The 20S core proteasome represents the key degradation machinery in this system and exists in various isoforms, depending on differential compositions among its catalytic subunits, that may be either those of the constitutive proteasome (β1, β2 and β5) or their immune counterparts, expressed upon pro-inflammatory cytokine stimuli (β1i/LMP2 encoded by PSMB9, β2i/MECL1 encoded by PSMB10, β5i/LMP7 encoded by PSMB8). These isoforms are found either as free particles or associated with regulatory particles (RP), such as 19S RP, PA28αβ, PA28γ, PA200 and PI31 that control substrate entrance and degradation (Chondrogianni et al., 2014, 2015; Marshall and Vierstra, 2019). Apart from the ubiquitin-dependent proteasome degradation by the 26S/30S complexes, many proteins can be cleaved by the 20S proteasome in an ATP-independent manner, without preceding ubiquitination. Under oxidative stress conditions, the 26S proteasome partially dissociates into 20S proteasomes, which can degrade non-ubiquitinated oxidized proteins. The ubiquitin-independent recognition of oxidized proteins by 20S proteasomes is initiated by the loss of their secondary structure (Abi Habib et al., 2020). This type of cleavage requires the presence of unstructured regions in the amino acid sequences that provide interaction with the 20S proteasome and significantly contributes to the regulation of protein homeostasis (Baugh et al., 2009).

The mechanisms regulating proteasome expression respond to distinct intracellular and extracellular circumstances such as proteasomal dysfunction or inhibition, nutrient conditions, inflammatory stimuli and oxidative stress (Marshall and Vierstra, 2019; Motosugi and Murata, 2019). A proteasome regulatory network was firstly described a few decades ago in yeast. Upon attenuated proteasome function, RPN4, a C2H2-type zinc finger TF, becomes stabilized and translocates to the nucleus, where it binds to a hexameric consensus nucleotide sequence [(A/G) GTGGC], known as the proteasome-associated control element (PACE), present in the promoters of most proteasome subunit genes, along with some proteasome assembly chaperone genes (Xie and Varshavsky, 2001). The RPN4 protein has an extremely short half-life and is constantly degraded by the proteasome. Accumulation of RPN4 due to impaired proteasome degradation stimulates proteasome expression in a negative feedback circuit in which the same protein enhances proteolysis and is destroyed by the newly assembled active proteasome. Interestingly, RPN4 regulates and is regulated by several TFs related to HSR (HSF1), pandrug resistance (PDRs) and oxidative stress (YAP1), combining multiple roles in proteostasis (Boos et al., 2019). An analogous regulatory circuit exists in mammalian cells, where proteasome dysfunction can trigger an upturn in the expression of proteasome subunits. Nuclear factor erythroid 2-related factor 1 (NRF1, also known as NFE2L1) is a membrane protein normally located at the ER. The ubiquitin ligases HRD1, FBX7 and β-TRCP and the AAA+ ATPase p97/valosin-containing protein mediate the continuous retro-translocation of NRF1 to the cytosol (Northrop et al., 2020). Retro-translocated NRF1 is rapidly degraded by the 26S proteasome. In the presence of dysfunctional proteasomes, NRF1 is stabilized during retro-translocation, where it is cleaved by the aspartic protease and ubiquitin shuttling factor DNA damaged inducible 1 homolog 2 (DDI2; DDI1/VSM1 in yeast), a critical regulator of NRF1 activation (Dirac-Svejstrup et al., 2020). The resulting active form of NRF1 is deglycosylated by PNG1/NGLY1 and then it translocates to the nucleus, where it binds AREs to activate the transcription of its target genes, including those encoding proteasome subunits (Baird et al., 2017; Radhakrishnan et al., 2010; Steffen et al., 2010). In addition, RUVBL1 and TIP60, two subunits of the epigenetic regulator TIP60 complex, serve as chromatin modifiers that expose NRF1 target promoter regions, therefore providing access to the transcriptional machinery (Vangala and Radhakrishnan, 2019).

The mechanistic target of rapamycin complex 1 (mTORC1) is activated under nutrient abundance or in response to growth factors and promotes anabolic metabolism (Saxton and Sabatini, 2017). Moreover, mTORC1 activation promotes protein degradation by upregulating proteasome gene expression to eventually increase the intracellular amino acid pool and to facilitate new protein synthesis (Zhang and Manning, 2015). The sterol regulatory element-binding protein SREBP-1, which regulates expression of lipogenic genes by growth factor signaling through mTORC1 activation, has been shown to induce NRF1 expression and thus proteasome transcription (Zhang and Manning, 2015). On the contrary, induction of catabolism during nutrient deprivation activates the family of Forkhead Box (Fox) O (FOXO) TFs. FOXOs control the regulation of different UPS components including ubiquitin ligases, like MUSA1 (required for muscle loss), proteasome subunits such as PSMD11 (necessary for its assembly and activity), ubiquitin C gene and the de-ubiquitinating enzyme USP14 (Milan et al., 2015). Importantly, FOXO1 directly binds to the promoter of the catalytic β5 subunit to upregulate its expression. Consequently, low insulin/insulin-like growth factor 1 (IGF1)-signaling (IIS) pathway results in enhanced proteasome activity in murine liver and brain, while FOXO1 knockout reduces proteasome function (Kapetanou et al., 2021). Accordingly, FOXO4 regulates proteasome activity in human embryonic stem cells (hESCs) and its loss results in reduced differentiation capacity (Vilchez et al., 2013).

Proteasome and immunoproteasome gene expression is also subjected to various inflammatory signals. Whether immunoproteasome efficiency in the ubiquitin-dependent protein clearance is higher than that of the constitutive subtype is still under debate. Immunoproteasome has been shown to have a higher degradation capacity compared to standard proteasomes (Seifert et al., 2010). However, these results were soon challenged by Nathan et al., showing that both subtypes share the same degrading capacity for ubiquitinated proteins (Nathan et al., 2013). Recent evidence turn the attention to intermediate proteasomes that contain both immune and constitutive catalytic subunits (Abi Habib et al., 2020). The expression of immunoproteasome and PA28αβ genes in such complexes is regulated through IFN (types I and II) signaling via the JAK-STAT pathway. Binding of IFN-γ to its receptor results in JAK1 and JAK2 phosphorylation and activation, which in turn phosphorylate the receptor to recruit STAT1. Thereafter, STAT1 dimerizes and translocates into the nucleus, where it induces the transcription of IFN regulatory factor-1 (IRF-1).

IRF-1 subsequently upregulates the expression of immunoproteasome subunits, PA28αβ and MHC class I genes (Kors et al., 2019).

Moreover, IFNs induce the expression of various genes via the AKT-mTOR pathway, which will increase the levels of newly synthetized proteins, but also that of misfolded proteins, known as defective ribosomal products (DriPs). Oxidized proteins due to IFN-γ-induced free radicals will further aggregate along with DriPs (van Deventer and Neefjes, 2010). 26S immunoproteasomes are necessary for the degradation of these DriPs which are partly polyubiquitinated, likely through the IFN-γ-induced upregulation of 20 E3 ligases and of the ubiquitin-conjugating enzyme E2 L6 (UBE2L6) (Seifert et al., 2010). Upon oxidative stress, the immunoproteasome subunits may also be upregulated through the NF-κB pathway (Johnston-Carey et al., 2015). Degradation of inhibitor κB (IκBα) results in the activation of NF-κB, which induces type I IFNs and thus immunoproteasomes (Moschonas et al., 2008).

As previously mentioned, elevated ROS levels and oxidized proteins trigger OxR. Nrf2, the master coordinator of the OxR, also upregulates the expression of several proteasome subunits by binding to AREs in their proximal promoter regions, thus protecting the cells from toxic protein overload (Pickering et al., 2012). Upregulation of the expression of multiple representative proteasome subunits via compounds that are characterized as Nrf2 activators is confirmed in cell cultures and animal models (Kapeta et al., 2010; Kwak et al., 2003; Mladenovic Djordjevic et al., 2020; Papaevgeniou et al., 2016). Moreover, Nrf2 activation upon adaptation to oxidative stress results in high expression of the PSMB1 (20S) and PA28α subunits (Pickering et al., 2012). Finally, Nrf2 controls the expression of the proteasome maturation protein (POMP), a proteasome chaperone, which in turn modulates the proliferation of self-renewing hESCs (Jang et al., 2014). Collectively, Nrf2 upregulates the expression of key components of the UPS and therefore contributes to proteostasis via multiple ways.

3.3.2 Autophagy

Autophagy, the other major intracellular catabolic and proteostatic system, depends on the coordinated function of lysosomes and autophagosomes. Although initially, it was a general consensus that autophagy was exclusively regulated by cytosolic processes, it is nowadays well established that the life cycle of both lysosomes and autophagosomes is tightly controlled by the nucleus, through multiple transcriptional programs. The first evidence supporting this new concept was that nitrogen starvation in yeast induces the upregulation of the essential autophagy gene Apg8p, the homologous of mammalian lc327 (Kirisako et al., 1999). A decade after, the transcription factor EB (TFEB) emerged as a key regulator of multiple autophagy-related genes that control glucose homeostasis, mitochondrial biogenesis and fatty acids β-oxidation (Settembre et al., 2011).

TFEB, a basic helix–loop–helix-leucine-zipper (bHLH-Zip) protein of the microphthalmia/TFE (MiT/TFE) family, regulates the expression of target genes involved in autophagy, lysosomal biogenesis, lysosomal exocytosis, lipophagy and mitophagy that contain the coordinated lysosomal expression and regulation (CLEAR) motif in their promoter region. These target genes are crucial in different steps of autophagy-mediated clearance, such as autophagy initiation, autophagosome membrane elongation, autophagosomes trafficking, fusion with lysosomes and substrate docking (Palmieri et al., 2011). Upon nutrient abundance, TFEB is mainly found in its inactive state that is established by mTORC1 and ERK2-mediated phosphorylation. The Mitf-Tfe family of bHLH-Zip TFs that also regulates autophagy and lysosomal biogenesis is similarly subjected to mTORC1 control (Martina and Puertollano, 2018; Ploper et al., 2015). When cells experience stress or under conditions of lysosomal dysfunction, TFEB is dephosphorylated and translocates to the nucleus to activate target gene expression (Palmieri et al., 2011). Interestingly, mTORC1 can also inhibit TFEB by modulating the zinc finger TFs harboring Kruppel-associated box (KRAB) and SCAN domain (ZKSCAN3) activity (Chauhan et al., 2013). ZKSCAN3 negatively regulates autophagosome formation and lysosomal biogenesis by repressing the expression of autophagy-related genes including Map1lc3b, Atg5, Atg12 and Atg3 (Hu et al., 2020). The orchestrated regulation of the autophagy–lysosomal system by TFEB/ZKSCAN3 highlights the importance of this pathway in cellular adaptation to environmental signals.

In addition, FOXO factors have a well-established role in autophagy upregulation.

When activated, the FOXO TFs translocate to the nucleus to induce the expression of a number of autophagy-related genes, including Atg4, Atg12, Becn1, Bnip3, Map1lc3B, Ulk1, Vps34 (Pik3c3 in human), Gabarapl1, Atg5, Atg10, Atg14, Pink1 and several mitophagy-related genes (Webb and Brunet, 2014). FOXOs were initially described as crucial factors for the adaptation to starvation in muscle, as they induce cellular nutrient recycling through autophagy in response to fasting or denervation (Mammucari et al., 2007; Zhao et al., 2007). FOXO1 and FOXO3 also activate autophagy in other differentiated cell types, such as cardiomyocytes (Mammucari et al., 2007), while FOXO3 promotes autophagy in hematopoietic stem cells (Warr et al., 2013). The upregulation of autophagy can be advantageous (e.g. HSCs) or detrimental (muscle) depending on the cell type. Additionally, FOXOs can induce autophagy in many tissues of invertebrate organisms in a non-cell-autonomous manner, with beneficial effects on longevity (Webb and Brunet, 2014).

While located in the nucleus, members of p53 (tumor-suppressor protein TP53) family are found in their active state and upregulate autophagy. Their activation occurs following stimuli such as DNA damage or the presence of activated oncogenes and leads to the expression of genes related to autophagy induction and autophagosome maturation (Atg2, Atg4, Atg7, Atg10, Gabarap, Lkb1 and Ulk1/2). Loss of p53 can be partly compensated by p63 and p73, which appear to share in part similar autophagy-related target genes. Moreover, p53 regulates both FOXO3a expression and activity and promotes TFEB/TFE3 nuclear translocation upon DNA damage (Jeong et al., 2018; Renault et al., 2011), thus controlling key upstream modulators of the autophagy pathway. Along with DNA damage, multiple other stimuli regulate the often-overlapping functions of p53 family members, such as dynamics of oxygen levels.

E2F1, an important co-regulator of p53, that is active in hypoxic conditions, is also a key transcriptional regulator of autophagy genes, such as Bnip3, Ulk1, At5 and/or Map1lc3a (Füllgrabe et al., 2016). BNIP3 disrupts the inhibitory binding of B-cell lymphoma 2 (BCL-2) to BECLIN1, a component of the phosphoinositide 3-kinase (PI3K) complex that promotes autophagosome biogenesis. Activated E2F1 upregulates BNIP3 transcription; however, NF-kB inhibits E2F1-mediated BNIP3 transactivation by keeping its promoter occupied. These two antagonists are connected in another way; E2F1 also induces the stabilization of IκB, the inhibitor of NF-κB. Although the role of those two factors is opposing regarding BNIP3 transcription, NF-κB in other circumstances, such as apoptosis inhibition, promotes the expression of other autophagy genes. The example of E2F1 is also indicative of how the degree of hypoxia appears to determine which TFs activate autophagy. Its target gene Bnip3, for instance, is also activated by HIF1α, in moderate hypoxia. The same TF, in severe hypoxia, leads to a response involving the activation of ATF4 (Pike et al., 2012) that transcriptionally upregulates autophagy genes directly or indirectly through transcription of DNA damage inducible transcript 3 (DDIT3) (Füllgrabe et al., 2016).

Various TFs inducing autophagy genes are characterized as nutrient-sensing regulators. For instance, during nutrient sufficiency, the farnesoid X receptor FXR blocks the transcription of autophagy-related genes, as it dimerizes with CREB. Upon fasting, CREB forms a functional complex with CREB-regulated transcription coactivator 2 (CRTC2) and induces the transcription of target genes. On the other hand, the nuclear receptor PPARα (peroxisome proliferator activated receptor alpha) shares the same specific DNA-binding sites with FXR (DR1 elements) in the promoters of many autophagy-related genes. Consequently, these two nuclear receptors compete for the binding to the same target genes. Notably, CREB induces the expression of TFEB to transcriptionally enhance PPARα activity within a dynamic regulatory circuit (Di Malta et al., 2019). Finally, upon amino acid starvation (or in response to various stressors), JNK-Jun pathway is activated and regulates the expression of Annexin A2 (ANXA2), responsible for vesicular trafficking and autophagy enhancement in a feedback loop (Moreau et al., 2015). Additionally, Jun directly induces BECN1 and MAP1LC3B expression (Füllgrabe et al., 2016). The TFs that participate in the regulation of members of the PN are summarized in Table 3.1.

3.4 CROSSTALK BETWEEN PROTEOSTATIC MECHANISMS

The PN dynamically adapts to meet cellular requirements and to prevent proteostasis imbalance through the orchestrated transcriptional

TABLE 3.1

List of TFs that participate in the up- (↑) or downregulation (↓) of members of the proteostasis network

TF	HSR, UPR, OXR	Autography	Proteasome
ATF-4	↑, Gardner et al., 2013	↑, Pike et al., 2012	↑, Conn and Qian, 2011
ATF-6	↑, Conn and Qian, 2011	↑, Baird et al., 2017	↑, Periz et al., 2015
ATFS-1	↑, Jensen and Jasper, 2014	↑, Nargund et al., 2015	N/A
C-JUN	↑, Kansanen et al., 2013	↑, Moreau et al., 2015	↑, Niforou et al., 2014
CREB	N/A	↑, Seok et al., 2014	N/A
E2F1	↑, Conn and Qian, 2011	↑, Shaw et al., 2008	↑, Conn and Qian, 2011
FOXK1	N/A	↓, Füllgrabe et al., 2016	N/A
FOXO1	↑, Hipp et al., 2019	↑, Zhao et al., 2010	↑, Kapetanou et. al, 2021
FOXO3A	↑, Jeong et al., 2018	↑, Audesse et al., 2019	↑, Milan et al., 2015
FOXO4	N/A	N/A	↑, Vilchez et al., 2013
FXR	N/A	↓, Seok et al., 2014	N/A
HSF1	↑, Kansanen et al., 2013	↑, Luo et al., 2016	↑, Boos et al., 2019
XBP1	↑, Hetz and Glimcher, 2009	↑, Audesse et al., 2019	N/A
NF-KB	↑, Yuan et al., 2017	↑, Pike et al., 2012	↑, Moschonas et al., 2008
NRF1	↑, Baird et al., 2017	N/A	↑, Baird et al., 2017; Radhakrishnan et al., 2010; Steffen et al., 2010
NRF2	↑, Park et al., 2019	↑, Pajares et al., 2017	↑, Kapeta et al., 2010
P53	↑, Renault et al., 2011	↑, Renault et al., 2011	↑, Renault et al., 2011
P63	N/A	↑, Füllgrabe et al., 2016	N/A
P73	↑, Kors et al., 2019	↑, Füllgrabe et al., 2016	↑, Kors et al., 2019
PPARA	↑, Di Malta et al., 2019	↑, Di Malta et al., 2019	N/A
SREBP-1/2	N/A	N/A	↑, Zhang and Manning, 2015
STAT-1	N/A	↓, Goldberg et al., 2017	↑, Pajares et al., 2017
TFEB	N/A	↑, Settembre et al., 2011	N/A
ZKSCAN3	N/A	↓, Hu et al., 2020	N/A

regulation of the UPR, UPS and autophagy genes (Figure 3.2).

In the case of UPR, PERK-dependent phosphorylation triggers the dissociation of Nrf2/Keap1 complexes and inhibits the reassociation of Nrf2/Keap1 complexes in vitro. Activation of PERK via agents that initiate the UPR is both necessary and sufficient for dissociation of cytoplasmic Nrf2/Keap1 and the subsequent Nrf2 nuclear import (Cullinan et al., 2003). Nrf2 subsequently upregulates its target genes involved in OxR, proteasome activation and autophagy regulation.

Upregulation of autophagy is observed upon proteasome dysfunction and amino acid starvation. More specifically, eIF2α kinase GCN2 is activated and mTOR is inactivated. Furthermore, upon proteasome impairment, autophagy is upregulated most likely by mTOR inhibition via amino acid starvation (Park and Cuervo, 2013; Wang et al., 2013). In such conditions, p53 acts as a TF for autophagy housekeeping genes such as the damage-regulated autophagy modifier (DRAM). Alternatively, increased p53 levels may activate AMPK-dependent autophagy by inhibiting the mTOR pathway (Ji and Kwon, 2017). FOXOs within MAPK8/JNK1 axis can be induced upon nutritional or oxidative stress to upregulate the transcription of both proteasome catalytic subunits and crucial autophagy-related genes (Audesse et al., 2019). As a surprising example,

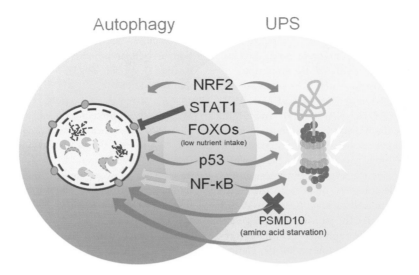

Figure 3.2 **Crosstalk between the UPS and autophagy.** Besides the transcription factors such as NRF2, STAT1, FOXOs, p53 and NF-κB that mediate the interaction and crosstalk among the two main proteolytic cellular systems, the changes in the activity of one proteolytic system may impact on the transcriptional regulation of the other system, as described in the text. Upregulation is marked with a green arrow, inhibition with a red hindrance arrow, dual role in activation or inhibition depending on conditions with a yellow fork and pathway blockage is marked with a red cross.

upon amino acid starvation, proteasome 26S sub-unit non-ATPase 10 (PSMD10) has recently been reported to translocate to the nucleus and to bind to HSF1 at the ATG7 promoter to induce its transcription (Luo et al., 2016).

Overall, the existence of different transcriptional regulators of these machineries also suggests a crosstalk that may assure proteostasis under different circumstances. STAT1 for instance, which induces the expression of immunoproteasome subunits, is suggested to function as an endogenous and stress-inducible repressor of the autophagy gene transcription (Moreau et al., 2015). Accordingly, Nrf1 and Nrf2 both bind to ARE sequences in the promoter regions of their target genes, which suggests that they have overlapping transcriptional activities (although they differ in their regulatory mechanisms and cellular localization) and may have a relevant role in tissues that support high levels of oxidative stress. For instance, oxidative stress-induced Nrf2 may function under nutrient-rich conditions to transcriptionally upregulate autophagy (Pajares et al., 2017), similarly to TFEB under starvation conditions. In such circumstances, inhibition of mTORC1 leads to nuclear translocation of TFEB and induction of the expression of autophagy genes. In summary, these overlapping TFs act as

interconnection hubs that link proteotoxic stimuli and protective transcriptional responses.

3.5 CONCLUSIONS AND PERSPECTIVES

Eukaryotic cells have evolved highly sophisticated systems to constantly monitor proteome fidelity and trigger transcriptional programs that act to promote and maintain proteostasis by curating the expression of molecular chaperones, proteolytic machineries and of other factors. Importantly, in metazoans, the robustness of the PN is crucial for the viability of the entire organism. Studies in model organisms have revealed many intimate connections between pathways that regulate PN expression and aging, suggesting that proteostasis maintenance is essential for longevity. The inhibition of IIS and dietary restriction represent two of the most effective ways to increase lifespan (Kenyon, 2010). IIS was the first pathway shown to influence aging in animals and to link the maintenance of proteostasis and stress resistance to longevity. IIS inhibition exerts beneficial effects on lifespan through the activation of FOXOs/DAF-16, HSF-1 and NRF1-NRF2/SKN-1 TFs, which in turn induce the expression of protective stress response pathways, such as the upregulation of chaperones, detoxification enzymes and proteolytic systems

(Kors et al., 2019). Furthermore, there is ample evidence that the effects of dietary restriction, which is the most robust intervention to increase animal lifespan, are mediated by alterations of regulators of proteostasis, such as FoxOs/DAF-16, HSF-1, NRF2/SKN-1 and TOR (Xie et al., 2020). Besides, FoxO/DAF-16, NRF2/SKN1 and HSF1 are required for the lifespan extension of *C. elegans* strain that overexpresses *pbs-5* proteasome subunit (β5 homolog) (Chondrogianni et al., 2015), while several lines of evidence reveal a continuous interplay between the integrity of proteasome function, senescence and cell survival (Chondrogianni et al., 2008). Essentially, the pathways that regulate both aging and the PN represent promising targets for future therapeutic interventions. Hence, a better understanding of proteostasis regulation and its alterations during aging is instrumental for the development of novel approaches to combat degenerative disorders and to increase the human healthspan.

REFERENCES

Abi Habib, J., De Plaen, E., Stroobant, V., et al. 2020. Efficiency of the four proteasome subtypes to degrade ubiquitinated or oxidized proteins. *Sci Rep*, 10(1), 15765. doi:10.1038/s41598-020-71550-5

Adams, C. J., Kopp, M. C., Larburu, N., et al. 2019. Structure and molecular mechanism of ER stress signaling by the unfolded protein response signal activator IRE1. *Front Mol Biosci*, 6(11). doi:10.3389/fmolb.2019.00011

Akerfelt, M., Morimoto, R. I., & Sistonen, L. 2010. Heat shock factors: integrators of cell stress, development and lifespan. *Nat Rev: Mol Cell Biol*, 11(8), 545–555. doi:10.1038/nrm2938

Amin-Wetzel, N., Saunders, R. A., Kamphuis, M. J., et al. 2017. A J-protein co-chaperone recruits BiP to monomerize IRE1 and repress the unfolded protein response. *Cell*, 171(7), 1625–1637.e1613. doi:10.1016/j.cell.2017.10.040

Audesse, A. J., Dhakal, S., Hassell, L.-A., et al. 2019. FOXO3 directly regulates an autophagy network to functionally regulate proteostasis in adult neural stem cells. *PLoS Genet*, 15(4), e1008097–e1008097. doi:10.1371/journal.pgen.1008097

Baird, L., Tsujita, T., Kobayashi, E. H., et al. 2017. A homeostatic shift facilitates endoplasmic reticulum proteostasis through transcriptional integration of proteostatic stress response pathways. *Mol Cell Biol*, 37, E00651–U00199. doi:10.1128/MCB.00439.16

Baugh, J. M., Viktorova, E. G., & Pilipenko, E. V. 2009. Proteasomes can degrade a significant proportion of cellular proteins independent of ubiquitination. *J Mol Biol*, 386(3), 814–827. doi:10.1016/j.jmb.2008.12.081

Boos, F., Kraemer, L., Groh, C., et al. 2019. Mitochondrial protein-induced stress triggers a global adaptive transcriptional programme. *Nat Cell Biol*, 21, 442–442. doi:10.1038/s41556-019-0294-5

Brehme, M., Sverchkova, A., & Voisine, C. 2019. Proteostasis network deregulation signatures as biomarkers for pharmacological disease intervention. *Curr Opin Syst Biol*, 15, 74–81. doi:10.1016/j.coisb.2019.03.008

Chauhan, S., Goodwin, J. G., Chauhan, S., et al. 2013. ZKSCAN3 is a master transcriptional repressor of autophagy. *Mol Cell*, 50(1), 16–28. doi:10.1016/j.molcel.2013.01.024

Chondrogianni, N., Trougakos, I. P., Kletsas, D., et al. 2008. Partial proteasome inhibition in human fibroblasts triggers accelerated M1 senescence or M2 crisis depending on p53 and Rb status. *Aging Cell*, 7(5), 717–732. doi:10.1111/j.1474-9726.2008.00425.x

Chondrogianni, N., Sakellari, M., Lefaki, M., et al. 2014. Proteasome activation delays aging in vitro and in vivo. *Free Radic Biol Med*, 71, 303–320. doi:10.1016/j.freeradbiomed.2014.03.031

Chondrogianni, N., Voutetakis, K., Kapetanou, M., et al. 2015. Proteasome activation: an innovative promising approach for delaying aging and retarding age-related diseases. *Ageing Res Rev*, 23(Pt. A), 37–55. doi:10.1016/j.arr.2014.12.003

Conn, C.S. and Qian, S.B., 2011. mTOR signaling in protein homeostasis: less is more? *Cell Cycle*. 10, 1940–7. doi:10.4161/cc.10.12.15858

Cullinan, S. B., Zhang, D., Hannink, M., et al. 2003. Nrf2 is a direct PERK substrate and effector of PERK-dependent cell survival. *Mol Cell Biol*, 23(20), 7198–7209. doi:10.1128/mcb.23.20.7198-7209.2003

Di Malta, C., Cinque, L., & Settembre, C. 2019. Transcriptional regulation of autophagy: Mechanisms and diseases. *Front Cell Dev Biol*, 7, 114–114. doi:10.3389/fcell.2019.00114

Dirac-Svejstrup, A. B., Walker, J., Faull, P., et al. 2020. DDI2 is a ubiquitin-directed endoprotease responsible for cleavage of transcription factor NRF1. *Mol Cell*, 79(2), 332–341.e337. doi:10.1016/j.molcel.2020.05.035

Füllgrabe, J., Ghislat, G., Cho, D. H., et al. 2016. Transcriptional regulation of mammalian autophagy at a glance. *J Cell Sci*, 129(16), 3059–3066. doi:10.1242/jcs.188920

Gardner, B. M., Pincus, D., Gotthardt, K., et al. 2013. Endoplasmic reticulum stress sensing in the unfolded protein response. *Cold Spring Harbor Perspect Biol*, 5(3), a013169–a013169. doi:10.1101/cshperspect.a013169

Goldberg, A.A., Nkengfac, B., Sanchez, A.M.J. et al., 2017. Regulation of ULK1 Expression and Autophagy by STAT1. *J Biol Chem*. 292, 1899–1909. doi:10.1074/jbc.M116.771584

Gonos, E. S., Kapetanou, M., Sereikaite, J., et al. 2018. Origin and pathophysiology of protein carbonylation, nitration and chlorination in age-related brain diseases and aging. *Aging*, 10(5), 868–901. doi:10.18632/aging.101450

Hetz, C. and Glimcher, L.H., 2009. Fine-tuning of the unfolded protein response: Assembling the IRE1alpha interactome. *Mol Cell*. 35, 551–61. doi:10.1016/j.molcel.2009.08.021.

Hietakangas, V., Ahlskog, J. K., Jakobsson, A. M., et al. 2003. Phosphorylation of serine 303 is a prerequisite for the stress-inducible SUMO modification of heat shock factor 1. *Mol Cell Biol*, 23(8), 2953–2968. doi:10.1128/mcb.23.8.2953-2968.2003

Hipp, M.S., Kasturi, P. and Hartl, F.U., 2019. The proteostasis network and its decline in ageing. *Nat Rev Mol Cell Biol*. 20, 421–435. doi:10.1038/s41580-019-0101-y

Hjerpe, R., Bett, J. S., Keuss, M. J., et al. 2016. UBQLN2 mediates autophagy-independent protein aggregate clearance by the proteasome. *Cell*, 166(4), 935–949. doi:10.1016/j.cell.2016.07.001

Holland, R., & Fishbein, J. C. 2010. Chemistry of the cysteine sensors in Kelch-like ECH-associated protein 1. *Antioxid Redox Signal*, 13(11), 1749–1761. doi:10.1089/ars.2010.3273

Hu, H., Ji, Q., Song, M., et al. 2020. ZKSCAN3 counteracts cellular senescence by stabilizing heterochromatin. *Nucl Acids Res*, 48(11), 6001–6018. doi:10.1093/nar/gkaa425

Jang, J., Wang, Y., Kim, H.-S., et al. 2014. Nrf2, a regulator of the proteasome, controls self-renewal and pluripotency in human embryonic stem cells. *Stem Cells (Dayton, Ohio)*, 32(10), 2616–2625. doi:10.1002/stem.1764

Jensen, M. B., & Jasper, H. 2014. Mitochondrial proteostasis in the control of aging and longevity. *Cell Metab*, 20(2), 214–225. doi:10.1016/j.cmet.2014.05.006

Jeong, E., Brady, O. A., Martina, J. A., et al. 2018. The transcription factors TFE3 and TFEB amplify p53 dependent transcriptional programs in response to DNA damage. *eLife*, 7 :e40856. doi: 10.7554/eLife.40856.

Ji, C. H., & Kwon, Y. T. 2017. Crosstalk and interplay between the ubiquitin-proteasome system and autophagy. *Mol Cells*, 40(7), 441–449. doi:10.14348/molcells.2017.0115

Johnston-Carey, H. K., Pomatto, L. C., & Davies, K. J. 2015. The immunoproteasome in oxidative stress, aging, and disease. *Crit Rev Biochem Mol Biol*, 51(4), 268–281. doi:10.3109/10409238.2016.1172554

Kansanen, E., Kuosmanen, S. M., Leinonen, H., et al. 2013. The Keap1-Nrf2 pathway: mechanisms of activation and dysregulation in cancer. *Redox Biol*, 1(1), 45–49. doi:10.1016/j.redox.2012.10.001

Kapeta, S., Chondrogianni, N., & Gonos, E. S. 2010. Nuclear erythroid factor 2-mediated proteasome activation delays senescence in human fibroblasts. *J Biol Chem*, 285(11), 8171–8184. doi:10.1074/jbc.M109.031575

Kapetanou, M., Nespital, T., Tain, L. S., et al. 2021. FoxO1 is a novel regulator of 20S proteasome subunits expression and activity. *Front Cell Dev Biol*, 9, 625715–625715. doi:10.3389/fcell.2021.625715

Kenyon, C. J. 2010. The genetics of ageing. *Nature*, 464(7288), 504–512. doi:10.1038/nature08980

Kirisako, T., Baba, M., Ishihara, N., et al. 1999. Formation process of autophagosome is traced with Apg8/Aut7p in yeast. *J Cell Biol*, 147(2), 435–446. doi:10.1083/jcb.147.2.435

Kmiecik, S. W., Le Breton, L., & Mayer, M. P. 2020. Feedback regulation of heat shock factor 1 (Hsf1) activity by Hsp70-mediated trimer unzipping and dissociation from DNA. *EMBO J*, 39(14), e104096. doi:10.15252/embj.2019104096

Kors, S., Geijtenbeek, K., Reits, E., et al. 2019. Regulation of proteasome activity by (post-)transcriptional mechanisms. *Front Mol Biosci*, 6(48). doi:10.3389/fmolb.2019.00048

Kwak, M.-K., Wakabayashi, N., Greenlaw, J. L., et al. 2003. Antioxidants enhance mammalian proteasome expression through the Keap1-Nrf2 signaling pathway. *Mol Cell Biol*, 23(23), 8786–8794. doi:10.1128/mcb.23.23.8786-8794.2003

Labbadia, J., & Morimoto, R. I. 2015. The biology of proteostasis in aging and disease. *Annu Rev Biochem*, 84, 435–464. doi:10.1146/annurev-biochem-060614-033955

Li, R., Jia, Z., & Zhu, H. 2019. Regulation of Nrf2 signaling. *React Oxyg Species (Apex)*, 8(24), 312–322.

Lingappan, K. 2018. NF-κB in oxidative stress. *Curr Opin Toxicol*, 7, 81–86. doi:10.1016/j.cotox.2017.11.002

Luo, T., Fu, J., Xu, A., et al. 2016. PSMD10/Gankyrin induces autophagy to promote tumor progression through cytoplasmic interaction with ATG7 and nuclear transactivation of ATG7 expression. *Autophagy*, 12(8), 1355–1371. doi:10.1080/15548627. 2015.1034405

Mammucari, C., Milan, G., Romanello, V., et al. 2007. FoxO3 controls autophagy in skeletal muscle in vivo. *Cell Metab*, 6(6), 458–471. doi:10.1016/j. cmet.2007.11.001

Marshall, R. S., & Vierstra, R. D. 2019. Dynamic regulation of the 26S proteasome: from synthesis to degradation. *Front Mol Biosci*, 6(40). doi:10.3389/ fmolb.2019.00040

Martina, J. A., & Puertollano, R. 2018. Protein phosphatase 2A stimulates activation of TFEB and TFE3 transcription factors in response to oxidative stress. *J Biol Chem*, 293(32), 12525–12534. doi:10.1074/jbc. RA118.003471

Milan, G., Romanello, V., Pescatore, F., et al. 2015. Regulation of autophagy and the ubiquitin-proteasome system by the FoxO transcriptional network during muscle atrophy. *Nat Commun*, 6. doi:10.1038/ncomms7670

Miles, J., Scherz-Shouval, R., & van Oosten-Hawle, P. 2019. Expanding the organismal proteostasis network: linking systemic stress signaling with the innate immune response. *Trends Biochem Sci*, 44(11), 927–942. doi:10.1016/j.tibs.2019.06.009

Mladenovic Djordjevic, A. N., Kapetanou, M., Loncarevic-Vasiljkovic, N., et al. 2020. Pharmacological intervention in a transgenic mouse model improves Alzheimer's-associated pathological phenotype: involvement of proteasome activation. *Free Radic Biol Med*. doi:10.1016/j. freeradbiomed.2020.11.038

Moreau, K., Ghislat, G., Hochfeld, W., et al. 2015. Transcriptional regulation of Annexin A2 promotes starvation-induced autophagy. *Nat Commun*, 6(1), 8045–8045. doi:10.1038/ncomms9045

Moschonas, A., Kouraki, M., Knox, P. G., et al. 2008. CD40 induces antigen transporter and immunoproteasome gene expression in carcinomas via the coordinated action of NF-kappaB and of NF-kappaB-mediated de novo synthesis of IRF-1. *Mol Cell Biol*, 28(20), 6208–6222. doi:10.1128/mcb.00611-08

Motosugi, R., & Murata, S. 2019. Dynamic regulation of proteasome expression. *Front Mol Biosci*, 6, 30. doi:10.3389/fmolb.2019.00030

Nargund, A. M., Fiorese, C. J., Pellegrino, M. W., et al. 2015. Mitochondrial and nuclear accumulation of the transcription factor ATFS-1 promotes OXPHOS recovery during the UPR(mt). *Mol Cell*, 58(1), 123–133. doi:10.1016/j.molcel.2015.02.008

Nathan, J. A., Spinnenhirn, V., Schmidtke, G., et al. 2013. Immuno- and constitutive proteasomes do not differ in their abilities to degrade ubiquitinated proteins. *Cell*, 152(5), 1184–1194. doi:10.1016/j. cell.2013.01.037

Niforou, K., Cheimonidou, C., & Trougakos, I. P. 2014. Molecular chaperones and proteostasis regulation during redox imbalance. *Redox Biol*, 2, 323–332. doi:10.1016/j.redox.2014.01.017

Northrop, A., Byers, H. A., & Radhakrishnan, S. K. 2020. Regulation of NRF1, a master transcription factor of proteasome genes: implications for cancer and neurodegeneration. *Mol Biol Cell*, 31(20), 2158–2163. doi:10.1091/mbc.E20-04-0238

Pajares, M., Cuadrado, A., & Rojo, A. I. 2017. Modulation of proteostasis by transcription factor NRF2 and impact in neurodegenerative diseases. *Redox Biol*, 11, 543–553. doi:10.1016/j.redox.2017.01.006

Palmieri, M., Impey, S., Kang, H., et al. 2011. Characterization of the CLEAR network reveals an integrated control of cellular clearance pathways. *Hum Mol Genet*, 20(19), 3852–3866. doi:10.1093/ hmg/ddr306

Papaevgeniou, N., Sakellari, M., Jha, S., et al. 2016. 18α-Glycyrrhetinic acid proteasome activator decelerates aging and Alzheimer's disease progression in *Caenorhabditis elegans* and neuronal cultures. *Antioxid Redox Signal*, 25(16), 855–869. doi:10.1089/ ars.2015.6494

Park, C., & Cuervo, A. M. 2013. Selective autophagy: talking with the UPS. *Cell Biochem Biophys*, 67(1), 3–13. doi:10.1007/s12013-013-9623-7

Park, J.-Y., Kim, S., Sohn, H.Y. et al., 2019. TFEB activates Nrf2 by repressing its E3 ubiquitin ligase DCAF11 and promoting phosphorylation of p62. Scientific Reports. 9, 14354. doi:10.1038/s41598-019-50877-8

Periz, G., Lu, J., Zhang, T. et al., 2015. Regulation of protein quality control by UBE4B and LSD1 through p53-mediated transcription. *PLoS Biol*. 13, e1002114. doi: 10.1371/journal.pbio.1002114.

Pickering, A. M., Linder, R. A., Zhang, H., et al. 2012. Nrf2-dependent induction of proteasome and Pa28αβ regulator are required for adaptation to oxidative stress. *J Biol Chem*, 287(13), 10021–10031. doi:10.1074/jbc.M111.277145

Pike, L. R. G., Singleton, D. C., Buffa, F., et al. 2012. Transcriptional up-regulation of ULK1 by ATF4 contributes to cancer cell survival. *Biochem J*, 449(2), 389–400. doi:10.1042/BJ20120972

Ploper, D., Taelman, V. F., Robert, L., et al. 2015. MITF drives endolysosomal biogenesis and potentiates Wnt signaling in melanoma cells. *Proc*

Natl Acad Sci USA, 112(5), E420–429. doi:10.1073/pnas.1424576112

Radhakrishnan, S. K., Lee, C. S., Young, P., et al. 2010. Transcription factor Nrf1 mediates the proteasome recovery pathway after proteasome inhibition in mammalian cells. Mol Cell, 38(1), 17–28. doi:10.1016/j.molcel.2010.02.029

Raghunath, A., Sundarraj, K., Nagarajan, R., et al. 2018. Antioxidant response elements: discovery, classes, regulation and potential applications. Redox Biol, 17, 297–314. doi:10.1016/j.redox.2018.05.002

Renault, V. M., Thekkat, P. U., Hoang, K. L., et al. 2011. The pro-longevity gene FoxO3 is a direct target of the p53 tumor suppressor. Oncogene, 30(29), 3207–3221. doi:10.1038/onc.2011.35

Rius-Pérez, S., Torres-Cuevas, I., Millán, I., et al. 2020. PGC-1α, inflammation, and oxidative stress: an integrative view in metabolism. Oxidat Med Cell Long, 2020, 1452696–1452696. doi:10.1155/2020/1452696

Samant, R. S., Livingston, C. M., Sontag, E. M., et al. 2018. Distinct proteostasis circuits cooperate in nuclear and cytoplasmic protein quality control. Nature, 563(7731), 407–411. doi:10.1038/s41586-018-0678-x

Saxton, R. A., & Sabatini, D. M. 2017. mTOR Signaling in growth, metabolism, and disease. Cell, 168(6), 960–976. doi:10.1016/j.cell.2017.02.004

Seok, S., Fu, T., Choi, S.E. et al., 2014. Transcriptional regulation of autophagy by an FXR-CREB axis. Nature. 516, 108–11. doi: 10.1038/nature13949.

Seifert, U., Bialy, L. P., Ebstein, F., et al. 2010. Immunoproteasomes preserve protein homeostasis upon interferon-induced oxidative stress. Cell, 142(4), 613–624. doi:10.1016/j.cell.2010.07.036

Settembre, C., Di Malta, C., Polito, V. A., et al. 2011. TFEB links autophagy to lysosomal biogenesis. Science, 332(6036), 1429–1433. doi:10.1126/science.1204592

Shaw, J., Yurkova, N., Zhang, T. et al., 2008. Antagonism of E2F-1 regulated Bnip3 transcription by NF-kappaB is essential for basal cell survival. Proc Natl Acad Sci U S A. 105, 20734–9. doi:10.1073/pnas.0807735105

Sims, J. D., McCready, J., & Jay, D. G. 2011. Extracellular heat shock protein (Hsp)70 and Hsp90α assist in matrix metalloproteinase-2 activation and breast cancer cell migration and invasion. PLoS One, 6(4), e18848. doi:10.1371/journal.pone.0018848

Steffen, J., Seeger, M., Koch, A., et al. 2010. Proteasomal degradation is transcriptionally controlled by TCF11 via an ERAD-dependent feedback loop. Mol Cell, 40(1), 147–158. doi:10.1016/j.molcel.2010.09.012

van Deventer, S., & Neefjes, J. 2010. The immunoproteasome cleans up after inflammation. Cell, 142(4), 517–518. doi:10.1016/j.cell.2010.08.002

Vangala, J. R., & Radhakrishnan, S. K. 2019. Nrf1-mediated transcriptional regulation of the proteasome requires a functional TIP60 complex. J Biol Chem, 294, 2036–2045. doi:10.1074/jbc.RA118.006290

Vilchez, D., Boyer, L., Lutz, M., et al. 2013. FOXO4 is necessary for neural differentiation of human embryonic stem cells. Aging Cell, 12, 518–522. doi:10.1111/acel.12067

Wang, X. J., Yu, J., Wong, S. H., et al. 2013. A novel crosstalk between two major protein degradation systems. Autophagy, 9(10), 1500–1508. doi:10.4161/auto.25573

Warr, M. R., Binnewies, M., Flach, J., et al. 2013. FOXO3A directs a protective autophagy program in haematopoietic stem cells. Nature, 494(7437), 323–327. doi:10.1038/nature11895

Webb, A. E., & Brunet, A. 2014. FOXO transcription factors: key regulators of cellular quality control. Trends Biochem Sci, 39(4), 159–169. doi:10.1016/j.tibs.2014.02.003

Westerheide, S. D., Anckar, J., Stevens, S. M., Jr., et al. 2009. Stress-inducible regulation of heat shock factor 1 by the deacetylase SIRT1. Science, 323(5917), 1063–1066. doi:10.1126/science.1165946

Xie, Y., & Varshavsky, A. 2001. RPN4 is a ligand, substrate, and transcriptional regulator of the 26S proteasome: a negative feedback circuit. Proc Natl Acad Sci, 98(6), 3056–3061. doi:10.1073/pnas.071022298

Xie, K., Kapetanou, M., Sidiropoulou, K., et al. 2020. Signaling pathways of dietary energy restriction and metabolism on brain physiology and in age-related neurodegenerative diseases. Mech Ageing Dev, 192, 111364. doi:10.1016/j.mad.2020.111364

Yuan, J., Tan, T., Geng, M. et al., 2017. Novel Small Molecule Inhibitors of Protein Kinase D Suppress NF-kappaB Activation and Attenuate the Severity of Rat Cerulein Pancreatitis. Front Physiol. 8, 1014. doi:10.3389/fphys.2017.01014

Zhang, Y., & Manning, B. D. 2015. mTORC1 signaling activates NRF1 to increase cellular proteasome levels. Cell Cycle, 14, 2011–2017. doi:10.1080/15384101.2015.1044188

Zhao, J., Brault, J. J., Schild, A., et al. 2007. FoxO3 coordinately activates protein degradation by the autophagic/lysosomal and proteasomal pathways in atrophying muscle cells. Cell Metab, 6(6), 472–483. doi:10.1016/j.cmet.2007.11.004

Zhao, Y., Yang, J., Liao, W. et al., 2010. Cytosolic FoxO1 is essential for the induction of autophagy and tumour suppressor activity. Nat Cell Biol. 12, 665–75. doi:10.1038/ncb2069

CHAPTER FOUR

MicroRNAs as Central Regulators of Adult Myogenesis and Proteostasis Loss in Skeletal Muscle Aging

Ioannis Kanakis, Ioanna Myrtziou, Aphrodite Vasilaki, Katarzyna Goljanek-Whysall

CONTENTS

4.1 INTRODUCTION

Skeletal muscle is a vital organ in mammals, which controls the vast majority of body movements. It is also considered as one of the most important and highly active metabolic tissues accounting for 40% of the total body mass. A 30–50% loss of muscle mass occurs between the ages of 50 and 80 years that impacts profoundly on the quality of life of older people resulting in a reduced ability to carry out everyday tasks and increased susceptibility to falling (Bortz, 2002; Espinoza and Walston, 2005; Fried et al., 2001). All individuals lose muscle mass and develop age-related muscle weakness (termed sarcopenia when it reaches clinically relevant severity). Several factors can lead to this dysfunction and degeneration of skeletal muscle including inflammation, dysregulation of motor neurons and oxidative stress (Narici and Maffulli,

2010). It has also been shown that aging causes the impairment of skeletal muscle stem cells leading to a decline in regeneration due to cell senescence (Cosgrove et al., 2014; Sousa-Victor et al., 2014). Furthermore, sarcopenic patients are more prone to develop other age-related pathologies such as diabetes, hypertension and complications of the cardiovascular system (J. C. Brown et al., 2016; Morley, 2008).

Skeletal muscle mass is dictated by the number and the size of muscle fibers. The decline in muscle mass and strength in people after the age of ~50 appears primarily due to loss of muscle fibers with weakening of the remaining fibers (Lexell et al., 1986). In addition, a switch from fast type IIa and IIx muscle fibers to slow type I has been observed during aging (Kosek et al., 2006; Nilwik et al., 2013). Data clearly indicate that in human and

DOI: 10.1201/9781003048138-4

45

rodents aging, loss of motor neurons also accompanies the loss of muscle fibers (Einsiedel and Luff, 1992; Larsson and Ansved, 1995; Lexell et al., 1988), with a 25–50% reduction in the number of α-motor neurons occurring with aging. Several studies have also reported loss of axons as well as the presence of swollen, segmental demyelinated and remyelinated axons in peripheral nerve of old animals and humans (Adinolfi et al., 1991) and such neuronal changes have been proposed to play a major role in the age-related loss of muscle mass and function (Delbono, 2003).

4.2 BIOGENESIS AND REGULATORY FUNCTION OF miRs

MicroRNAs (miRNAs, miRs) are a part of the small non-coding RNAs (sncRNAs) and are 19–22 nucleotides long. miRs control muscle development, disease and aging (Brown and Goljanek-Whysall, 2015; Goljanek-Whysall et al., 2012) through regulation of post-transcriptional gene expression. MiR genes are transcribed into pri-miRs by RNA polymerase II or III and incorporated into pre-miR hairpins by the RNase III endonuclease enzyme Drosha (Diebel et al., 2014; J. Han et al., 2006). Subsequently, exportin-5 translocates pre-miRs outside the nucleus, where Dicer cleaves the pre-miR haiprins into mature, double-stranded short miRs (Winter et al., 2009). The miR duplex strands are separated and one strand of the mature miR (guide strand) together with Argonaute protein and co-factors forms the RNA-induced silencing complex (RISC) protein complex (Schraivogel and Meister, 2014). The other strand (passenger strand) is subsequently degraded. After generation of mature miRs, miR binding to target mRNA triggers translational repression or mRNA degradation (Y. W. Kong et al., 2008). The effectiveness of miR-mediated translational inhibition largely depends on the binding capacity to the target mRNA (Brodersen and Voinnet, 2009; Hu and Bruno, 2011; Kim et al., 2016). Generally, mRNAs binding sites of miRs are located in the 3′-untranslated region (UTR) and less frequently in the 5′-UTR, although many mRNA molecules have multiple binding sites (Hu and Bruno, 2011). Two types of binding have been identified; the first type refers to the perfect complementarity between the 3′-UTR of the target mRNA and the 5′-end of the miR, which is named as seed site of the miR; the other

type is incomplete match between the 3′-UTR sequence of mRNA and the seed region of miR (Quattrocelli and Sampaolesi, 2015). Both types of RISC binding, with or without mismatches, lead to mRNA suppression of expression either by degradation of the mRNA transcripts via Argonaute-2 (Ago-2) cleaving activity or destabilization of the mRNA molecule causing direct repression of protein translation.

Despite the knowledge that has been gained during the last two decades for the miR-mRNA interactions, there are some mechanistic and regulatory questions that still need to be addressed. To this point, the development of specific software, like TargetScan and miRWalk (Dweep et al., 2011; Lewis et al., 2005), and the extended use of bioinformatic tools have largely contributed. Since miRs have multiple mRNA targets and one gene can be targeted by several miRs, it has been well established that miRs regulate different cell functions and are involved in the interplay between different cell signaling pathways. Therefore, dysregulation of miR expression has been linked to cancer, neurodegenerative disorders and cardiomyopathies (Calderon-Dominguez et al., 2020; Juzwik et al., 2019; Slack and Chinnaiyan, 2019).

4.3 MYOGENESIS IN ADULTHOOD AND SARCOPENIA

The development of skeletal muscle or myogenesis involves defined steps that are well characterized. In brief, myoblasts, originated from activated muscle stem cells, known as satellite cells, proliferate and differentiate into myocytes, which undergo fusion to form myofibers. Myogenesis is finely regulated by a complex regulatory gene network of expression which involves various myogenic regulatory transcription factors (MRFs), such as myogenic differentiation (MyoD), myogenin, myogenic factor 5 (Myf5) and myogenic regulatory factor 4 (MRF4) (Asfour et al., 2018). Satellite cells, which are able to self-renew and regenerate the tissue, express the paired box transcription factors, Pax3 and Pax7, and not MyoD, which maintain them in a quiescent status, but upon activation, they co-express Pax7, MyoD and Myf5 (Schmidt et al., 2019). During myogenesis, Pax expression is repressed, and this is followed by enhanced expression of MRFs. On the other hand, Pax7 expression is required during skeletal muscle regeneration in adulthood (von Maltzahn

PROTEOSTASIS AND PROTEOLYSIS

et al., 2013). In the aged skeletal muscle, satellite cells lose their ability to preserve their quiescent status with simultaneous deregulation of self-renewal and regenerative capacity that contributes to sarcopenia (Snijders and Parise, 2017; Sousa-Victor et al., 2014).

4.4 MiRs ROLE IN ADULT MYOGENESIS

4.4.1 Skeletal Muscle-Specific miRs – myomiRs

Myogenesis regulation by miRs is mainly mediated through modulation of MRFs expression. A certain group of miRs, known as myomiRs, including miR-1, miR-133a, miR-133b, miR-206, miR-208, miR-208b, miR-486 and miR-499, are abundantly expressed in skeletal muscle tissue (Horak et al., 2016; van Rooij et al., 2009). These myomiRs have been shown to control the size of muscle fibers as well as the response to aging and exercise with regard to muscle fiber type switch. miR-499, miR-208 and miR-208b are reported as regulators of myosin expression in skeletal muscle as well as of muscle functionality (van Rooij et al., 2009). Importantly, conditional knockout of Dicer, only in skeletal muscle of mice using MyoD-Cre mice, results in reduced expression of miR-1, miR-133 and miR-206, decreased skeletal muscle mass, and alteration of morphological characteristics of myofibers, indicating that the biogenesis and expression levels of specific myomiRs are vital for normal skeletal muscle development and function. However, when Dicer is depleted during adulthood and for lifelong, although miR expression is reduced, skeletal muscle mass is preserved (Vechetti et al., 2019), but muscle regeneration is affected (Oikawa et al., 2019). It is important to note that skeletal muscle mass reduction is observed only when Dicer is inactivated during embryonic development, highlighting the essential role for Dicer in skeletal muscle and the requirement of miRs for embryonic myogenesis (O'Rourke et al., 2007).

4.4.2 miR Effects on Satellite Cells

During myogenesis in adulthood, satellite cell activation as well as the proliferation of myoblasts are also regulated by specific miRs, which target transcription factor genes participating in various myogenic signaling pathways. For example, miR-1, miR-206, miR-27b, miR-486 and miR-133b suppress the expression of Pax3/7 and, thus, satellite cells are activated (Crist et al., 2009; Cui et al., 2019; Hirai et al., 2010). In addition, miR-1 and miR-206 overexpression enhances the activation of satellite cells, inhibits myoblast proliferation (Chen et al., 2010) and results in timely myogenin expression by targeting Pax3 in myoblasts (Goljanek-Whysall et al., 2011). Another miR that has been implicated in the downregulation of Pax3, with no effect on Pax7, is miR-27b. Mice injected with anti-miR-27b antagomirs, in a model of muscle injury, showed altered levels of Pax3 and delayed regeneration of injured muscle (Crist et al., 2009), suggesting that miRs could be used as therapeutic interventions for damaged muscles.

4.4.3 myomiRs-1, -133 and -206 Effects on Skeletal Muscle Cells

Among the known myomiRs, miR-1, miR-133 and miR-206 have specific effects on skeletal muscle cells. Recently, it has been reported that miR-206 targets the glucose-6-phosphate dehydrogenase (G6PD) gene and suppresses its expression resulting in the inhibition of muscle cell proliferation and cell cycle arrest in G0/G1 phase (Jiang et al., 2019).

Both miR-1 and miR-133 are transcribed from the same loci on chromosomes 18 and 20 but inhibit the expression of different genes. MiR-1 supports muscle development by repressing histone deacetylase 4 (HDAC4) expression, which inhibits myocyte differentiation through suppression of myocyte enhancer factor 2 (MEF2) (Chen et al., 2010). The role of miR-133 is still debatable since results from different studies are contradictory. It is reported that miR-133a inhibits the fusion of myoblasts by targeting serum response factor (SRF) gene expression, which supports the growth and differentiation of muscle cells through metastasis-associated lung adenocarcinoma transcript 1 (Malat1) modulation (X. Han et al., 2015). Furthermore, miR-133 decreases Cyclin D1 expression and inhibits myoblast proliferation by G1 phase arrest stimulation and suppression of the transcription factor Sp1 (D. Zhang et al., 2012). It has been also demonstrated that miR-133 may reduce ERK1/2 kinases expression and promote myoblast proliferation (Feng et al., 2013). It is also of great importance that the Wnt/β-catenin signaling cascade is involved in skeletal muscle gene regulation by myomiRs. In particular, Wnt3 has been proven to enhance miR-133b and miR-206 expression, which inhibit Pax7, but not miR-1 and miR-133a (Cui et al., 2019).

4.4.4 miR Involvement in Signaling Pathways

On the other hand, there are also miRs that induce myogenesis by interacting with molecules involved in several cell-signaling pathways. For example, miR-26a and miR-214 are upregulated during myogenic differentiation and suppress the histone methyltransferase enhancer of zeste homolog 2 (Ezh2), a known myogenic inhibitor (Juan et al., 2009; Wong and Tellam, 2008). Furthermore, selected miRs have been demonstrated to inhibit the myogenic differentiation acting on transforming growth factor-beta (TGFβ) and Wnt/β-catenin signaling. In addition, miR-499 induces the proliferation of C2C12 cells by decreasing TGFβ-receptor 1 expression, while Smad7 is a target for miR-216a altering muscle cell differentiation (Wu et al., 2019; Z. Yang et al., 2019). It is also known that miR-26a, exhibiting elevated expression levels in skeletal muscle, suppresses Smad1 and Smad4 and causes alterations in TGFβ-signaling pathway which has an effect on the expression of MyoD and myogenin (Dey et al., 2012). Furthermore, miR-675-3p and miR-675-5p target the Smad transcription factors Smad1 and Smad5 as well as the DNA replication initiation factor Cdc6, resulting in muscle differentiation and regeneration (Dey et al., 2014). Other miRs enhance myogenesis by targeting

the NF-κB signaling pathway. miR-29 targets yin yang 1 (YY1) transcription factor and its binding protein Rybp, which repress genes involved in myogenic differentiation. The formation of the complex Rybp/YY1/Ezh2/HDAC4 leads to suppression of miR-29 expression, but, during myogenesis, MyoD/SRF dislocates the complex and induces miR-29 expression (Wang et al., 2008; Zhou et al., 2012). miR-17 and miR-20a can also promote C2C12 proliferation and primary bovine satellite cell activation by targeting Ccnd2, Jak1 and Rhoc genes that are essential regulators of cell proliferation and fusion (Kong et al., 2019). Further studies have suggested that miR-143-3p expression is elevated during C2C12 myoblast proliferation and its overexpression decreases the expression levels of MyoD, MyoG, myf5 and MyHC, by inhibition of Wnt5a, LRP5, Axin2 and β-catenin (Du et al., 2016).

Finally, miRs have been implicated in the insulin-like growth factor/insulin-like growth factor 1 receptor (IGF/IGF1R) signaling pathway. miR-125b, miR-133 and miR-199a-3p decrease cell differentiation and muscle regeneration potential through IGF-1/AKT/mTOR-signaling pathway (Ge et al., 2011; Jia et al., 2013). Importantly, miR-143 regulates the insulin growth factor-binding protein 5 (Igfbp5) in primary myoblasts and its expression in satellite cells from old mice is dysregulated. Thus, suppression of miR-143 in the aged skeletal muscle could contribute to myogenesis as a compensatory mechanism (Soriano-Arroquia, McCormick, et al., 2016). The main miRs affecting the different stages of skeletal myogenesis are summarized in Figure 4.1.

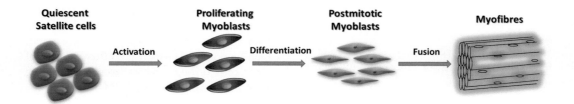

Quiescence	Cell proliferation	Myogenic differentiation
miR-31, miR-195, miR-489	*Inducers:* miR-27a, miR-133, miR-199, miR-208b, miR-499, miR-351	*Inducers:* miR-1, miR-206, miR-26a, miR-29, miR-27b, miR-133, miR-181, miR-486, miR-34c
Activation		
miR-1, miR-206, miR-27b, miR-486	*Inhibitors:* miR-1, miR-27b, miR-29, miR-34c, miR-128a, miR-206, miR-323-5p, miR-669	*Inhibitors:* miR-124, miR-125b, miR-155, miR-143-3p, miR-499, miR-323-5p. miR-23a

Figure 4.1 A schematic summary of miRs implicated in skeletal muscle myogenesis.

4.5 miRs IN SKELETAL MUSCLE AGING AND SARCOPENIA

4.5.1 Differential Expression of miRs in Young and Old Skeletal Muscle

A large number of comparative studies of miR expression levels between young and old muscle tissues of rodents, monkeys and humans have demonstrated that miRs play an important role in the impairment of skeletal muscle homeostasis during aging (Drummond et al., 2008; Hamrick et al., 2010; Mercken et al., 2013). Precursor miR expression measurements using adult healthy skeletal muscle biopsies revealed that miR-1, miR-133a and miR-206 were upregulated in aged individuals (Drummond et al., 2008), but this difference was alleviated when the expression of the corresponding mature miRs was analyzed, highlighting the importance in miR form and balance in aging. Another study focused on miR-206 and miR-21 reported that these two miRs are essential in muscle atrophy by targeting transcription factor YY1 and the translational initiator factor eIF4E3 (Soares et al., 2014). The levels of miR-let-7b and -7e were found higher in the elderly in comparison to young human subjects (Drummond et al., 2011). It has been suggested that the miR-let-7 family targets CDK6, CDC25A and CDC34 resulting in the inhibition of satellite cell activation in aged skeletal muscle (Drummond et al., 2011).

4.5.2 miR-181a Regulates Skeletal Muscle Aging

Small RNA-sequencing data have reported differential expression of several miRs after comparison between young and aged skeletal muscles in animal models and human samples (Hamrick et al., 2010; Mercken et al., 2013; Zheng et al., 2018). Many studies have explored the role of miR-181a in the aged muscle tissue. During skeletal muscle differentiation, miR-181a is upregulated and decreases the expression of the homeobox protein Hox-A11, which inhibits MyoD, and leads to high expression of myogenin (Naguibneva et al., 2006). In contrast, during aging miR-181a levels are decreased, leading to the upregulation of TGFβ-related activin receptor type IIA (ActRIIA) which is associated with inhibition of satellite cell proliferation through phosphorylation of Smad2 and Smad3 (Zacharewicz et al., 2013). Furthermore, miR-181a has been also implicated in inflammatory pathways during aging and has

been found to regulate the expression of pro-inflammatory cytokines including IL-6-and -8, IL-1β and TNF-α resulting in inflammaging (Xie et al., 2013). Finally, miR-181a restoration in aged skeletal muscle promotes mitochondrial dynamics by regulating p62/SQSTM1, parkin and protein deglycase DJ-1 (PARK7) (Borja-Gonzalez et al., 2020; Goljanek-Whysall et al., 2020).

4.5.3 miR-Dependent Regulation of Longevity-Related SIRT1 in Aged Skeletal Muscle

Recent studies have highlighted miR-181a as crucial miR in aging due to its interaction with sirtuins, mainly Sirtuin1 (SIRT1). Sirtuins are known histone deacetylases which promote longevity by inhibition of cellular senescence via interactions with IGF1, AMPK and FOXO-signaling pathways and support skeletal muscle mass preservation via regulation of autophagy (Carnio et al., 2014; Garcia-Prat et al., 2016; Lapierre et al., 2015; Masiero and Sandri, 2010). It is known that miR-181a down-regulation leads to SIRT1 increase in aged mouse skeletal (Soriano-Arroquia, House, et al., 2016). Another miR, miR-195, has been shown to reduce SIRT1 and telomerase reverse transcriptase (TERT) levels in aged skeletal myogenic cells (Kondo et al., 2016), while miR-29 upregulation increases senescence-associated β-galactosidase levels by targeting IGF-1 and the p85α regulatory subunit of PI3K (Hu et al., 2014). In vitro studies suggest that miR-431 may support myogenesis in aged mouse myoblasts by inhibition of Smad4 (Lee et al., 2015) while Notch1 suppresses miR-155 expression, and Notch1 loss induces miR-155 overexpression, an inflammation-related miR (Onodera et al., 2018). The elevated level of oxidative stress is a known factor that contributes to age-related skeletal muscle dysfunction and muscle cell apoptosis. miR-434-3p expression is reduced in aged mice, leading to mitochondrial apoptosis via the increase of eukaryotic translation initiation factor 5A1 (eIF5A1), a factor demonstrated to induce apoptosis via the mitochondrial apoptotic pathway (Pardo et al., 2017). Finally, one more miR that may play a central role in sarcopenia is miR-34, which also targets SIRT1 as well as vascular endothelial growth factor (VEGF) and has been found increased in elderly people with sarcopenia, along with miR-449b-5p and miR-424-5p (Connolly et al., 2018; Yamakuchi et al., 2008; Zheng et al., 2018).

4.6 PROTEOSTASIS LOSS AND AUTOPHAGY REGULATION BY miRs IN AGED SKELETAL MUSCLE

The disruption of the balance between protein synthesis and degradation leads to loss of proteostasis and impairment of the interplay between these two procedures. Skeletal muscle is hugely affected by this dysregulation as a tissue that mainly consists of protein but also actively participates in protein homeostasis through basic metabolism (Attaix et al., 2005; Fernando et al., 2019). Evidence from a number of studies suggest that the aged muscle fails to respond to external anabolic stimuli, e.g. exercise resistance, and reduces the rates of protein synthesis (Breen and Phillips, 2011; Deutz et al., 2014). Furthermore, the protein degradation machinery is also affected in skeletal muscle during aging (Bowen et al., 2015; Strucksberg et al., 2010; Wohlgemuth et al., 2010) and especially the ubiquitin-proteasome pathway (UPS), which induces muscle atrophy (Chai et al., 2003; Sishi et al., 2011). Muscle RING finger 1 (MuRF1) and muscle atrophy F-box (MAFbx) are E3 ligases which are key mediators of protein ubiquitination in skeletal muscle (Hartmann-Petersen and Gordon, 2004; Ratti et al., 2015). It has been demonstrated that MuRF1 and MAFbx are upregulated in atrophic muscle and, reversely, their inhibition results in the decrease of muscle loss (Clavel et al., 2006; Eddins et al., 2011). miR-23a has been implicated in the regulation of MuRF1 and MAFbx and overexpression of miR-23a in mice results in reduced muscle loss, which is caused by glucocorticoid treatment (Wada et al., 2011). Furthermore, miR-1 myomiR is increased in the mouse model of dexamethasone-induced muscle atrophy, which leads to upregulation of MuRF1 and MAFbx expression via the HSP70/PKB/Akt/FOXO3 signaling pathway (Kukreti et al., 2013). Furthermore, miR-199/214 can also regulate the UPS via twist basic helix–loop–helix transcription factor 1 (TWIST1) (Baumgarten et al., 2013). Recent studies have also shown that caloric restriction improves mitochondrial proteostasis and may prevent sarcopenia through the action of miRs (Rhoads et al., 2020; R. Zhang et al., 2019). Overall, miRs are considered as central regulators of the UPS activation and further studies are needed to unravel the mechanisms of miR-mediated proteostasis loss in sarcopenia.

Another important catabolic biological process that contributes to age-related sarcopenia is autophagy. Reactive oxygen species (ROS) accumulation during aging, which results in sarcopenia onset, is related to autophagy dysregulation and excessive activity of autophagosomes causing cellular stress (Fan et al., 2016; Terman and Brunk, 2006). miR-34 is upregulated during aging and is known to modulate autophagy-associated proteins (Yamakuchi et al., 2008; J. Yang et al., 2013). Recently, it was suggested that miR-378, a highly abundant miR in skeletal muscle, promotes normal muscle homeostasis by coordinating autophagy and apoptosis via FOXO signaling and Caspase 9 regulation (Li et al., 2018). The roles of specific miRs on skeletal muscle aging are illustrated in Figure 4.2.

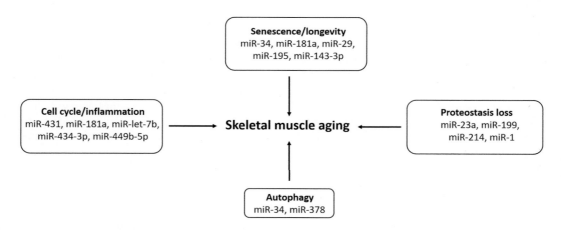

Figure 4.2 miRs involved in post-transcriptional regulation of skeletal muscle aging.

4.7 CONCLUSIONS AND FUTURE PERSPECTIVES

The evolution of miRs has been remarkably rapid since their discovery almost three decades ago. It is now known that miRs are very important regulators of post-transcriptional modifications and can have essential effects on a variety of tissues. Skeletal muscle is inevitably affected by a large number of miRs, which seem to control major biological processes including myogenesis, homeostasis and disease development such as sarcopenia. Specific miRs play central roles by targeting a plethora of genes, which regulate skeletal muscle functions via cell signaling pathways. Therefore, modulation of miR expression can potentially lead to new therapeutic approaches and interventions to counteract age-related muscle loss. In support of miR-mediated treatments, a recent study reported the therapeutic potential of an intramuscular adenoviral delivery of miR-376c-3p against age-related muscle atrophy by targeting Atrogin-1 in mice (Shin et al., 2020).

REFERENCES

Adinolfi, A. M., Yamuy, J., Morales, F. R., et al. 1991. Segmental demyelination in peripheral nerves of old cats. Neurobiol Aging, 12(2), 175–179. doi:10.1016/0197-4580(91)90058-r

Asfour, H. A., Allouh, M. Z., & Said, R. S. 2018. Myogenic regulatory factors: the orchestrators of myogenesis after 30 years of discovery. Exp Biol Med (Maywood), 243(2), 118–128. doi:10.1177/1535370217749494

Attaix, D., Ventadour, S., Codran, A., et al. 2005. The ubiquitin-proteasome system and skeletal muscle wasting. Essays Biochem, 41, 173–186. doi:10.1042/EB0410173

Baumgarten, A., Bang, C., Tschirner, A., et al. 2013. TWIST1 regulates the activity of ubiquitin proteasome system via the miR-199/214 cluster in human end-stage dilated cardiomyopathy. Int J Cardiol, 168(2), 1447–1452. doi:10.1016/j.ijcard.2012.12.094

Borja-Gonzalez, M., Casas-Martinez, J. C., McDonagh, B., et al. 2020. Aging science talks: the role of miR-181a in age-related loss of muscle mass and function. Transl Med Aging, 4, 81–85. doi:10.1016/j.tma.2020.07.001

Bortz, W. M., 2nd. 2002. A conceptual framework of frailty: a review. J Gerontol A Biol Sci Med Sci, 57(5), M283–288. doi:10.1093/gerona/57.5.m283

Bowen, T. S., Schuler, G., & Adams, V. 2015. Skeletal muscle wasting in cachexia and sarcopenia: molecular pathophysiology and impact of exercise training. J Cachexia Sarcopenia Muscle, 6(3), 197–207. doi:10.1002/jcsm.12043

Breen, L., & Phillips, S. M. 2011. Skeletal muscle protein metabolism in the elderly: interventions to counteract the 'anabolic resistance' of ageing. Nutr Metab (Lond), 8, 68. doi:10.1186/1743-7075-8-68

Brodersen, P., & Voinnet, O. 2009. Revisiting the principles of microRNA target recognition and mode of action. Nat Rev Mol Cell Biol, 10(2), 141–148. doi:10.1038/nrm2619

Brown, D. M., & Goljanek-Whysall, K. 2015. microRNAs: modulators of the underlying pathophysiology of sarcopenia? Ageing Res Rev, 24(Pt B), 263–273. doi:10.1016/j.arr.2015.08.007

Brown, J. C., Harhay, M. O., & Harhay, M. N. 2016. Sarcopenia and mortality among a population-based sample of community-dwelling older adults. J Cachexia Sarcopenia Muscle, 7(3), 290–298. doi:10.1002/jcsm.12073

Calderon-Dominguez, M., Belmonte, T., Quezada-Feijoo, M., et al. 2020. Emerging role of microRNAs in dilated cardiomyopathy: evidence regarding etiology. Transl Res, 215, 86–101. doi:10.1016/j.trsl.2019.08.007

Carnio, S., LoVerso, F., Baraibar, M. A., et al. 2014. Autophagy impairment in muscle induces neuromuscular junction degeneration and precocious aging. Cell Rep, 8(5), 1509–1521. doi:10.1016/j.celrep.2014.07.061

Chai, J., Wu, Y., & Sheng, Z. Z. 2003. Role of ubiquitin-proteasome pathway in skeletal muscle wasting in rats with endotoxemia. Crit Care Med, 31(6), 1802–1807. doi:10.1097/01.CCM.0000069728.49939.E4

Chen, J. F., Tao, Y., Li, J., et al. 2010. microRNA-1 and microRNA-206 regulate skeletal muscle satellite cell proliferation and differentiation by repressing Pax7. J Cell Biol, 190(5), 867–879. doi:10.1083/jcb.200911036

Clavel, S., Coldefy, A. S., Kurkdjian, E., et al. 2006. Atrophy-related ubiquitin ligases, atrogin-1 and MuRF1 are up-regulated in aged rat tibialis anterior muscle. Mech Ageing Dev, 127(10), 794–801. doi:10.1016/j.mad.2006.07.005

Connolly, M., Paul, R., Farre-Garros, R., et al. 2018. miR-424-5p reduces ribosomal RNA and protein synthesis in muscle wasting. J Cachexia Sarcopenia Muscle, 9(2), 400–416. doi:10.1002/jcsm.12266

Cosgrove, B. D., Gilbert, P. M., Porpiglia, E., et al. 2014. Rejuvenation of the muscle stem cell population restores strength to injured aged muscles. *Nat Med*, 20(3), 255–264. doi:10.1038/nm.3464

Crist, C. G., Montarras, D., Pallafacchina, G., et al. 2009. Muscle stem cell behavior is modified by microRNA-27 regulation of Pax3 expression. *Proc Natl Acad Sci U S A*, 106(32), 13383–13387. doi:10.1073/pnas.0900210106

Cui, S., Li, L., Mubarokah, S. N., et al. 2019. Wnt/beta-catenin signaling induces the myomiRs miR-133b and miR-206 to suppress Pax7 and induce the myogenic differentiation program. *J Cell Biochem*, 120(8), 12740–12751. doi:10.1002/jcb.28542

Delbono, O. 2003. Neural control of aging skeletal muscle. *Aging Cell*, 2(1), 21–29. doi:10.1046/j.1474-9728.2003.00011.x

Deutz, N. E., Bauer, J. M., Barazzoni, R., et al. 2014. Protein intake and exercise for optimal muscle function with aging: recommendations from the ESPEN Expert Group. *Clin Nutr*, 33(6), 929–936. doi:10.1016/j.clnu.2014.04.007

Dey, B. K., Gagan, J., Yan, Z., et al. 2012. miR-26a is required for skeletal muscle differentiation and regeneration in mice. *Genes Dev*, 26(19), 2180–2191. doi:10.1101/gad.198085.112

Dey, B. K., Pfeifer, K., & Dutta, A. 2014. The H19 long noncoding RNA gives rise to microRNAs miR-675-3p and miR-675-5p to promote skeletal muscle differentiation and regeneration. *Genes Dev*, 28(5), 491–501. doi:10.1101/gad.234419.113

Diebel, K. W., Claypool, D. J., & van Dyk, L. F. 2014. A conserved RNA polymerase III promoter required for gammaherpesvirus TMER transcription and microRNA processing. *Gene*, 544(1), 8–18. doi:10.1016/j.gene.2014.04.026

Drummond, M. J., McCarthy, J. J., Fry, C. S., et al. 2008. Aging differentially affects human skeletal muscle microRNA expression at rest and after an anabolic stimulus of resistance exercise and essential amino acids. *Am J Physiol Endocrinol Metab*, 295(6), E1333–1340. doi:10.1152/ajpendo.90562.2008

Drummond, M. J., McCarthy, J. J., Sinha, M., et al. 2011. Aging and microRNA expression in human skeletal muscle: a microarray and bioinformatics analysis. *Physiol Genomics*, 43(10), 595–603. doi:10.1152/physiolgenomics.00148.2010

Du, J., Zhang, Y., Shen, L., et al. 2016. Effect of miR-143-3p on C2C12 myoblast differentiation. *Biosci Biotechnol Biochem*, 80(4), 706–711. doi:10.1080/09168451.2015.1123604

Dweep, H., Sticht, C., Pandey, P., et al. 2011. miRWalk-database: prediction of possible miRNA binding sites by "walking" the genes of three genomes.

J Biomed Inform, 44(5), 839–847. doi:10.1016/j.jbi.2011.05.002

Eddins, M. J., Marblestone, J. G., Suresh Kumar, K. G., et al. 2011. Targeting the ubiquitin E3 ligase MuRF1 to inhibit muscle atrophy. *Cell Biochem Biophys*, 60(1–2), 113–118. doi:10.1007/s12013-011-9175-7

Einsiedel, L. J., & Luff, A. R. 1992. Alterations in the contractile properties of motor units within the ageing rat medial gastrocnemius. *J Neurol Sci*, 112(1–2), 170–177. doi:10.1016/0022-510x(92)90147-d

Espinoza, S., & Walston, J. D. 2005. Frailty in older adults: insights and interventions. *Cleve Clin J Med*, 72(12), 1105–1112. doi:10.3949/ccjm.72.12.1105

Fan, J., Kou, X., Jia, S., et al. 2016. Autophagy as a potential target for sarcopenia. *J Cell Physiol*, 231(7), 1450–1459. doi:10.1002/jcp.25260

Feng, Y., Niu, L. L., Wei, W., et al. 2013. A feedback circuit between miR-133 and the ERK1/2 pathway involving an exquisite mechanism for regulating myoblast proliferation and differentiation. *Cell Death Dis*, 4, e934. doi:10.1038/cddis.2013.462

Fernando, R., Drescher, C., Nowotny, K., et al. 2019. Impaired proteostasis during skeletal muscle aging. *Free Radic Biol Med*, 132, 58–66. doi:10.1016/j.freeradbiomed.2018.08.037

Fried, L. P., Tangen, C. M., Walston, J., et al. 2001. Frailty in older adults: evidence for a phenotype. *J Gerontol A Biol Sci Med Sci*, 56(3), M146–156. doi:10.1093/gerona/56.3.m146

Garcia-Prat, L., Martinez-Vicente, M., Perdiguero, E., et al. 2016. Autophagy maintains stemness by preventing senescence. *Nature*, 529(7584), 37–42. doi:10.1038/nature16187

Ge, Y., Sun, Y., & Chen, J. 2011. IGF-II is regulated by microRNA-125b in skeletal myogenesis. *J Cell Biol*, 192(1), 69–81. doi:10.1083/jcb.201007165

Goljanek-Whysall, K., Sweetman, D., Abu-Elmagd, M., et al. 2011. MicroRNA regulation of the paired-box transcription factor Pax3 confers robustness to developmental timing of myogenesis. *Proc Natl Acad Sci U S A*, 108(29), 11936–11941. doi:10.1073/pnas.1105362108

Goljanek-Whysall, K., Pais, H., Rathjen, T., et al. 2012. Regulation of multiple target genes by miR-1 and miR-206 is pivotal for C2C12 myoblast differentiation. *J Cell Sci*, 125(Pt 15), 3590–3600. doi:10.1242/jcs.101758

Goljanek-Whysall, K., Soriano-Arroquia, A., McCormick, R., et al. 2020. miR-181a regulates p62/SQSTM1, parkin, and protein DJ-1 promoting mitochondrial dynamics in skeletal muscle aging. *Aging Cell*, 19(4), e13140. doi:10.1111/acel.13140

Hamrick, M. W., Herberg, S., Arounleut, P., et al. 2010. The adipokine leptin increases skeletal muscle mass and significantly alters skeletal muscle miRNA expression profile in aged mice. *Biochem Biophys Res Commun*, 400(3), 379–383. doi:10.1016/j.bbrc.2010.08.079

Han, J., Lee, Y., Yeom, K. H., et al. 2006. Molecular basis for the recognition of primary microRNAs by the Drosha-DGCR8 complex. *Cell*, 125(5), 887–901. doi:10.1016/j.cell.2006.03.043

Han, X., Yang, F., Cao, H., et al. 2015. Malat1 regulates serum response factor through miR-133 as a competing endogenous RNA in myogenesis. *FASEB J*, 29(7), 3054–3064. doi:10.1096/fj.14-259952

Hartmann-Petersen, R., & Gordon, C. 2004. Proteins interacting with the 26S proteasome. *Cell Mol Life Sci*, 61(13), 1589–1595. doi:10.1007/s00018-004-4132-x

Hirai, H., Verma, M., Watanabe, S., et al. 2010. MyoD regulates apoptosis of myoblasts through microRNA-mediated down-regulation of Pax3. *J Cell Biol*, 191(2), 347–365. doi:10.1083/jcb.201006025

Horak, M., Novak, J., & Bienertova-Vasku, J. 2016. Muscle-specific microRNAs in skeletal muscle development. *Dev Biol*, 410(1), 1–13. doi:10.1016/j.ydbio.2015.12.013

Hu, Z., & Bruno, A. E. 2011. The influence of 3′ UTRs on microRNA function inferred from human SNP data. *Comp Funct Genomics*, 2011, 910769. doi:10.1155/2011/910769

Hu, Z., Klein, J. D., Mitch, W. E., et al. 2014. MicroRNA-29 induces cellular senescence in aging muscle through multiple signaling pathways. *Aging (Albany NY)*, 6(3), 160–175. doi:10.18632/aging.100643

Jia, L., Li, Y. F., Wu, G. F., et al. 2013. MiRNA-199a-3p regulates C2C12 myoblast differentiation through IGF-1/AKT/mTOR signal pathway. *Int J Mol Sci*, 15(1), 296–308. doi:10.3390/ijms15010296

Jiang, A., Dong, C., Li, B., et al. 2019. MicroRNA-206 regulates cell proliferation by targeting G6PD in skeletal muscle. *FASEB J*, 33(12), 14083–14094. doi:10.1096/fj.201900502RRRR

Juan, A. H., Kumar, R. M., Marx, J. G., et al. 2009. Mir-214-dependent regulation of the polycomb protein Ezh2 in skeletal muscle and embryonic stem cells. *Mol Cell*, 36(1), 61–74. doi:10.1016/j.molcel.2009.08.008

Juzwik, C. A., Drake, S. S., Zhang, Y., et al. 2019. microRNA dysregulation in neurodegenerative diseases: a systematic review. *Prog Neurobiol*, 182, 101664. doi:10.1016/j.pneurobio.2019.101664

Kim, Y. K., Kim, B., & Kim, V. N. 2016. Re-evaluation of the roles of DROSHA, Export in 5, and DICER in microRNA biogenesis. *Proc Natl Acad Sci U S A*, 113(13), E1881–1889. doi:10.1073/pnas.1602532113

Kondo, H., Kim, H. W., Wang, L., et al. 2016. Blockade of senescence-associated microRNA-195 in aged skeletal muscle cells facilitates reprogramming to produce induced pluripotent stem cells. *Aging Cell*, 15(1), 56–66. doi:10.1111/acel.12411

Kong, Y. W., Cannell, I. G., de Moor, C. H., et al. 2008. The mechanism of micro-RNA-mediated translation repression is determined by the promoter of the target gene. *Proc Natl Acad Sci U S A*, 105(26), 8866–8871. doi:10.1073/pnas.0800650105

Kong, D., He, M., Yang, L., et al. 2019. MiR-17 and miR-19 cooperatively promote skeletal muscle cell differentiation. *Cell Mol Life Sci*, 76(24), 5041–5054. doi:10.1007/s00018-019-03165-7

Kosek, D. J., Kim, J. S., Petrella, J. K., et al. 2006. Efficacy of 3 days/wk resistance training on myofiber hypertrophy and myogenic mechanisms in young vs. older adults. *J Appl Physiol (1985)*, 101(2), 531–544. doi:10.1152/japplphysiol.01474.2005

Kukreti, H., Amuthavalli, K., Harikumar, A., et al. 2013. Muscle-specific microRNA1 (miR1) targets heat shock protein 70 (HSP70) during dexamethasone-mediated atrophy. *J Biol Chem*, 288(9), 6663–6678. doi:10.1074/jbc.M112.390369

Lapierre, L. R., Kumsta, C., Sandri, M., et al. 2015. Transcriptional and epigenetic regulation of autophagy in aging. *Autophagy*, 11(6), 867–880. doi:10.1080/15548627.2015.1034410

Larsson, L., & Ansved, T. 1995. Effects of ageing on the motor unit. *Prog Neurobiol*, 45(5), 397–458. doi:10.1016/0301-0082(95)98601-z

Lee, K. P., Shin, Y. J., & Kwon, K. S. 2015. microRNA for determining the age-related myogenic capabilities of skeletal muscle. *BMB Rep*, 48(11), 595–596. doi:10.5483/bmbrep.2015.48.11.211

Lewis, B. P., Burge, C. B., & Bartel, D. P. 2005. Conserved seed pairing, often flanked by adenosines, indicates that thousands of human genes are microRNA targets. *Cell*, 120(1), 15–20. doi:10.1016/j.cell.2004.12.035

Lexell, J., Downham, D., & Sjostrom, M. 1986. Distribution of different fibre types in human skeletal muscles: fibre type arrangement in m. vastus lateralis from three groups of healthy men between 15 and 83 years. *J Neurol Sci*, 72(2-3), 211–222. doi:10.1016/0022-510x(86)90009-2

Lexell, J., Taylor, C. C., & Sjostrom, M. 1988. What is the cause of the ageing atrophy? Total number, size and proportion of different fiber types studied in whole vastus lateralis muscle from 15- to 83-year-old men. *J Neurol Sci*, 84(2–3), 275–294. doi:10.1016/0022-510x(88)90132-3

Li, Y., Jiang, J., Liu, W., et al. 2018. microRNA-378 promotes autophagy and inhibits apoptosis in skeletal muscle. *Proc Natl Acad Sci U S A*, 115(46), E10849–E10858. doi:10.1073/pnas.1803377115

Masiero, E., & Sandri, M. 2010. Autophagy inhibition induces atrophy and myopathy in adult skeletal muscles. *Autophagy*, 6(2), 307–309. doi:10.4161/auto.6.2.11137

Mercken, E. M., Majounie, E., Ding, J., et al. 2013. Age-associated miRNA alterations in skeletal muscle from rhesus monkeys reversed by caloric restriction. *Aging (Albany NY)*, 5(9), 692–703. doi:10.18632/aging.100598

Morley, J. E. 2008. Sarcopenia: diagnosis and treatment. *J Nutr Health Aging*, 12(7), 452–456. doi:10.1007/BF02982705

Naguibneva, I., Ameyar-Zazoua, M., Polesskaya, A., et al. 2006. The microRNA miR-181 targets the homeobox protein Hox-A11 during mammalian myoblast differentiation. *Nat Cell Biol*, 8(3), 278–284. doi:10.1038/ncb1373

Narici, M. V., & Maffulli, N. 2010. Sarcopenia: characteristics, mechanisms and functional significance. *Br Med Bull*, 95, 139–159. doi:10.1093/bmb/ldq008

Nilwik, R., Snijders, T., Leenders, M., et al. 2013. The decline in skeletal muscle mass with aging is mainly attributed to a reduction in type II muscle fiber size. *Exp Gerontol*, 48(5), 492–498. doi:10.1016/j.exger.2013.02.012

O'Rourke, J. R., Georges, S. A., Seay, H. R., et al. 2007. Essential role for Dicer during skeletal muscle development. *Dev Biol*, 311(2), 359–368. doi:10.1016/j.ydbio.2007.08.032

Oikawa, S., Lee, M., & Akimoto, T. 2019. Conditional deletion of Dicer in adult mice impairs skeletal muscle regeneration. *Int J Mol Sci*, 20(22). doi:10.3390/ijms20225686

Onodera, Y., Teramura, T., Takehara, T., et al. 2018. Inflammation-associated miR-155 activates differentiation of muscular satellite cells. *PLoS One*, 13(10), e0204860. doi:10.1371/journal.pone.0204860

Pardo, P. S., Hajira, A., Boriek, A. M., et al. 2017. MicroRNA-434-3p regulates age-related apoptosis through eIF5A1 in the skeletal muscle. *Aging (Albany NY)*, 9(3), 1012–1029. doi:10.18632/aging.101207

Quattrocelli, M., & Sampaolesi, M. 2015. The mesmiRizing complexity of microRNAs for striated muscle tissue engineering. *Adv Drug Deliv Rev*, 88, 37–52. doi:10.1016/j.addr.2015.04.011

Ratti, F., Ramond, F., Moncollin, V., et al. 2015. Histone deacetylase 6 is a FoxO transcription factor-dependent effector in skeletal muscle atrophy. *J Biol Chem*, 290(7), 4215–4224. doi:10.1074/jbc.M114.600916

Rhoads, T. W., Clark, J. P., Gustafson, G. E., et al. 2020. Molecular and functional networks linked to sarcopenia prevention by caloric restriction in rhesus monkeys. *Cell Syst*, 10(2), 156–168 e155. doi:10.1016/j.cels.2019.12.002

Schmidt, M., Schuler, S. C., Huttner, S. S., et al. 2019. Adult stem cells at work: regenerating skeletal muscle. *Cell Mol Life Sci*, 76(13), 2559–2570. doi:10.1007/s00018-019-03093-6

Schraivogel, D., & Meister, G. 2014. Import routes and nuclear functions of Argonaute and other small RNA-silencing proteins. *Trends Biochem Sci*, 39(9), 420–431. doi:10.1016/j.tibs.2014.07.004

Shin, Y. J., Kwon, E. S., Lee, S. M., et al. 2020. A subset of microRNAs in the Dlk1-Dio3 cluster regulates age-associated muscle atrophy by targeting Atrogin-1. *J Cachexia Sarcopenia Muscle*, 11(5), 1336–1350. doi:10.1002/jcsm.12578

Sishi, B., Loos, B., Ellis, B., et al. 2011. Diet-induced obesity alters signalling pathways and induces atrophy and apoptosis in skeletal muscle in a prediabetic rat model. *Exp Physiol*, 96(2), 179–193. doi:10.1113/expphysiol.2010.054189

Slack, F. J., & Chinnaiyan, A. M. 2019. The role of non-coding RNAs in oncology. *Cell*, 179(5), 1033–1055. doi:10.1016/j.cell.2019.10.017

Snijders, T., & Parise, G. 2017. Role of muscle stem cells in sarcopenia. *Curr Opin Clin Nutr Metab Care*, 20(3), 186–190. doi:10.1097/MCO.0000000000000360

Soares, R. J., Cagnin, S., Chemello, F., et al. 2014. Involvement of microRNAs in the regulation of muscle wasting during catabolic conditions. *J Biol Chem*, 289(32), 21909–21925. doi:10.1074/jbc.M114.561845

Soriano-Arroquia, A., House, L., Tregilgas, L., et al. 2016. The functional consequences of age-related changes in microRNA expression in skeletal muscle. *Biogerontology*, 17(3), 641–654. doi:10.1007/s10522-016-9638-8

Soriano-Arroquia, A., McCormick, R., Molloy, A. P., et al. 2016. Age-related changes in miR-143-3p:Igfbp5 interactions affect muscle regeneration. *Aging Cell*, 15(2), 361–369. doi:10.1111/acel.12442

Sousa-Victor, P., Gutarra, S., Garcia-Prat, L., et al. 2014. Geriatric muscle stem cells switch reversible quiescence into senescence. *Nature*, 506(7488), 316–321. doi:10.1038/nature13013

Strucksberg, K. H., Tangavelou, K., Schroder, R., et al. 2010. Proteasomal activity in skeletal muscle: a matter of assay design, muscle type, and age. *Anal Biochem*, 399(2), 225–229. doi:10.1016/j.ab.2009.12.026

Terman, A., & Brunk, U. T. 2006. Oxidative stress, accumulation of biological 'garbage', and aging. *Antioxid Redox Signal*, 8(1–2), 197–204. doi:10.1089/ars.2006.8.197

van Rooij, E., Quiat, D., Johnson, B. A., et al. 2009. A family of microRNAs encoded by myosin genes governs myosin expression and muscle performance. *Dev Cell*, 17(5), 662–673. doi:10.1016/j.devcel.2009.10.013

Vechetti, I. J., Jr., Wen, Y., Chaillou, T., et al. 2019. Life-long reduction in myomiR expression does not adversely affect skeletal muscle morphology. *Sci Rep*, 9(1), 5483. doi:10.1038/s41598-019-41476-8

von Maltzahn, J., Jones, A. E., Parks, R. J., et al. 2013. Pax7 is critical for the normal function of satellite cells in adult skeletal muscle. *Proc Natl Acad Sci U S A*, 110(41), 16474–16479. doi:10.1073/pnas.1307680110

Wada, S., Kato, Y., Okutsu, M., et al. 2011. Translational suppression of atrophic regulators by microRNA-23a integrates resistance to skeletal muscle atrophy. *J Biol Chem*, 286(44), 38456–38465. doi:10.1074/jbc.M111.271270

Wang, H., Garzon, R., Sun, H., et al. 2008. NF-kappaB-YY1-miR-29 regulatory circuitry in skeletal myogenesis and rhabdomyosarcoma. *Cancer Cell*, 14(5), 369–381. doi:10.1016/j.ccr.2008.10.006

Winter, J., Jung, S., Keller, S., et al. 2009. Many roads to maturity: microRNA biogenesis pathways and their regulation. *Nat Cell Biol*, 11(3), 228–234. doi:10.1038/ncb0309-228

Wohlgemuth, S. E., Seo, A. Y., Marzetti, E., et al. 2010. Skeletal muscle autophagy and apoptosis during aging: effects of calorie restriction and life-long exercise. *Exp Gerontol*, 45(2), 138–148. doi:10.1016/j.exger.2009.11.002

Wong, C. F., & Tellam, R. L. 2008. MicroRNA-26a targets the histone methyltransferase enhancer of Zeste homolog 2 during myogenesis. *J Biol Chem*, 283(15), 9836–9843. doi:10.1074/jbc.M709614200

Wu, J., Yue, B., Lan, X., et al. 2019. MiR-499 regulates myoblast proliferation and differentiation by targeting transforming growth factor beta receptor 1. *J Cell Physiol*, 234(3), 2523–2536. doi:10.1002/jcp.26903

Xie, W., Li, M., Xu, N., et al. 2013. MiR-181a regulates inflammation responses in monocytes and macrophages. *PLoS One*, 8(3), e58639. doi:10.1371/journal.pone.0058639

Yamakuchi, M., Ferlito, M., & Lowenstein, C. J. 2008. miR-34a repression of SIRT1 regulates apoptosis. *Proc Natl Acad Sci U S A*, 105(36), 13421–13426. doi:10.1073/pnas.0801613105

Yang, J., Chen, D., He, Y., et al. 2013. MiR-34 modulates Caenorhabditis elegans lifespan via repressing the autophagy gene atg9. *Age (Dordr)*, 35(1), 11–22. doi:10.1007/s11357-011-9324-3

Yang, Z., Song, C., Jiang, R., et al. 2019. Micro-ribonucleic acid-216a regulates bovine primary muscle cells proliferation and differentiation via targeting SMAD nuclear interacting protein-1 and Smad7. *Front Genet*, 10, 1112. doi:10.3389/fgene.2019.01112

Zacharewicz, E., Lamon, S., & Russell, A. P. 2013. MicroRNAs in skeletal muscle and their regulation with exercise, ageing, and disease. *Front Physiol*, 4, 266. doi:10.3389/fphys.2013.00266

Zhang, D., Li, X., Chen, C., et al. 2012. Attenuation of p38-mediated miR-1/133 expression facilitates myoblast proliferation during the early stage of muscle regeneration. *PLoS One*, 7(7), e41478. doi:10.1371/journal.pone.0041478

Zhang, R., Wang, X., Qu, J. H., et al. 2019. Caloric restriction induces microRNAs to improve mitochondrial proteostasis. *iScience*, 17, 155–166. doi:10.1016/j.isci.2019.06.028

Zheng, Y., Kong, J., Li, Q., et al. 2018. Role of miR-NAs in skeletal muscle aging. *Clin Interv Aging*, 13, 2407–2419. doi:10.2147/CIA.S169202

Zhou, L., Wang, L., Lu, L., et al. 2012. A novel target of microRNA-29, Ring1 and YY1-binding protein (Rybp), negatively regulates skeletal myogenesis. *J Biol Chem*, 287(30), 25255–25265. doi:10.1074/jbc.M112.357053

CHAPTER FIVE

mRNA Granules and Proteostasis in Aging and Age-Related Diseases

Fivos Borbolis and Popi Syntichaki

CONTENTS

5.1 INTRODUCTION

All eukaryotic cells possess several nuclear and cytoplasmic non-membrane-bound compartments that consist of heterogeneous mixtures of proteins and RNAs and allow the creation of distinct microenvironments that influence various cellular processes. Such membrane-less organelles include nucleoli, nuclear speckles, paraspeckles, nuclear gems, Cajal bodies and promyelocytic leukemia (PLM) bodies in the nucleus, and processing bodies or P-bodies (PBs), stress granules (SGs), germ granules and transport granules in the cytoplasm (Gomes and Shorter, 2019; van Leeuwen and Rabouille, 2019). These cytoplasmic condensates of ribonucleoprotein complexes (or mRNA granules for short) have emerged as important modulators of post-transcriptional and epigenetic gene expression. PBs and SGs are found in somatic cells, germ granules in embryos and transport granules in neurons. The composition and functions of PBs and SGs are analyzed in the next sections. Germ granules are involved in the storage and localization of nascent maternal transcripts and determine cell fate in oocytes and spatial patterning in embryos of lower metazoans (Seydoux, 2018). Transport granules suppress translation during mRNA transport in axons and dendrites and promote local translation at synapses upon activation by extracellular signals, two important features for synaptic plasticity, learning and memory (Hutten et al., 2014).

Cytoplasmic mRNA granules (PBs, SGs, germ and transport granules) have both distinct and overlapping constituents and exert diverse cellular functions, related to mRNA transport, translation, silencing, storage and decay. They are highly dynamic, able to exchange their materials with the surrounding cytoplasm and interact with each other (Anderson and Kedersha, 2009). Several studies suggest that mRNA granule assembly is promoted by multivalent interactions between their components (Tauber et al., 2020). The first observations of germ granules in the nematode *Caenorhabditis elegans* and subsequent data from *in vitro* and *in vivo* studies support that many of the membrane-less compartments assemble by phase separation, resembling the two phases

DOI: 10.1201/9781003048138-5

that are spontaneously formed in a mixture of oil and water. This liquid-like behavior enables rapid exchange of their components (Alberti and Dormann, 2019; Gomes and Shorter, 2019). In fact, this phase separation has been linked with the wide range of biological functions exerted by mRNA granules, providing spatiotemporal control or even tuning some biochemical reactions (Gomes and Shorter, 2019; Wang et al., 2018). Several key components of these granules are evolutionarily conserved and their manipulation by genetic or pharmacological means impacts development, cellular viability and organismal fitness (Buchan, 2014). It is now well established that mRNA granules could serve as key modulators of all aspects of RNA function, having a significant influence on cellular signaling and homeostasis.

Proteostasis (i.e. protein homeostasis) is critical for cellular and organismal viability as it functions to preserve proteome integrity of a living organism, in a constantly changing cellular environment. Proteostasis requires the precise control of protein synthesis, folding, transport and degradation, which is accomplished via a complex and adaptive network (Klaips et al., 2018). Cells engage this protein quality control (PQC) network under various forms of stress to counteract adverse effects on proteostasis. Stress-induced assembly of mRNA granules is part of the PQC mechanisms that regulate localization, storage and turnover of mRNAs under potentially harmful conditions. Moreover, imbalanced proteostasis can result in unfolded, misfolded or aggregated proteins, a phenomenon common in aging and age-related diseases (Taylor and Dillin, 2011). The aging process is characterized by a gradual failure to maintain proteostasis, mostly due to progressive decline of PQC mechanisms and uncontrollable aggregation of misfolded proteins. Protein dyshomeostasis is exacerbated in the presence of toxic or aggregation-prone proteins that are associated with neurodegeneration. Intriguingly, recent findings support a role of mRNA granules, and specifically SGs, in the formation of pathological protein aggregates. Impaired proteostasis can drive aberrant phase separation of cytoplasmic mRNA granules, promoting their transition from liquid droplets to solid, amyloid-like aggregates (Alberti and Dormann, 2019). These, in turn, might serve as a platform for more aggregates, which exacerbate proteostasis

failure and result in pathological inclusions, present in many protein-misfolding disorders (Ciryam et al., 2015). Disassembly and clearance of mRNA granules upon stress relief rely on major PQC mechanisms like chaperones, proteasome and autophagy. Here, we shortly overview the basic characteristics, dynamics, functions and regulation of cytoplasmic RNA granules, with emphasis on PBs and SGs, and discuss their connection with proteostasis. Given the established link between proteostasis control and aging, this connection highlights the importance of mRNA granules in human pathology and neurodegenerative disorders, offering new promising therapeutic avenues.

5.2 DEFINITIONS AND COMPONENTS OF PBs AND SGs

5.2.1 P-Bodies

PBs are condensed foci of ribonucleoprotein complexes, constitutively formed in the cytoplasm, and their number and size can vary in response to various stress signals. They contain untranslated mRNAs and proteins that are involved in 5′–3′ mRNA decay, translation repression, miRNA silencing or mRNA surveillance pathways (Borbolis and Syntichaki, 2015) (Table 5.1). Since PBs host mRNA decay intermediates and 5′–3′ mRNA decay factors, such as the mRNA deadenylase and decapping complexes, it was thought until recently that bulk mRNA degradation takes place into these granules (Decker and Parker, 2012). Subsequent studies, however, showed that PBs are not required for mRNA decay, which can occur co-translationally, but rather are shelters for mRNAs that escape degradation and can reenter the translation machinery (Ivanov et al., 2019). Recent data on purified constitutive PBs from human cells further support a major role in the stabilization and storage of specific, mostly AU-rich mRNAs, encoding regulatory factors of chromatin remodeling, transcription regulation, RNA processing, protein modifications and other cellular processes (Hubstenberger et al., 2017; Wang et al., 2018). Several stresses elicit the formation of PBs, resulting in the accumulation of translationally repressed mRNAs. Nevertheless, PB assembly may not be just a consequence of translation attenuation, as it can be dictated by distinct signaling pathways, depending on

TABLE 5.1

Composition of PBs and SGs

Name	Functions	Key components
Processing bodies (PBs)	Storing of translationally repressed mRNAs 5′–3′ mRNA degradation miRNA-mediated silencing mRNA surveillance	Deadenylation factors (CCR4, PAN3) mRNA decapping complex (DCP1/2) Decapping activators (LSM1-7, EDC3/4, PATL1, RCK/p54) Translation repressors (eIF4E-T, DDX3, CPEB1) miRNA function (GW182, AGO1-4) Nonsense-mediated decay (UPF1, SMG5/7) Ribonuclease (XRN1)
Stress granules (SGs)	Storing of translationally repressed mRNAs Sorting of mRNAs for reinitiation or decay	40S ribosomal subunit Translation initiation factors Translation/mRNA processing (PABP, ATAXIN-2, DDX3, FUS, TDP-43) Translation repression/mRNA stability (TIA-1, TIAR, RCK/p54) Scaffolding proteins (G3BP, RACK1) miRNA function (FMRP, AGO1-4) Ribonuclease (XRN1, G3BP)

SOURCE: Data derived from Becker et al. (2017) and references within.

the encountered stress (Kilchert et al., 2010). Indeed, yeast PBs contain distinct mRNA species in response to specific types of stress, which are either degraded or protected from decay (Wang et al., 2018). PB formation is affected by a network of multivalent RNA–protein and protein–protein interactions, as well as several post-translational protein modifications, like phosphorylation of the main decapping enzyme DCP2, ubiquitylation of its essential cofactor DCP1 and arginine methylation of the decapping activator LSM4 (Luo et al., 2018). Quantitative analysis of the major yeast PB resident proteins revealed that only a small subset is highly abundant and forms a stable core, whereas members of a second less ample group rapidly exchange between PBs and the cytoplasm (Xing et al., 2020).

5.2.2 Stress granules

SGs are complex membrane-less cytoplasmic assemblies that are transiently formed when cells are exposed to adverse environmental conditions, such as oxidative stress, heat shock, osmotic stress, nutrient deprivation, UV irradiation and viral infection (Decker and Parker, 2012; Ivanov et al., 2019). Many of these stressors induce phosphorylation of the eIF2α translation initiation factor, resulting in the repression of overall translation and the disassembly of polysomes both of which stimulate SG assembly. Accordingly, SGs can be generated upon

treatment of cells with various compounds, like proteasome inhibitors, ER stressors and mitochondrial poisons, or when some key protein components of SGs are overexpressed (Mahboubi and Stochaj, 2017). Typical components of SGs include stalled pre-initiation complexes, containing mRNAs, translation initiation factors and 40S ribosomal proteins, and a diverse group of proteins that include RNA-binding proteins related to mRNA processing, translation or stability (Table 5.1). Evidence suggests that SGs are composed of stable cores surrounded by a phase-separated shell (Jain et al., 2016; Wheeler et al., 2016). Transcriptomics and proteomics studies reveal that SGs are enriched in long and poorly translated mRNAs, and that their composition varies in vivo, depending on the type of tissue and stress (Jain et al., 2016; Khong et al., 2017; Markmiller et al., 2018; Namkoong et al., 2018). SGs can also recruit noncoding RNAs and signaling molecules, such as AMPK that controls biogenesis of SGs, while SG-dependent sequestration of mTOR, Raptor or RACK1 prevents stress-induced apoptosis (Arimoto et al., 2008; Mahboubi et al., 2015; Takahara and Maeda, 2012; Thedieck et al., 2013). Assembly and disassembly of SGs can be regulated by a wide range of post-translational modifications in resident proteins, as for example phosphorylation of the RNA-binding proteins FUS (fused in sarcoma) and G3BP1 (Ras GTPase-activating protein-binding protein 1) (Mahboubi and Stochaj, 2017).

5.3 PROTEOSTATIC FUNCTIONS OF PBs AND SGs

5.3.1 In Stress: The More Stable, the Less Worrying

Cells are constantly exposed to environmental fluctuations and a number of conserved cellular mechanisms have evolved that allow organisms to adapt and survive under harsh conditions. Inhibition of global protein synthesis is among the first cellular responses to various forms of stress, in order to preserve energy and promote selective synthesis of stress-adaptive proteins (Wek et al., 2006). Stress-induced formation of both PBs and SGs is tightly and inversely connected to the translational status of the cell and provides temporal regulation of translation regulation in response to cellular stress (Decker and Parker, 2012; Ivanov et al., 2019). Besides non-translating mRNAs, a complex network of interactions between their components shapes these mRNA granules. It is now known that transcript length or specific motifs in the 3′ untranslated region and 3′ AU-rich elements determine the localization of some mRNAs into PBs and SGs during stress (Namkoong et al., 2018; Wang et al., 2018).

PB and SG formation in response to stress is thought to protect cellular homeostasis by recruiting translationally silenced mRNAs, and temporarily storing mRNAs and proteins away from degradation machineries. Interestingly, a yeast mutant lacking both PBs and SGs exhibits potential defects in proteostasis, increased levels of protein misfolding and diminished ability to survive long term in stationary phase (Nostramo et al., 2019). Upon cessation of stress, SGs and PBs disassemble and mRNAs trapped into them can resume active translation, thus facilitating the rapid recovery of cells (Cherkasov et al., 2013; Riback et al., 2017). Moreover, both types of granules can confer cytoprotection by sequestrating or even regulating the activity of stress-related factors and function as transient signaling hubs for cellular responses to a variety of insults (Mahboubi and Stochaj, 2017). A recent study showed that SG formation is triggered by starvation and dietary restriction (DR) via AMPK signaling in *C. elegans* and is required for DR-dependent longevity (Kuo et al., 2020). In agreement with the crucial role of functional PBs and SGs in preserving proteostasis under adverse conditions, genetic alterations of key protein components that result in impaired formation or function of these granules affect cellular adaptive responses and reduce fitness of diverse organisms, such as yeast, *C. elegans* and *Drosophila* (Borbolis et al., 2017, 2020; Cornes et al., 2015; Mazzoni et al., 2005; Nostramo et al., 2019; Rousakis et al., 2014).

5.3.2 In Aging and Disease: The More Unstable, the Less Worrying

During the aging process, proteins tend to naturally aggregate mostly due to transcriptome deterioration and progressive loss of both translation control and proteostasis (Ayyadevara et al., 2016; David et al., 2010; Tanase et al., 2016; Walther et al., 2015). Although little is known about the effect aging has on PB and SG function, increased accumulation of PBs (Rieckher et al., 2018; Rousakis et al., 2014) and highly insoluble aggregates of SG-related RNA-binding proteins have been reported in aged *C. elegans* (Lechler et al., 2017). Whether the age-dependent aggregation of mRNA granules is required for proper regulation of mRNA metabolism and other cellular functions in aged cells or it is a consequence of uncontrolled protein synthesis and degradation remains to be clarified. Cytoplasmic protein aggregation is also a hallmark of Alzheimer's (AD), Parkinson's (PD) and Huntington's diseases (HD), frontotemporal dementia (FTD) or neuromuscular disorders, such as amyotrophic lateral sclerosis (ALS). Although diverse in their symptoms, all these diseases share a common factor, the accumulation of insoluble cytoplasmic aggregates in patients' neurons (Dobra et al., 2018). Interestingly, several RNA-binding proteins that are connected to these neurodegenerative diseases have been found to misfold and accumulate into SGs when mutated (Wolozin and Ivanov, 2019). These include FUS, TDP-43, ATAXIN-2, hnRNPA1 and TIA1 proteins that are associated with ALS and FTD pathology, tau protein that is linked to AD as well as ALS-associated variants of SOD1 (Mateju et al., 2017). These findings provided the first link between the formation of persistent SG (analyzed below) and the pathophysiology of neurodegenerative diseases, while SGs have been further implicated in other neurological diseases, inflammatory diseases, viral infection and in some forms of cancer (Wolozin and Ivanov, 2019).

SGs contain many RNA-binding proteins that harbor self-interacting domains that contribute to the rapid and transient assembly of SGs under stress (Decker and Parker, 2012) but can also drive aberrant aggregation in the case of disease. Indeed, seminal studies have shown that such proteins

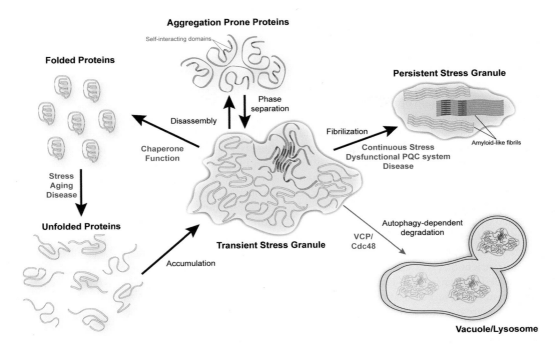

Figure 5.1 Model depicting the interplay of misfolded and intrinsically aggregation-prone proteins with SGs and PQC systems.

can form liquid droplets *in vitro*, but they turn into a solid-like state over time (Conicella et al., 2016; Kato et al., 2012; Patel et al., 2015). Missense mutations, repeat expansion or aberrant post-transcriptional modifications in these proteins accelerate this transition *in vitro* and *in vivo*, leading to the formation of stable amyloid-like fibrils (Patel et al., 2015). The conversion of transient to persistent alters the dynamics of SGs which may act as seeds for extensive aggregation of nonfunctional proteins, mRNAs or miRNAs, thereby contributing to cellular toxicity (Wolozin and Ivanov, 2019). Thus, the emerging model is that accumulation of intrinsically aggregation-prone proteins above a critical threshold can induce SG phase transition to a less dynamic or more stable state (Figure 5.1). Furthermore, the progressive failure of proteostasis that occurs during aging can not only impair the dynamic state of mRNA granules but also reduce their solubility and clearance by PQC mechanisms.

5.4 REGULATION OF mRNA GRANULES BY PROTEOSTATIC MECHANISMS

To counteract protein aggregation, cells employ a set of chaperones and protein degradation systems, and several observations argue that these

systems can regulate mRNA granules disassembly too. Protein chaperones of the HSP70 and HSP40 family, or small HSPs, colocalize with SGs in yeast, fruitflies and mammalian cells, and they are required for SG disassembly and restoration of translation activity during stress recovery (Cherkasov et al., 2013; Ganassi et al., 2016; Kroschwald et al., 2015; Walters et al., 2015). This chaperone-mediated disassembly is considered to be the preferred pathway of SG clearance but other PQC systems are often engaged. A genetic screen in *Saccharomyces cerevisiae* aiming to identify mutations that affect PB and SG assembly and disassembly revealed that both types of mRNA granules can be targeted for degradation by autophagy, in a process termed "granulophagy" (Buchan et al., 2013). Granulophagy requires the AAA-ATPase Cdc48, the yeast ortholog of valosin-containing protein (VCP), which is a ubiquitin segregase and loss of its function in yeast and mammalian cells results in significantly reduced SG clearance (Buchan et al., 2013). Also, compound (MG132)-mediated inhibition of the ubiquitin-proteasome system activity induces SG assembly in human cells (Mazroui et al., 2007), but inhibition of autophagy, lysosomes or VCP blocks stress or MG132 induced SG formation

(Seguin et al., 2014). Interestingly, mutations in VCP or other PQC components, such as ubiquilin 2, optineurin, sequestosome 1 and dynactin, which affect autophagic clearance of aggregated proteins, have been all linked to ALS (Ramesh and Pandey, 2017). Inhibition of autophagy in cultured neurons expressing an ALS-linked FUS mutation results in increased accumulation of FUS-positive SGs that contain the autophagy receptor p62 (Ryu et al., 2014). Likewise, SGs containing ALS-linked variants of SOD1 are transported along microtubules to the aggresome, a perinuclear protein inclusion that sequesters misfolded proteins and is degraded by autophagy (Mateju et al., 2017).

These findings support that misfolded proteins generated due to stress or a pathology accumulate in transient SGs that can be disassembled by chaperone function and/or autophagy. However, under continuous stress or after a PQC malfunction, SGs aggregate into persistent and potentially harmful mRNA granules (Figure 5.1). When the PQC machinery is compromised, several highly aggregation-prone proteins can be misdirected and accumulated into SGs, altering the composition, dynamics and disassembly of these granules. One of the main sources of aggregation-prone proteins in cells is the heterogenous group of proteins named defective ribosomal products (DRiPs), consisting of nascent- and prematurely terminated polypeptides that are released from disassembling polysomes. Normally, they are recognized by the HSP70 chaperone family and VCP that assist their degradation by the proteasome. However, when the PQC is impaired, DRiPs can accumulate in mammalian SGs and promote their conversion from liquid-like into solid-like granules that are connected to pathological aggregates (Ganassi et al., 2016; Mateju et al., 2017). Overall, these findings further reinforce the notion that disaggregation or clearance of SGs is intimately linked to PQC machineries and proteostasis control.

5.5 mRNA GRANULES-BASED POTENTIAL INTERVENTIONS IN AGING AND AGE-RELATED DISEASES

Several studies have revealed that proper formation, function and dynamism of mRNA granules, such as PBs and SGs, are required for normal chronological lifespan in yeast (Nostramo et al., 2019) and for both many forms of lifespan

extension in worms (Kuo et al., 2020; Rousakis et al., 2014). Moreover, overexpression of DCP1, a key PB component, in worms and flies extends lifespan (Borbolis et al., 2020), supporting a modulatory role of such granules in the aging process. A large body of data has also highlighted that lifespan-extending strategies can mitigate toxicity and confer neuroprotection in age-related diseases in model organisms. The involvement of SGs in age-related neurodegenerative disorders or other human diseases advocates that approaches reducing accumulation of disease-associated RNA-binding proteins into mRNA granules, or inhibiting aberrant SG phase transition, could be of therapeutic value. Genetic studies in animal models showed that reducing the levels of TIA1, a defining marker of SGs that promotes tau misfolding and insolubility, could inhibit degeneration in cultured hippocampal neurons and prolong survival in mice (Apicco et al., 2018). Likewise, decreasing the expression of ATAXIN-2, another SG component, either genetically or with antisense oligonucleotides, reduced TDP-43 toxicity and increased survival in yeast, flies and mouse models of ALS (Becker et al., 2017; Elden et al., 2010). Genetic blocking of SG formation in human sarcoma cells was also shown to inhibit tumor invasion and prevent lung metastasis in mouse models (Somasekharan et al., 2015). On the other hand, sequestration of translation factors into PBs or SGs can lead to protein synthesis inhibition (Rieckher et al., 2018), which in turn is overall protective in response to stress, aging and disease. In agreement with this, inhibiting elevated phosphorylation of eIF2α, by genetic or pharmacological means, was shown to ameliorate disease progression in several systems, such as in prion-diseased mice (Moreno et al., 2012), in TDP-43-induced neurotoxicity in *Drosophila* and primary rat neurons (Kim et al., 2014), and in mouse models of FTD (Radford et al., 2015) or AD (Ma et al., 2013).

Additionally, cellular mechanisms that enhance clearance of mRNA granules and disrupt harmful aggregates may have a potential as targets to treat related pathologies (Klaips et al., 2018). These mechanisms include ATP-driven chaperones, RNA helicases, ubiquitin segregases, the ubiquitin-proteasome system and the autophagic-lysosomal system. Overexpression of such chaperones and induction of autophagy or the proteasome system with pharmacological or genetic means have resulted in neuroprotection in many experimental systems, and even increased lifespan in model organisms (Papaevgeniou and Chondrogianni, 2018; Ryu et al., 2014). Thus, small-molecule enhancers of chaperones or de novo–designed chaperone proteins have been proposed as potential approaches for targeting neurodegenerative diseases (Shorter, 2017). Such therapeutic interventions that boost proteolytic mechanisms or chaperones could prevent age- or disease-associated toxic aggregations of mRNA granules, providing protection against various pathologies.

5.6 CONCLUDING REMARKS

Advanced technologies and progress in many research areas in the last years have provided detailed information about the assembly of cellular mRNA granules and their properties. The importance of cytoplasmic mRNA granules, not only in post-transcriptional regulation of gene expression but also in proteostasis control during stress, aging or disease, is becoming increasingly recognized. The transient sequestration of protein-bound mRNAs and formation of mRNA granules, such as PBs and SGs, in response to developmental stimuli or environmental insults, is required for proper regulation of developmental events or to combat stress. Conversely, a persistent expansion and accumulation of aberrant mRNA granules during aging and age-associated diseases can be detrimental for cellular viability and organismal health. Protein misfolding and aggregation in such aberrant granules are exacerbated by dysregulation of PQC mechanisms, which are also involved in the surveillance and clearance of these granules. Deciphering the functional link of different mRNA granule components, the regulators of their dynamic and reversible interactions as well as the exact mechanisms that cause their age- or disease-related aggregation is a great challenge.

REFERENCES

Alberti, S., Dormann, D. 2019. Liquid-liquid phase separation in disease. Annu Rev Genet, 53, 171–194. doi:10.1146/annurev-genet-112618-043527

Anderson, P., Kedersha, N. 2009. RNA granules: post-transcriptional and epigenetic modulators of gene expression. Nat Rev Mol Cell Biol, 10(6), 430–436. doi:10.1038/nrm2694

Apicco, D. J., Ash, P. E. A., Maziuk, B., et al. 2018. Reducing the RNA binding protein TIA1 protects against tau-mediated neurodegeneration in vivo. Nat Neurosci, 21(1), 72–80. doi:10.1038/s41593-017-0022-z

Arimoto, K., Fukuda, H., Imajoh-Ohmi, S., et al. 2008. Formation of stress granules inhibits apoptosis by suppressing stress-responsive MAPK pathways. Nat Cell Biol, 10(11), 1324–1332. doi:10.1038/ncb1791

Ayyadevara, S., Balasubramaniam, M., Suri, P., et al. 2016. Proteins that accumulate with age in human skeletal-muscle aggregates contribute to declines in muscle mass and function in Caenorhabditis elegans. Aging (Albany NY), 8(12), 3486–3497. doi:10.18632/aging.101141

Becker, L. A., Huang, B., Bieri, G., et al. 2017. Therapeutic reduction of ataxin-2 extends lifespan and reduces pathology in TDP-43 mice. Nature, 544(7650), 367–371. doi:10.1038/nature22038

Borbolis, F., Syntichaki, P. 2015. Cytoplasmic mRNA turnover and ageing. Mech Ageing Dev, 152, 32–42. doi:10.1016/j.mad.2015.09.006

Borbolis, F., Flessa, C. M., Roumelioti, F., et al. 2017. Neuronal function of the mRNA decapping complex determines survival of Caenorhabditis elegans at high temperature through temporal regulation of heterochronic gene expression. Open Biol, 7(3). doi:10.1098/rsob.160313

Borbolis, F., Rallis, J., Kanatouris, G., et al. 2020. mRNA decapping is an evolutionarily conserved modulator of neuroendocrine signaling that controls development and ageing. Elife, 9. doi:10.7554/eLife.53757

Buchan, J. R. 2014. mRNP granules: assembly, function, and connections with disease. RNA Biol, 11(8), 1019–1030. doi:10.4161/15476286.2014.972208

Buchan, J. R., Kolaitis, R. M., Taylor, J. P., et al. 2013. Eukaryotic stress granules are cleared by autophagy and Cdc48/VCP function. Cell, 153(7), 1461–1474. doi:10.1016/j.cell.2013.05.037

Cherkasov, V., Hofmann, S., Druffel-Augustin, S., et al. 2013. Coordination of translational control and protein homeostasis during severe heat stress. Curr Biol, 23(24), 2452–2462. doi:10.1016/j.cub.2013.09.058

Ciryam, P., Kundra, R., Morimoto, R. I., et al. 2015. Supersaturation is a major driving force for protein aggregation in neurodegenerative diseases. Trends Pharmacol Sci, 36(2), 72–77. doi:10.1016/j.tips.2014.12.004

Conicella, A. E., Zerze, G. H., Mittal, J., et al. 2016. ALS mutations disrupt phase separation mediated by alpha-helical structure in the TDP-43 low-complexity C-terminal domain. Structure, 24(9), 1537–1549. doi:10.1016/j.str.2016.07.007

Cornes, E., Porta-De-La-Riva, M., Aristizabal-Corrales, D., et al. 2015. Cytoplasmic LSM-1 protein regulates stress responses through the insulin/IGF-1 signaling pathway in Caenorhabditis elegans. RNA, 21(9), 1544–1553. doi:10.1261/rna.052324.115

David, D. C., Ollikainen, N., Trinidad, J. C., et al. 2010. Widespread protein aggregation as an inherent part of aging in C. elegans. PLoS Biol, 8(8), e1000450. doi:10.1371/journal.pbio.1000450

Decker, C. J., Parker, R. 2012. P-bodies and stress granules: possible roles in the control of translation and mRNA degradation. Cold Spring Harb Perspect Biol, 4(9), a012286. doi:10.1101/cshperspect.a012286

Dobra, I., Pankivskyi, S., Samsonova, A., et al. 2018. Relation between stress granules and cytoplasmic protein aggregates linked to neurodegenerative diseases. Curr Neurol Neurosci Rep, 18(12), 107. doi:10.1007/s11910-018-0914-7

Elden, A. C., Kim, H. J., Hart, M. P., et al. 2010. Ataxin-2 intermediate-length polyglutamine expansions are associated with increased risk for ALS. Nature, 466(7310), 1069–1075. doi:10.1038/nature09320

Ganassi, M., Mateju, D., Bigi, I., et al. 2016. A surveillance function of the HSPB8-BAG3-HSP70 chaperone complex ensures stress granule integrity and dynamism. Mol Cell, 63(5), 796–810. doi:10.1016/j.molcel.2016.07.021

Gomes, E., Shorter, J. 2019. The molecular language of membraneless organelles. J Biol Chem, 294(18), 7115–7127. doi:10.1074/jbc.TM118.001192

Hubstenberger, A., Courel, M., Benard, M., et al. 2017. P-body purification reveals the condensation of repressed mRNA regulons. Mol Cell, 68(1), 144–157 e145. doi:10.1016/j.molcel.2017.09.003

Hutten, S., Sharangdhar, T., Kiebler, M. 2014. Unmasking the messenger. RNA Biol, 11(8), 992–997. doi:10.4161/rna.32091

Ivanov, P., Kedersha, N., Anderson, P. 2019. Stress granules and processing bodies in translational control. Cold Spring Harb Perspect Biol, 11(5). doi:10.1101/cshperspect.a032813

Jain, S., Wheeler, J. R., Walters, R. W., et al. 2016. ATPase-modulated stress granules contain a diverse proteome and substructure. Cell, 164(3), 487–498. doi:10.1016/j.cell.2015.12.038

Kato, M., Han, T. W., Xie, S., et al. 2012. Cell-free formation of RNA granules: low complexity sequence domains form dynamic fibers within hydrogels. Cell, 149(4), 753–767. doi:10.1016/j.cell.2012.04.017

Khong, A., Matheny, T., Jain, S., et al. 2017. The stress granule transcriptome reveals principles of mRNA accumulation in stress granules. Mol Cell, 68(4), 808–820.e805. doi:10.1016/j.molcel.2017.10.015

Kilchert, C., Weidner, J., Prescianotto-Baschong, C., et al. 2010. Defects in the secretory pathway and high Ca^{2+} induce multiple P-bodies. Mol Biol Cell, 21(15), 2624–2638. doi:10.1091/mbc.E10-02-0099

Kim, H. J., Raphael, A. R., LaDow, E. S., et al. 2014. Therapeutic modulation of eIF2alpha phosphorylation rescues TDP-43 toxicity in amyotrophic lateral sclerosis disease models. Nat Genet, 46(2), 152–160. doi:10.1038/ng.2853

Klaips, C. L., Jayaraj, G. G., Hartl, F. U. 2018. Pathways of cellular proteostasis in aging and disease. J Cell Biol, 217(1), 51–63. doi:10.1083/jcb.201709072

Kroschwald, S., Maharana, S., Mateju, D., et al. 2015. Promiscuous interactions and protein disaggregases determine the material state of stress-inducible RNP granules. Elife, 4, e06807. doi:10.7554/eLife.06807

Kuo, C. T., You, G. T., Jian, Y. J., et al. 2020. AMPK-mediated formation of stress granules is required for dietary restriction-induced longevity in Caenorhabditis elegans. Aging Cell, 19(6), e13157. doi:10.1111/acel.13157

Lechler, M. C., Crawford, E. D., Groh, N., et al. 2017. Reduced insulin/IGF-1 signaling restores the dynamic properties of key stress granule proteins during Aging. Cell Rep, 18(2), 454–467. doi:10.1016/j.celrep.2016.12.033

Luo, Y., Na, Z., Slavoff, S. A. 2018. P-bodies: composition, properties, and functions. Biochemistry, 57(17), 2424–2431. doi:10.1021/acs.biochem.7b01162

Ma, T., Trinh, M. A., Wexler, A. J., et al. 2013. Suppression of eIF2alpha kinases alleviates Alzheimer's disease-related plasticity and memory deficits. Nat Neurosci, 16(9), 1299–1305. doi:10.1038/nn.3486

Mahboubi, H., Stochaj, U. 2017. Cytoplasmic stress granules: dynamic modulators of cell signaling and disease. Biochim Biophys Acta Mol Basis Dis, 1863(4), 884–895. doi:10.1016/j.bbadis.2016.12.022

Mahboubi, H., Barise, R., Stochaj, U. 2015. 5′-AMP-activated protein kinase alpha regulates stress granule biogenesis. Biochim Biophys Acta, 1853(7), 1725–1737. doi:10.1016/j.bbamcr.2015.03.015

Markmiller, S., Soltanieh, S., Server, K. L., et al. 2018. Context-dependent and disease-specific diversity in protein interactions within stress granules. Cell, 172(3), 590–604.e513. doi:10.1016/j.cell.2017.12.032

Mateju, D., Franzmann, T. M., Patel, A., et al. 2017. An aberrant phase transition of stress granules triggered by misfolded protein and prevented by chaperone function. EMBO J, 36(12), 1669–1687. doi:10.15252/embj.201695957

Mazroui, R., Di Marco, S., Kaufman, R. J., et al. 2007. Inhibition of the ubiquitin-proteasome system induces stress granule formation. Mol Biol Cell, 18(7), 2603–2618. doi:10.1091/mbc.e06-12-1079

Mazzoni, C., Herker, E., Palermo, V., et al. 2005. Yeast caspase 1 links messenger RNA stability to apoptosis in yeast. EMBO Rep, 6(11), 1076–1081. doi:10.1038/sj.embor.7400514

Moreno, J. A., Radford, H., Peretti, D., et al. 2012. Sustained translational repression by eIF2alpha-P mediates prion neurodegeneration. Nature, 485(7399), 507–511. doi:10.1038/nature11058

Namkoong, S., Ho, A., Woo, Y. M., et al. 2018. Systematic characterization of stress-induced RNA granulation. Mol Cell, 70(1), 175–187.e178. doi:10.1016/j.molcel.2018.02.025

Nostramo, R., Xing, S., Zhang, B., et al. 2019. Insights into the role of P-bodies and stress granules in protein quality control. Genetics, 213(1), 251–265. doi:10.1534/genetics.119.302376

Papaevgeniou, N., Chondrogianni, N. 2018. Anti-aging and anti-aggregation properties of polyphenolic compounds in C. elegans. Curr Pharm Des, 24(19), 2107–2120. doi:10.2174/1381612824666180515145652

Patel, A., Lee, H. O., Jawerth, L., et al. 2015. A liquid-to-solid phase transition of the ALS protein FUS accelerated by disease mutation. Cell, 162(5), 1066–1077. doi:10.1016/j.cell.2015.07.047

Radford, H., Moreno, J. A., Verity, N., et al. 2015. PERK inhibition prevents tau-mediated neurodegeneration in a mouse model of frontotemporal dementia. Acta Neuropathol, 130(5), 633–642. doi:10.1007/s00401-015-1487-z

Ramesh, N., Pandey, U. B. 2017. Autophagy dysregulation in ALS: when protein aggregates get out of hand. Front Mol Neurosci, 10, 263. doi:10.3389/fnmol.2017.00263

Riback, J. A., Katanski, C. D., Kear-Scott, J. L., et al. 2017. Stress-triggered phase separation is an adaptive, evolutionarily tuned response. Cell, 168(6), 1028–1040.e1019. doi:10.1016/j.cell.2017.02.027

Rieckher, M., Markaki, M., Princz, A., et al. 2018. Maintenance of proteostasis by P-body-mediated regulation of eIF4E availability during aging in Caenorhabditis elegans. Cell Rep, 25(1), 199–211.e196. doi:10.1016/j.celrep.2018.09.009

Rousakis, A., Vlanti, A., Borbolis, F., et al. 2014. Diverse functions of mRNA metabolism factors in stress defense and aging of Caenorhabditis elegans. PLoS One, 9(7), e103365. doi:10.1371/journal.pone.0103365

Ryu, H. H., Jun, M. H., Min, K. J., et al. 2014. Autophagy regulates amyotrophic lateral sclerosis-linked fused in sarcoma-positive stress granules in neurons. Neurobiol Aging, 35(12), 2822–2831. doi:10.1016/j.neurobiolaging.2014.07.026

Seguin, S. J., Morelli, F. F., Vinet, J., et al. 2014. Inhibition of autophagy, lysosome and VCP function impairs stress granule assembly. Cell Death Differ, 21(12), 1838–1851. doi:10.1038/cdd.2014.103

Seydoux, G. 2018. The P granules of C. elegans: a genetic model for the study of RNA-protein condensates. J Mol Biol, 430(23), 4702–4710. doi:10.1016/j.jmb.2018.08.007

Shorter, J. 2017. Designer protein disaggregases to counter neurodegenerative disease. Curr Opin Genet Dev, 44, 1–8. doi:10.1016/j.gde.2017.01.008

Somasekharan, S. P., El-Naggar, A., Leprivier, G., et al. 2015. YB-1 regulates stress granule formation and tumor progression by translationally activating G3BP1. J Cell Biol, 208(7), 913–929. doi:10.1083/jcb.201411047

Takahara, T., Maeda, T. 2012. Transient sequestration of TORC1 into stress granules during heat stress. Mol Cell, 47(2), 242–252. doi:10.1016/j.molcel.2012.05.019

Tanase, M., Urbanska, A. M., Zolla, V., et al. 2016. Role of carbonyl modifications on aging-associated protein aggregation. Sci Rep, 6, 19311. doi:10.1038/srep19311

Tauber, D., Tauber, G., Parker, R. 2020. Mechanisms and regulation of RNA condensation in RNP granule formation. Trends Biochem Sci, 45(9), 764–778. doi:10.1016/j.tibs.2020.05.002

Taylor, R. C., Dillin, A. 2011. Aging as an event of proteostasis collapse. Cold Spring Harb Perspect Biol, 3(5). doi:10.1101/cshperspect.a004440

Thedieck, K., Holzwarth, B., Prentzell, M. T., et al. 2013. Inhibition of mTORC1 by astrin and stress granules prevents apoptosis in cancer cells. Cell, 154(4), 859–874. doi:10.1016/j.cell.2013.07.031

van Leeuwen, W., Rabouille, C. 2019. Cellular stress leads to the formation of membraneless stress assemblies in eukaryotic cells. Traffic, 20(9), 623–638. doi:10.1111/tra.12669

Walters, R. W., Muhlrad, D., Garcia, J., et al. 2015. Differential effects of Ydj1 and Sis1 on Hsp70-mediated clearance of stress granules in Saccharomyces cerevisiae. RNA, 21(9), 1660–1671. doi:10.1261/rna.053116.115

Walther, D. M., Kasturi, P., Zheng, M., et al. 2015. Widespread proteome remodeling and aggregation in aging C. elegans. Cell, 161(4), 919–932. doi:10.1016/j.cell.2015.03.032

Wang, C., Schmich, F., Srivatsa, S., et al. 2018. Context-dependent deposition and regulation of mRNAs in P-bodies. Elife, 7. doi:10.7554/eLife.29815

Wek, R. C., Jiang, H. Y., Anthony, T. G. 2006. Coping with stress: eIF2 kinases and translational control. Biochem Soc Trans, 34(Pt 1), 7–11. doi:BST20060007 [pii] 10.1042/BST20060007

Wheeler, J. R., Matheny, T., Jain, S., et al. 2016. Distinct stages in stress granule assembly and disassembly. Elife, 5. doi:10.7554/eLife.18413

Wolozin, B., Ivanov, P. 2019. Stress granules and neurodegeneration. Nat Rev Neurosci, 20(11), 649–666. doi:10.1038/s41583-019-0222-5

Xing, W., Muhlrad, D., Parker, R., et al. 2020. A quantitative inventory of yeast P-body proteins reveals principles of composition and specificity. Elife, 9. doi:10.7554/eLife.56525

CHAPTER SIX

Phospholipids and the Unfolded Protein Response

Ilias Gkikas and Nektarios Tavernarakis

CONTENTS

6.1 INTRODUCTION

Eukaryotic membranes, encircling cells and organelles, consist of proteins and lipid molecules such as phospholipids and sterols. The dynamic interplay between these constituents underpins membrane integrity and fluidity, and ultimately cellular protein homeostasis, or proteostasis. The endoplasmic reticulum (ER) and mitochondria possess a unique composition of proteins and lipids which allows maintenance of organelle structure and function. The ER is a highly regulated organelle forming large and dynamic membranous networks. This complex membrane architecture is essential for more than one-third of cellular protein synthesis, folding and secretion (Pendin et al., 2011). In addition, most abundant phospholipids such as glycerophospholipids (GPLs) including phosphatidic acid (PA), phosphatidylcholine (PC), phosphatidylethanolamine (PE), phosphatidylinositol (PI) and phosphatidylserine (PS) are predominantly synthesized in the ER (Ridgway, 2016).

Similarly, mitochondria are double-membrane organelles forming tubular networks with diverse size, number and position within cells (Kuhlbrandt, 2015). Beyond their crucial role in energy production, mitochondria are also involved in the synthesis of specialized phospholipids. While ER-synthesized phospholipids can

DOI: 10.1201/9781003048138-6

be transported to mitochondria, a string of enzymatic activities within mitochondria is required for phosphatidylglycerol (PG), cardiolipin (CL) and PE synthesis (Horvath and Daum, 2013).

Interestingly, alternation of phospholipid composition within membranes can cause ER and mitochondrial proteotoxic stress (Halbleib et al., 2017; Ho et al., 2018). To maintain proteostasis, the load of newly synthesized and the degradation of unwanted proteins should be kept in balance. From yeast to mammals, compartmentalized protein quality control mechanisms, such as the unfolded protein response (UPR) and autophagy, assure the integrity of these processes (Senft and Ronai, 2015). In this chapter, we discuss recent findings on UPR activation by proteotoxic and phospholipid bilayer stress. Emphasis will also be placed on the role of UPR signaling in phospholipid biosynthesis (hereafter we will refer to both the ER and mitochondrial UPRs as UPR unless otherwise noted).

6.2 REGULATION OF ENDOPLASMIC RETICULUM AND MITOCHONDRIAL UNFOLDED PROTEIN RESPONSE

6.2.1 The UPR^{ER} Signaling

The ER proteostasis is confronted with diverse insults often leading to lipid disequilibrium, accumulation of unfolded proteins, depletion of Ca^{2+} and glycosylation defects among others, which in turn activate the UPR^{ER} (Hetz and Papa, 2018; Ho et al., 2018; Metcalf et al., 2020). The UPR^{ER} signal is emanated from three discrete ER transmembrane proteins, namely the inositol-requiring kinase 1 (IRE1), activating transcription factor 6 (ATF6), and double-stranded RNA-activated protein kinase (PKR)-like ER kinase (PERK). While IRE1 is well conserved across species, PERK and ATF6 are only present in metazoans (Hetz and Papa, 2018). A member of the HSP70 chaperone family of proteins, the binding immunoglobin protein (BiP), determines their activity. Beyond its classical ER chaperone role, BiP binds to ER luminal domain (LD) of IRE1, PERK and ATF6. Accumulation of unfolded proteins results in BiP release from IRE1, PERK and ATF6 LDs and subsequently activates UPR^{ER}. Several models have been proposed to rationalize the transient binding of either misfolded proteins or ER stress sensors to BiP (Kopp et al., 2019). Nevertheless, the exact mechanism underpinning the inverse correlation between its role as an ER chaperone and as a UPR^{ER} activator is not fully understood.

6.2.1.1 Sensing by IRE1

Structurally and functionally conserved, IRE1 is the only UPR^{ER} transducer expressed from lower to higher eukaryotic cells. Unlike yeast, the mammalian IRE1 has two isoforms, IRE1α and IRE1β, exhibiting tissue-specific expression. Through their cytosolic regions, both exert Ser/Thr kinase and endoribonuclease activities, and sense ER stress through their N-terminal ER LDs. Under physiological conditions, IRE1α remains monomeric and inactive. Mammalian IRE1α is activated upon either BiP uncoupling or direct binding to unfolded proteins. The latter is accompanied by IRE1α oligomerization and/or dimerization, and trans-autophosphorylation (Hetz and Papa, 2018). Activated IRE1α catalyzes the unconventional splicing of X-box binding protein 1 (XBP1) mRNA more effectively than IRE1β (Imagawa et al., 2008). Concurrently, the spliced and functional isoform of XBP1 triggers the UPR^{ER} response by orchestrating the transcription of genes involved in protein folding and quality control (Uemura et al., 2009). Apart from XBP1 splicing, activities of both IRE1α and IRE1β endonucleases have been involved in the degradation of several ER-targeted and cytosolic mRNAs via a process known as IRE1-dependent decay (RIDD). Under physiological conditions, RIDD sequence-specific cleavage is essential for sustaining the normal load of newly synthesized proteins targeting the ER. Interestingly, cleavage sites recognized by IRE1α and IRE1β appear to be similar for both XBP1 and RIDD substrates. Notably, substrate recognition is a highly selective process. As such, activation of IRE1α favors XBP1 mRNA splicing to preserve ER homeostasis, whereas IRE1β preferentially recognizes RIDD substrates to induce cell death signaling upon irremediable ER stress (Maurel et al., 2014).

6.2.1.2 Sensing by PERK

Under normal conditions, PERK exists in its monomeric inactive form bound by BiP. In response to ER stress, BiP is released while

PERK LD and kinase domains facilitate its oligomerization. Concomitantly, autophosphorylation of its kinase domain results in activation of PERK-mediated UPR^{ER} signaling followed by cytosolic protein synthesis inhibition. Crystalized PERK in dimeric and tetrameric forms has been also reported to trigger UPR^{ER} signaling (Carrara et al., 2015). Whether and how dimer-tetramer transition correlates with a more or less active state of PERK remains elusive. Furthermore, direct binding of misfolded proteins to PERK-LD is indispensable for PERK activation (Wang et al., 2018). Intriguingly, a by-product of choline metabolism, trimethylamine N-oxide (TMAO), has been also reported to bind exclusively PERK-LD and thus induce UPR^{ER} (Chen et al., 2019). Once activated, stimulation of PERK kinase domain phosphorylates eukaryotic translation initiation factor 2 alpha (eIF2α) at Ser51 and inactivates it, eventually attenuating global translation (Harding et al., 1999). While limited, the phosphorylated eIF2α favors the translation of several mRNAs possessing upstream open reading frames (uORFs) in their 5′-untranslated region (5′-UTR) (Lee et al., 2009). Among them, activated transcription factor 4 (ATF4) has a central pro-survival and pro-death activity depending on the origin, severity and duration of ER stress. Upon activation, ATF4 controls the expression of genes involved in protein homeostasis, autophagy, apoptosis, amino acid and redox metabolism (Carrara et al., 2015; Han et al., 2013). Particularly, ATF4 exerts its pro-apoptotic role by controlling the expression of C/EBP homology protein (CHOP) and growth-arrest and DNA-damage-induced transcript 34 (GADD34). Expression of GADD34 restores protein synthesis by promoting the dephosphorylation of eIF2a (Hetz and Papa, 2018).

6.2.1.3 Sensing by ATF6

From the two isoforms found in mammals, ATF6α has been studied to a greater extent than ATF6β. ER-located ATF6α exists as a mixed population of monomers, dimers and oligomers, linked by disulfide bridges formed between ATF6α-LD cysteine residues (Nadanaka et al., 2007). In unstressed cells, BiP binding to ATF6α-LD cancels its engagement with coat protein complex II (COPII) transport vesicles at the ER–Golgi interface. In response to superfluous or misfolded proteins, BiP is released from ATF6α-LD which in turn promotes ATF6α ER-to-Golgi translocation via COPII vesicles (Sato et al., 2011). Activation of ATF6α relies on Golgi translocation of the ATF6α precursor, therein it is processed by site-1 and site-2 proteases (S1P and S2P), that sequentially remove the LD and the transmembrane anchor. Subsequently, the liberated N-terminal fragment of 50 kDa (p50ATF6α) can enter the nucleus and activate UPR^{ER} downstream transcriptional targets (Ye et al., 2000). Recently, it has been shown that ER-resident thrombospondin 4 (Thbs4) binding to ATF6α-LD is sufficient for COPII-mediated transport to the Golgi apparatus. As such, nuclear entry of p50ATF6a fragment and transactivation of its UPR^{ER} target genes allows restoration of ER proteostasis (Lynch et al., 2012). Interestingly, upon ER stress, oxidoreductase activity of protein disulfide isomerase A5 (PDIA5) promotes accessibility of ATF6α into COPII vesicles in a BiP-independent manner (Higa et al., 2014). More recently, an additional ER oxidoreductase, ERp18, has been associated with ATF6α in response to ER stress. In this context, moderate ATF6α activation through improved cleavage by S1P/S2P and controlled ER-to-Golgi transport by ERp18 has been documented (Oka et al., 2019). Despite the effort, how BiP binding and release occur and how other ATF6 interacting partners ultimately mediate ATF6 response are not fully understood.

6.2.2 The UPR^{mt} Signaling

When excess misfolded and/or unfolded proteins accumulate within mitochondria, the UPR^{mt}-mediated transcriptional response emerges. Since its discovery in mammals, only few sensing and transducing molecules have been uncovered. Unequivocally, UPR^{mt} activation occurs through a mitochondria-to-nucleus signaling, initiated upon failure of mitochondrial tricarboxylic acid cycle (TCA) activity, electron transport chain (ETC) activity, fatty acid beta-oxidation (FAO), translation, protein import system etc., leading to proteotoxic stress (Nargund et al., 2012; Rolland et al., 2019). In turn, specific transcription factors allow adaptive transcriptional changes and restore collapsed mitochondrial proteostasis. Alongside chaperones and proteases, the UPR^mt-mediated transcriptional response favors glycolytic gene expression while it dampens ETC target genes (Nargund et al., 2015). The latter suggests that alternative energy sources are triggered to reduce the pressure on ETC function. Moreover, adequate evidence of how mitochondrial misfolded or unfolded proteins are recognized and how signaling is transduced across the mitochondrial membrane is missing.

The mammalian bZIP transcription factor ATF5 and its *Caenorhabditis elegans* homolog ATFS-1 are the master regulators of UPR^mt signaling. In unstressed cells, ATFS-1 is targeted predominantly to mitochondrial matrix, where degradation of ATFS-1 is facilitated by AAA^+-protease LON. As reported recently, the mitochondrial targeting sequence (MTS) of ATFS-1 is able to sense changes in mitochondrial membrane potential and thus initiate UPR^mt signaling (Rolland et al., 2019). In response to mitochondrial stress, mitigated import of ATFS-1 precursor into mitochondria favors its nuclear translocation. Nuclear accumulation of ATFS-1 triggers the expression of mitochondrial chaperone- and protease-encoding genes. Subsequently, nuclear ATFS-1 mitigates the expression of ETC transcripts to a sustainable level of bioenergetics, where less but still efficient ETC complexes are formed (Nargund et al., 2012). Interestingly, paraquat-induced (ETC inhibitor) expression of mitochondrial chaperones such as mtHSP70 and HSP10, and the protease LONP1, requires ATF5 activity in mammalian cells (Fiorese et al., 2016). It has been reported that, contrary to wild type, oligomycin- (inhibitor of ATP synthase) or doxycycline (mitochondrial translation inhibitor)-treated ATF5 knockout mice failed to activate UPR^mt as a protective mechanism against cardiac ischemia-reperfusion injury (Wang et al., 2019).

6.3 PHOSPHOLIPID BIOSYNTHESIS IN THE ENDOPLASMIC RETICULUM AND MITOCHONDRIA

Phospholipids are amphipathic molecules possessing a polar phosphate group (hydrophilic heads) and two non-polar long fatty acid (FA) chains (hydrophobic tails). Two FA chains can be attached either to glycerol or a sphingosine backbone, forming GPLs and/or

sphingophospholipids (SPLs), respectively. Here, we review GPL biosynthesis (referred to as phospholipid from now on) derived from a glycerol backbone. Infrequently found in cells, PA is the metabolic precursor of all phospholipids composed of a glycerol backbone, two FA chains and a phosphate group. Glycerol contains two hydroxyl groups esterified by FAs while esterification or not of the third glycerol hydroxyl group (sn3 position) with a phosphate defines PA or diacylglycerol (DAG) formations (Blanco and Blanco, 2017). Given that phospholipids are amphipathic molecules, their spontaneous self-assembly forms a continuous bilayer (Ridgway, 2016). Therefore, membrane formation of all mammalian cells relies primarily on the proper synthesis and distribution of these phospholipid classes (Figure 6.1).

6.3.1 Phosphatidylcholine

Mostly synthesized in the liver, PC appears to be the major phospholipid in biliary lipids, secreted lipoproteins and lipid droplet membranes (Van der Veen et al., 2017). PC synthesis requires the activity of several ER- and cytosolic-resident enzymes of the cytidine diphosphate (CDP) choline pathway, alternatively named as the Kennedy pathway. Choline is the main substrate for PC synthesis. Cytosolic choline kinase (CK) enzymatic activity promptly phosphorylates choline to phosphocholine which in turn is catalyzed by CTP:phosphocholine cytidyltransferase (CCT; encoded by *Pcyt1*), to form CDP-choline. Next, the ER-embedded 1,2-DAG choline phosphotransferase (CPT) catalyzes the formation of PC by attaching a phosphocholine moiety from CDP-choline to the third hydroxyl group of DAG (Ridgway, 2016). Alternatively, PE can be used as a substrate for PC synthesis. In this context, the ER- and mitochondrial-associated ER membrane (MAM)-embedded PE N-methyltransferase (PEMT) catalyzes the sequential methylation of PE to PC (Vance, 2014).

6.3.2 Phosphatidylethanolamine

PE is the second most abundant phospholipid in mammalian cells. Interestingly, coordinated *de novo* synthesis of PE in both ER and mitochondria has been reported. Of note, partial inhibition of mitochondrial PE synthesis activates the

PROTEOSTASIS AND PROTEOLYSIS

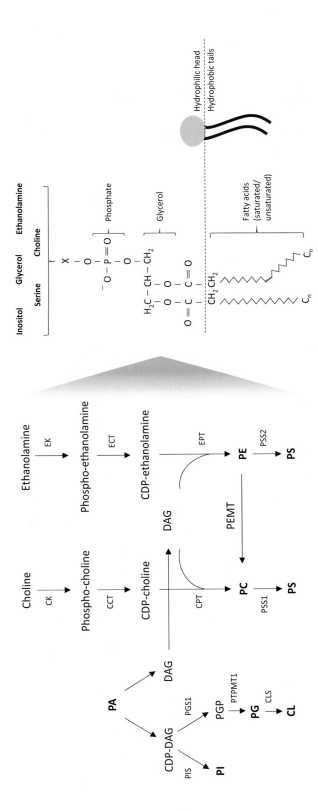

Figure 6.1 **Overview of phospholipid biosynthesis and structure in mammalian cells.** The enzymatic activities required for phospholipid biosynthesis are discussed in the text. Abbreviations: CK: Choline kinase, CCT: CTP:phosphocholine cytidyltransferase, CPT: 1,2-DAG choline phosphotransferase, EK: ethanolamine kinase, ECT: CTP:phosphoethanolamine cytidylyltransferase, EPT: 1,2-DAG ethanolamine phosphotransferase, PSS1: PS synthase 1, PSS2: PS synthase 2, PIS: PI synthase, PGS1: PGP synthase, PTPMT1: PTP localized to the mitochondrion 1, CLS: cardiolipin synthase, C_n: number of carbons.

CDP-ethanolamine pathway to restore imbalance of cellular PE levels (Steenbergen et al., 2005). ER-synthesized PE relies on the CDP-ethanolamine pathway. Initially, the cytosolic ethanolamine kinase (EK) catalyzes the phosphorylation of ethanolamine to phosphoethanolamine. Next, the rate-limiting CTP:phosphoethanolamine cytidylyltransferase (ECT; encoded by Pcyt2) converts phosphoethanolamine to CDP-ethanolamine. The last step involves the ER-localized enzymatic activity of 1,2-DAG ethanolamine phosphotransferase (EPT) which catalyzes the transfer of the phosphoethanolamine moiety of CDP-ethanolamine to the free hydroxyl group of DAG, to form PE (Gibellini and Smith, 2010). Mitochondrial synthesis of PE requires the proper trafficking of its precursor PS from MAMs of the ER to the inner mitochondrial membrane (IMM). Therein, PS decarboxylase (PSD) converts PS to PE, which is either assimilated into mitochondria or exported to the ER (Patel and Witt, 2017).

6.3.3 Phosphatidylserine

Contrary to high quantities of PC and PE, PS represents a minor phospholipid in mammalian cells. Its synthesis occurs in the MAMs and requires both PS synthase 1 (PSS1) and PS synthase 2 (PSS2). Based on intracellular calcium levels, PSS1 and PSS2 catalyze the replacement of either choline or ethanolamine polar heads from PC and PE, respectively, with L-serine (Vance, 2008). In yeast, a single PSS enzyme exists which catalyzes PS formation, in the expense of CDP-DAG and L-serine (Vance and Tasseva, 2013). Interestingly, PSS1 possesses distinct but complementary activity, since lack of PSS2 exhibits relatively normal amounts of PS and PE attributed to PSS1 activity (Arikketh et al., 2008).

6.3.4 Phosphatidylinositol

In eukaryotes, PI is found in relatively small amounts compared to other phospholipids. While DAG derived from PA dephosphorylation is the intermediate for PC and PE synthesis, conversion of PA to CDP-DAG provides the major precursor for PI synthesis (Yang et al., 2018). Particularly, the mammalian CDP-DAG synthases 1 and 2 (CDS1 and CDS2) as well as translocator assembly and maintenance homolog protein 41 (TAMM41)

catalyze the formation of CDP-DAG from PA and CTP, in the ER and mitochondria, respectively (Jennings and Epand, 2020). Despite this, mitochondrial CDS1 and CDS2 activity cannot be excluded. Subsequently, utilizing inositol and CDP-DAG as substrates, ER-embedded PI synthase (PIS) catalyzes the formation of PI (Blunsom and Cockcroft, 2020).

6.3.5 Phosphatidylglycerol

PG is also found in small quantities, accounting for approximately 1% of total mammalian phospholipid content. Of note, PG is significantly enriched in pulmonary surfactant comprising approximately 15% of phospholipid mass, and it prevents alveolar collapse. Although sharing a similar biosynthetic route with PI, PG is exclusively synthesized in mitochondria. From substrate incorporation to final PG synthesis, most of the associated enzymes are well conserved between yeast and mammals. Once PA is transported into IMM, it serves as a substrate for TAMM41 CDS activity required for CDP-DAG formation (Blunsom and Cockcroft, 2020). Next, a condensation reaction of CDP-DAG and glycerol-3-phosphate through PGP synthase 1 (PGS1) results in phosphatidylglycerol-phosphate (PGP) formation. Subsequently, PTP localized to the mitochondrion 1 (PTPMT1) phosphatase activity dephosphorylates PGP to PG (Ridgway, 2016).

6.3.6 Cardiolipin

In mammals, synthesized CL is predominantly assimilated into either the IMM or the OMM (outer mitochondrial membrane), where it accounts for approximately the 20% and 3% of total phospholipid mass, respectively. Contrary to other phospholipids, CL possesses a unique molecular structure which is constituted of two phosphate and three glycerol moieties as well as four acyl chains (Dudek, 2017). In mammalian and yeast cells, CL synthesis is amenable to a common biosynthetic pathway utilizing mitochondrial PA, CDP-DAG and PG precursors. When PG is synthesized, it rapidly reacts with CDP-DAG to form premature CL through the activity of CL synthase (CLS). Finally, premature to mature remodeling of CL requires the transacylase activity of taffazin (TAZ) (Dudek, 2017).

6.4 PHOSPHOLIPIDS AND UNFOLDED PROTEIN RESPONSE SIGNALING

Contrary to the prevailing notion that UPR is activated upon sensing of unfolded protein accumulation, aberrant phospholipid composition has also been shown to trigger UPR signaling. For instance, alterations in acyl chain length and/or degree of saturation of various phospholipid classes result in phospholipid bilayer stress, which in turn activates UPR. Perturbation of phospholipid homeostasis has been associated with various metabolic disorders including obesity and diabetes, among others. Conversely, in response to ER or mitochondrial proteotoxic stress, UPR activation has been shown to control phospholipid abundance. Here, our current understanding of phospholipids and UPR mutual regulation will be discussed.

6.4.1 Unfolded Protein Response as a Sensor of Phospholipid Bilayer Stress

Phospholipid biosynthesis in mammalian and yeast cells is primarily compartmentalized in the ER and mitochondria (Ridgway, 2016). In addition, locally synthesized phospholipids are convoyed to a designated cellular membrane in time of need. Once they have reached their destination, phospholipids are organized and self-assembled into bilayers, or used as a substrate for the synthesis of other phospholipid classes. Regardless of where they are synthesized or transported, their bilayer heterogeneity determines membrane fluidity and provides a versatile environment for integral membrane proteins to be embedded, folded, modified and recruited (Harayama and Riezman, 2018). Specifically, membrane fluidity relies on the acyl chain composition of various phospholipid classes which is unique among cell types and subcellular organelles (Ridgway, 2016). Moreover, the presence of saturated and unsaturated acyl chains in various phospholipid classes renders membranes as non-fluidic and fluidic, respectively (Harayama and Riezman, 2018; Manni et al., 2018). Importantly, the acyl chain ordering and length affects linearly the thickness of the PC bilayer in the presence of sterols. Together, acyl chain length and degree of saturation regulate PC bilayer thickness and engender proper self-assembly of membrane-bound proteins (Anbazhagan and Schneider, 2010). In addition, membrane protein conformation and function necessitate the proper matching of the hydrophobic core of membranes with the hydrophobic span of transmembrane helix. Similarly, to ER-synthesized phospholipids, acyl chain remodeling of CL affects mitochondrial membrane properties, and the function of the embedded ETC proteins (Pennington et al., 2019). Although membrane-bound proteins account for approximately 30% of the human proteome, the interplay between phospholipid bilayers and incorporated proteins is still not fully understood. Interestingly, it has been shown that phospholipid bilayer stress, as a result of aberrant phospholipid composition and remodeling, mitigates the harnessing of membrane properties while it promotes ER stress and proteostasis collapse (Shyu et al., 2019). Thus, it is not surprising that changes in the phospholipid composition in the ER and mitochondrial membranes can have profound effects on the activity of UPR transducers. Expectedly, disruption of phospholipid and protein homeostasis has also been associated with numerous pathologies including cancer, obesity, type II diabetes, liver and heart failure (Wang and Tontonoz, 2019).

Toward this direction, muscle cells from mice deficient of Lipin1 (a phosphatidate phosphatase enzyme catalyzing the dephosphorylation of phosphatidate to DAG) exhibit increased phospholipid biosynthesis which coincides with high levels of several phospholipid substrates. In addition to phospholipids, the levels of triglycerides (TAGs) and their DAGs precursors are also augmented. This accumulation of divergent lipid classes has been ascribed to higher expression of various lipogenic genes controlled by the sterol regulatory element-binding protein 1c/2 (SREBP1c/2) when Lipin1 is depleted. Ultimately, sarcoplasmic reticulum stress, resulting from dysregulation of lipid composition, is accompanied by active XBP1 and ATF6 forms (Rashid et al., 2019). Moreover, obese mice with high liver PC/PE ratio display irregular ER Ca^{2+} signaling and UPR[ER] activation (Fu et al., 2011). Likewise, increased phospholipid acyl chain saturation in mouse liver due to lysophosphatidylcholine acyltranferase 3 (Lpcat3) deficiency is accompanied by heightened UPR[ER] signaling (Rong et al., 2013). In addition, stearoyl-CoA desaturase 1 (SCD1)-deficient mice exhibit more saturated fatty acids (SFAs) and less monounsaturated fatty acids (MUFAs) of phospholipids followed by activation of CHOP, GRP78 and increased of spliced XBP1 transcripts (Ariyama et al., 2010). Since SCD1 catalyzes the biosynthesis of MUFAs (e.g. palmitate) from SFAs (e.g. oleate), when palmitate is supplemented, activation of UPR^{ER} is expected. In agreement, incubation of rat insulinoma cell line (INS1-β cells) with palmitate initiates UPR^{ER} signaling, which is reversed upon oleate supplementation (Sommerweiss et al., 2013). It is noteworthy that in INS1 β-cells, palmitate-induced ER stress is also accompanied by altered phospholipid composition (Moffitt et al., 2005). Furthermore, exposing hepatic cells to palmitate leads to ER membrane integrity collapse and accumulation of saturated phospholipids, followed by increased CHOP expression. These adverse conditions could be reversed upon oleate supplementation (Leamy et al., 2014). As evidenced in yeast and mammalian cells, IRE1 and PERK1 responsiveness in increased acyl chain saturation, resulting by phospholipid perturbation, has also been reported. Of note, IRE1 and PERK LDs, required for sensing unfolded proteins, are dispensable for UPR^{ER} activation in response to phospholipid bilayer stress. As expected, direct sensing of bilayer stress predominantly relies on their transmembrane domains (Ho et al., 2020; Volmer et al., 2013). Despite this, whether phospholipid bilayer stress and misfolded proteins act in parallel or not, for UPR^{ER} signaling to be activated, is far from understood. Further mechanistic insight into IRE1 oligomerization and activation showed that an amphipathic helix region within IRE1 transmembrane helix controls its responsiveness to both phospholipid bilayer and proteotoxic stress (Halbleib et al., 2017).

Interestingly, yeast cells lacking ubiquitin-like (UBX)-domain-containing protein 2 (UBX2) display more saturated than unsaturated phospholipid acyl chains resulting in UPR^{ER} activation. It is noteworthy that while total phospholipid saturation is increased, high discrepancies between cone-shaped PE and cylinder-shaped PC saturation have been reported (Surma et al., 2013). Most likely, high levels of saturated PE compensate for low levels of saturated PC to sustain FA composition at an optimal range, which in turn

favors bilayer-forming propensity (Basu Ball et al., 2018). In addition, knockout of acetyl-CoA synthase Fat1 in yeast increases abundance of very long-chain fatty acids (VLCFAs) with compensatory activation of UPR^{ER}. Specifically, lipidomic analysis of Fat1-deficient cells showed accumulation of PC species including FA chains with 32 and 34 carbons. Importantly, the ratio of di-unsaturated/monounsaturated PC species appeared higher in Fat1 mutants compared to wild type. Both increased acyl chain length and saturation of PC reflect membrane disruption and are necessary for UPR^{ER} activation (Micoogullari et al., 2020).

Attenuation of PC synthesis by knocking down ptm-2 in worms also induces UPR^{ER} signaling. Intriguingly, tunicamycin-treated (N-glycosylation inhibitor) and PTM-2 deficient worms have differential expression of UPR-related transcripts (Koh et al., 2018). Moreover, worms deficient of the mediator complex subunit 15 (MDT-15) display compromised unsaturated over saturated phospholipids ratio and reduced membrane fluidity, leading to UPR^{ER} activation, independent of proteotoxic stress. Despite the ER membrane disequilibrium and UPR^{ER} activation, it is surprising that MDT-15 deficient worms do not display altered CL abundance and FA composition and fail to activate the UPR^{mt} (Hou et al., 2014). Nevertheless, worms with reduced CL or PE levels upon inhibition of the CRLS-1 or PSD-1, respectively, have been shown to engender activation of UPR^{mt} (Rolland et al., 2019). Notably, worms deficient of the S-adenosyl methionine synthase 1 (SAMS-1) also exhibit compromised PC synthesis and increased UPR^{ER} signaling (Ehmke et al., 2014; Ho et al., 2020). Collectively, both the UPR^{ER} and UPR^{mt} act as sensors of imbalanced phospholipid biosynthesis and remodeling.

6.4.2 UPR as a Regulator of Phospholipid Abundance

Perturbation of phospholipids comes with a unique UPR transcriptional outcome, yet this is not a unidirectional interaction. Activation of UPR has been also shown to promote membrane biogenesis, as it regulates lipid metabolism, including phospholipids biosynthesis. Among UPR^{ER} transducers, the well-conserved IRE1 has been primarily identified in yeast cells to mediate inositol synthesis, which serves as a substrate for PI biosynthesis (Nikawa and Yamashita, 1992). Further characterization of the UPR[ER] transcriptional response in yeast identified various genes involved in phospholipid biosynthesis to be activated upon tunicamycin or dithiothreitol (DTT) treatment (a strong disulfide-reducing agent) (Travers et al., 2000). Through UPR[ER] signaling, increased expression of OPI3 and INO1, involved in PC and PI synthesis, respectively, has also been documented. This mutual activation of UPR[ER] signaling and phospholipid biosynthesis is necessary for proper membrane expansion upon ER stress (Schuck et al., 2009).

Following XBP1 activation in mammals, the enzymatic activity of several constituents of the CDP-choline pathway is also increased. Specifically, it has been suggested that increased PC synthesis contributes to the ER membrane biogenesis (Sriburi et al., 2004). Similarly, palmitate-exposed murine macrophages exhibit increased phospholipid content and enlarged ER, in an XBP1-dependent manner (Kim et al., 2015). Notably, chemically induced IRE1, ATF4 and PERK in the mouse lung result in elevated expression of lipogenic proteins such as SDC1, SREBP1c and FA synthase. The latter is evidenced by increased TAG and phospholipid contents, and a higher degree of PC saturation. When SCD1 is inhibited, defective UPR[ER] activation with increased expression of profibrotic factors comes with compromised phospholipid composition. On the contrary, elevated activity of SREBP1c accompanied by increased neutral- and phospholipid abundance alleviates both ER stress and expression of pro-fibrotic factors in lung epithelial cells (Romero et al., 2018).

Similarly, UPR[mt] activation is accompanied by changes in mitochondrial phospholipid content. Previous studies have shown that perturbation of mitochondrial proteostasis activates a mitochondrial-to-cytosolic stress response (MCSR) which functions in concert with the UPR[mt] and the cytosolic heat shock response (HSR). Interestingly, worms deficient of ATFS-1 failed to activate MCSR suggesting that UPR[mt] is indispensable for MCSR activation. In addition, it has been shown that MCSR and UPR[mt] activation is accompanied by increased levels of various phospholipid classes including mitochondrial PG which is a precursor of CL. Importantly, attenuation of CL synthesis in CRSL-1 deficient worms failed to activate MCRS (Kim et al., 2016). These findings further support the notion that phospholipid biosynthesis and UPR signaling are not mutually exclusive (Figure 6.2).

6.5 CONCLUSIONS

Compartmentalized in the ER and mitochondria, phospholipid biosynthesis and remodeling play an essential role in protein synthesis and secretion. When phospholipid or protein homeostasis collapses, the UPR signaling facilitates a specialized transcriptional response. Beyond its role in sensing phospholipid bilayer and proteotoxic stress, UPR signaling has also been shown to control phospholipid abundance in response to ER or mitochondrial stress. A vicious cycle linking phospholipid biosynthesis and UPR signaling therefore exists, which is responsive to both phospholipid bilayer and proteotoxic stress. Recent findings on the impact of bilayer stress and unfolded proteins on UPR signaling come with several questions regarding their mutual regulation. Notably, whether bilayer stress-mediated UPR activation relies on unfolded protein accumulation or not remains enigmatic. Moreover, whether the existence of distinct transcriptional outcomes upon UPR activation can provide a cross-protection response remains uncertain. In addition, how the composition and ratio of saturated versus unsaturated phospholipids regulate the UPR signaling is unclear. Since new biosynthetic enzymes and proteostatic factors have been identified, our understanding of their contribution to cellular homeostasis will continue to improve. Dissecting the molecular mechanisms underpinning co-regulation of phospholipids and UPR will open the path to a new era of resolving various severe metabolic disorders.

ACKNOWLEDGMENTS

We gratefully acknowledge the contributions of numerous investigators that we could not include in this chapter, owing to space limitations. This work was supported by grants from the European Research Council (ERC-GA695190-MANNA) and the General Secretariat for Research and Technology (GSRT) of the Greek Ministry of Development and Investments.

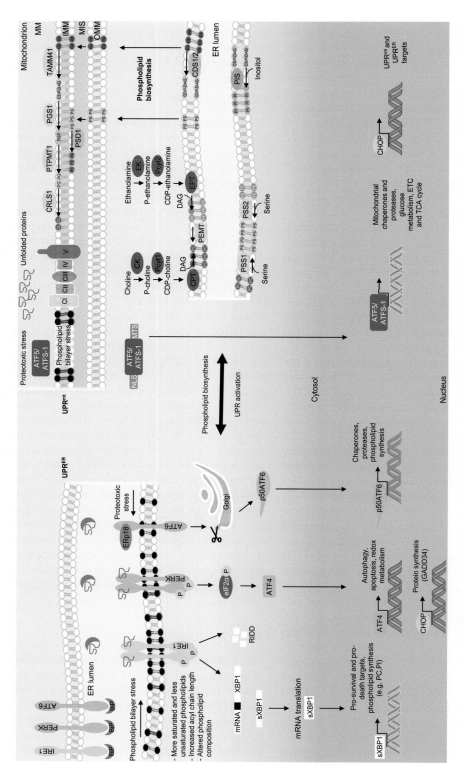

Figure 6.2 **Interplay between UPR signaling and biosynthetic pathways of phospholipids.** In mammals, sensing of ER proteotoxic and phospholipid bilayer stress relies on IRE1, PERK and ATF6 signaling. In C. elegans and mammals, sensing of mitochondria proteotoxic and phospholipid bilayer stress necessitates ATFS-1 and ATF5 activity, respectively. Intriguingly, UPR signaling proteins also regulate phospholipid abundance. While phospholipid biosynthesis is compartmentalized, trafficking of phospholipids between ER and mitochondrial also occurs. Abbreviations: MM: Mitochondrial matrix, IMM: inner mitochondrial membrane, MIS: mitochondrial intermembrane space, OMM: outer mitochondrial membrane.

REFERENCES

Anbazhagan, V., & Schneider, D. 2010. The membrane environment modulates self-association of the human GpA TM domain — implications for membrane protein folding and transmembrane signaling. *Biochim Biophys Acta*, 1798(10), 1899–1907. doi:10.1016/j.bbamem.2010.06.027

Arikketh, D., Nelson, R., & Vance, J. E. 2008. Defining the importance of phosphatidylserine synthase-1 (PSS1): unexpected viability of PSS1-deficient mice. *J Biol Chem*, 283(19), 12888–12897. doi:10.1074/jbc.M800714200

Ariyama, H., Kono, N., Matsuda, S., et al. 2010. Decrease in membrane phospholipid unsaturation induces unfolded protein response. *J Biol Chem*, 285(29), 22027–22035. doi:10.1074/jbc.M110.126870

Basu Ball, W., Baker, C. D., Neff, J. K., et al. 2018. Ethanolamine ameliorates mitochondrial dysfunction in cardiolipin-deficient yeast cells. *J Biol Chem*, 293(28), 10870–10883. doi:10.1074/jbc.RA118.004014

Blanco, A., & Blanco, G. 2017. Lipids. In A. Blanco & G. Blanco (Eds.), *Medical Biochemistry*, Academic Press, Elsevier Inc., San Diego, USA, (pp. 99–119). doi:10.1016/B978-0-12-803550-4.00005-7

Blunsom, N. J., & Cockcroft, S. 2020. CDP-diacylglycerol synthases (CDS): gateway to phosphatidylinositol and cardiolipin synthesis. *Front Cell Dev Biol*, 8, 63. doi:10.3389/fcell.2020.00063

Carrara, M., Prischi, F., Nowak, P. R., et al. 2015. Crystal structures reveal transient PERK luminal domain tetramerization in endoplasmic reticulum stress signaling. *EMBO J*, 34(11), 1589–1600. doi:10.15252/embj.201489183

Chen, S., Henderson, A., Petriello, M. C., et al. 2019. Trimethylamine N-oxide binds and activates PERK to promote metabolic dysfunction. *Cell Metab*, 30(6), 1141–1151 e1145. doi:10.1016/j.cmet.2019.08.021

Dudek, J. 2017. Role of cardiolipin in mitochondrial signaling pathways. *Front Cell Dev Biol*, 5, 90. doi:10.3389/fcell.2017.00090

Ehmke, M., Luthe, K., Schnabel, R., et al. 2014. S-adenosyl methionine synthetase 1 limits fat storage in Caenorhabditis elegans. *Genes Nutr*, 9(2), 386. doi:10.1007/s12263-014-0386-6

Fiorese, C. J., Schulz, A. M., Lin, Y. F., et al. 2016. The transcription factor ATF5 mediates a mammalian mitochondrial UPR. *Curr Biol*, 26(15), 2037–2043. doi:10.1016/j.cub.2016.06.002

Fu, S., Yang, L., Li, P., et al. 2011. Aberrant lipid metabolism disrupts calcium homeostasis causing liver endoplasmic reticulum stress in obesity. *Nature*, 473(7348), 528–531. doi:10.1038/nature09968

Gibellini, F., & Smith, T. K. 2010. The Kennedy pathway — de novo synthesis of phosphatidylethanolamine and phosphatidylcholine. *IUBMB Life*, 62(6), 414–428. doi:10.1002/iub.337

Halbleib, K., Pesek, K., Covino, R., et al. 2017. Activation of the unfolded protein response by lipid bilayer stress. *Mol Cell*, 67(4), 673–684.e678. doi:10.1016/j.molcel.2017.06.012

Han, J., Back, S. H., Hur, J., et al. 2013. ER-stress-induced transcriptional regulation increases protein synthesis leading to cell death. *Nat Cell Biol*, 15(5), 481–490. doi:10.1038/ncb2738

Harayama, T., & Riezman, H. 2018. Understanding the diversity of membrane lipid composition. *Nat Rev: Mol Cell Biol*, 19(5), 281–296. doi:10.1038/nrm.2017.138

Harding, H. P., Zhang, Y., & Ron, D. 1999. Protein translation and folding are coupled by an endoplasmic-reticulum-resident kinase. *Nature*, 397(6716), 271–274. doi:10.1038/16729

Hetz, C., & Papa, F. R. 2018. The unfolded protein response and cell fate control. *Mol Cell*, 69(2), 169–181. doi:10.1016/j.molcel.2017.06.017

Higa, A., Taouji, S., Lhomond, S., et al. 2014. Endoplasmic reticulum stress-activated transcription factor ATF6alpha requires the disulfide isomerase PDIA5 to modulate chemoresistance. *Mol Cell Biol*, 34(10), 1839–1849. doi:10.1128/MCB.01484-13

Ho, N., Xu, C., & Thibault, G. 2018. From the unfolded protein response to metabolic diseases — lipids under the spotlight. *J Cell Sci*, 131(3). doi:10.1242/jcs.199307

Ho, N., Yap, W. S., Xu, J., et al. 2020. Stress sensor Ire1 deploys a divergent transcriptional program in response to lipid bilayer stress. *J Cell Biol*, 219(7). doi:10.1083/jcb.201909165

Horvath, S. E., & Daum, G. 2013. Lipids of mitochondria. *Prog Lipid Res*, 52(4), 590–614. doi:10.1016/j.plipres.2013.07.002

Hou, N. S., Gutschmidt, A., Choi, D. Y., et al. 2014. Activation of the endoplasmic reticulum unfolded protein response by lipid disequilibrium without disturbed proteostasis in vivo. *Proc Natl Acad Sci U S A*, 111(22), E2271–2280. doi:10.1073/pnas.1318262111

Imagawa, Y., Hosoda, A., Sasaka, S., et al. 2008. RNase domains determine the functional difference between IRE1alpha and IRE1beta. *FEBS Lett*, 582(5), 656–660. doi:10.1016/j.febslet.2008.01.038

Jennings, W., & Epand, R. M. 2020. CDP-diacylglycerol, a critical intermediate in lipid metabolism. *Chem*

Phys Lipids, 230, 104914. doi:10.1016/j.chemphyslip. 2020.104914

Kim, S. K., Oh, E., Yun, M., et al. 2015. Palmitate induces cisternal ER expansion via the activation of XBP-1/CCTalpha-mediated phospholipid accumulation in RAW 264.7 cells. *Lipids Health Dis*, 14, 73. doi:10.1186/s12944-015-0077-3

Kim, H. E., Grant, A. R., Simic, M. S., et al. 2016. Lipid biosynthesis coordinates a mitochondrial-to-cytosolic stress response. *Cell*, 166(6), 1539–1552. e1516. doi:10.1016/j.cell.2016.08.027

Koh, J. H., Wang, L., Beaudoin-Chabot, C., et al. 2018. Lipid bilayer stress-activated IRE-1 modulates autophagy during endoplasmic reticulum stress. *J Cell Sci*, 131(22). doi:10.1242/jcs.217992

Kopp, M. C., Larburu, N., Durairaj, V., et al. 2019. UPR proteins IRE1 and PERK switch BiP from chaperone to ER stress sensor. *Nat Struct Mol Biol*, 26(11), 1053–1062. doi:10.1038/s41594-019-0324-9

Kuhlbrandt, W. 2015. Structure and function of mitochondrial membrane protein complexes. *BMC Biol*, 13, 89. doi:10.1186/s12915-015-0201-x

Leamy, A. K., Egnatchik, R. A., Shiota, M., et al. 2014. Enhanced synthesis of saturated phospholipids is associated with ER stress and lipotoxicity in palmitate treated hepatic cells. *J Lipid Res*, 55(7), 1478–1488. doi:10.1194/jlr.M050237

Lee, Y. Y., Cevallos, R. C., & Jan, E. 2009. An upstream open reading frame regulates translation of GADD34 during cellular stresses that induce eIF2alpha phosphorylation. *J Biol Chem*, 284(11), 6661–6673. doi:10.1074/jbc.M806735200

Lynch, J. M., Maillet, M., Vanhoutte, D., et al. 2012. A thrombospondin-dependent pathway for a protective ER stress response. *Cell*, 149(6), 1257–1268. doi:10.1016/j.cell.2012.03.050

Manni, M. M., Tiberti, M. L., Pagnotta, S., et al. 2018. Acyl chain asymmetry and polyunsaturation of brain phospholipids facilitate membrane vesiculation without leakage. *eLife*, 7. doi:10.7554/ eLife.34394

Maurel, M., Chevet, E., Tavernier, J., et al. 2014. Getting RIDD of RNA: IRE1 in cell fate regulation. *Trends Biochem Sci*, 39(5), 245–254. doi:10.1016/j. tibs.2014.02.008

Metcalf, M. G., Higuchi-Sanabria, R., Garcia, G., et al. 2020. Beyond the cell factory: homeostatic regulation of and by the UPR(ER). *Sci Adv*, 6(29), eabb9614. doi:10.1126/sciadv.abb9614

Micoogullari, Y., Basu, S. S., Ang, J., et al. 2020. Dysregulation of very-long-chain fatty acid metabolism causes membrane saturation and induction of the unfolded protein response. *Mol Biol Cell*, 31(1), 7–17. doi:10.1091/mbc.E19-07-0392

Moffitt, J. H., Fielding, B. A., Evershed, R., et al. 2005. Adverse physicochemical properties of tripalmitin in beta cells lead to morphological changes and lipotoxicity in vitro. *Diabetologia*, 48(9), 1819–1829. doi:10.1007/s00125-005-1861-9

Nadanaka, S., Okada, T., Yoshida, H., et al. 2007. Role of disulfide bridges formed in the luminal domain of ATF6 in sensing endoplasmic reticulum stress. *Mol Cell Biol*, 27(3), 1027–1043. doi:10.1128/ MCB.00408-06

Nargund, A. M., Pellegrino, M. W., Fiorese, C. J., et al. 2012. Mitochondrial import efficiency of ATFS-1 regulates mitochondrial UPR activation. *Science*, 337(6094), 587–590. doi:10.1126/science.1223560

Nargund, A. M., Fiorese, C. J., Pellegrino, M. W., et al. 2015. Mitochondrial and nuclear accumulation of the transcription factor ATFS-1 promotes OXPHOS recovery during the UPR(mt). *Mol Cell*, 58(1), 123–133. doi:10.1016/j.molcel.2015.02.008

Nikawa, J., & Yamashita, S. 1992. IRE1 encodes a putative protein kinase containing a membrane-spanning domain and is required for inositol phototrophy in Saccharomyces cerevisiae. *Mol Microbiol*, 6(11), 1441–1446. doi:10.1111/j.1365-2958.1992.tb00864.x

Oka, O. B., van Lith, M., Rudolf, J., et al. 2019. ERp18 regulates activation of ATF6alpha during unfolded protein response. *EMBO J*, 38(15), e100990. doi:10.15252/embj.2018100990

Patel, D., & Witt, S. N. 2017. Ethanolamine and phosphatidylethanolamine: partners in health and disease. *Oxidat Med Cell Longev*, 2017, 4829180. doi:10.1155/2017/4829180

Pendin, D., McNew, J. A., & Daga, A. 2011. Balancing ER dynamics: shaping, bending, severing, and mending membranes. *Curr Opin Cell Biol*, 23(4), 435–442. doi:10.1016/j.ceb.2011.04.007

Pennington, E. R., Funai, K., Brown, D. A., et al. 2019. The role of cardiolipin concentration and acyl chain composition on mitochondrial inner membrane molecular organization and function. *Biochim Biophys Acta Mol Cell Biol Lipids*, 1864(7), 1039–1052. doi:10.1016/j.bbalip.2019.03.012

Rashid, T., Nemazanyy, I., Paolini, C., et al. 2019. Lipin1 deficiency causes sarcoplasmic reticulum stress and chaperone-responsive myopathy. *EMBO J*, 38(1). doi:10.15252/embj.201899576

Ridgway, N. 2016. Phospholipid Synthesis in Mammalian Cells. In N. D. Ridgway & R. S. McLeod (Eds.), *Biochemistry of Lipids, Lipoproteins and Membranes* (6th ed), Elsevier B.V., Amsterdam,

The Netherlands, (pp. 209–236). dio: 10.1016/B978-0-444-63438-2.00007-9

Rolland, S. G., Schneid, S., Schwarz, M., et al. 2019. Compromised mitochondrial protein import acts as a signal for UPR(mt). *Cell Rep*, 28(7), 1659–1669. e1655. doi:10.1016/j.celrep.2019.07.049

Romero, F., Hong, X., Shah, D., et al. 2018. Lipid synthesis is required to resolve endoplasmic reticulum stress and limit fibrotic responses in the lung. *Am J Respir Cell Mol Biol*, 59(2), 225–236. doi:10.1165/rcmb.2017-0340OC

Rong, X., Albert, C. J., Hong, C., et al. 2013. LXRs regulate ER stress and inflammation through dynamic modulation of membrane phospholipid composition. *Cell Metab*, 18(5), 685–697. doi:10.1016/j.cmet.2013.10.002

Sato, Y., Nadanaka, S., Okada, T., et al. 2011. Luminal domain of ATF6 alone is sufficient for sensing endoplasmic reticulum stress and subsequent transport to the Golgi apparatus. *Cell Struct Funct*, 36(1), 35–47. *doi:10.1247/csf.10010*

Schuck, S., Prinz, W. A., Thorn, K. S., et al. 2009. Membrane expansion alleviates endoplasmic reticulum stress independently of the unfolded protein response. *J Cell Biol*, 187(4), 525–536. doi:10.1083/jcb.200907074

Senft, D., & Ronai, Z. A. 2015. UPR, autophagy, and mitochondria crosstalk underlies the ER stress response. *Trends Biochem Sci*, 40(3), 141–148. doi:10.1016/j.tibs.2015.01.002

Shyu, P., Jr., Ng, B. S. H., Ho, N., et al. 2019. Membrane phospholipid alteration causes chronic ER stress through early degradation of homeostatic ER-resident proteins. *Sci Rep*, 9(1), 8637. doi:10.1038/s41598-019-45020-6

Sommerweiss, D., Gorski, T., Richter, S., et al. 2013. Oleate rescues INS-1E beta-cells from palmitate-induced apoptosis by preventing activation of the unfolded protein response. *Biochem Biophys Res Commun*, 441(4), 770–776. doi:10.1016/j.bbrc.2013.10.130

Sriburi, R., Jackowski, S., Mori, K., et al. 2004. XBP1: a link between the unfolded protein response, lipid biosynthesis, and biogenesis of the endoplasmic reticulum. *J Cell Biol*, 167(1), 35–41. doi:10.1083/jcb.200406136

Steenbergen, R., Nanowski, T. S., Beigneux, A., et al. 2005. Disruption of the phosphatidylserine decarboxylase gene in mice causes embryonic lethality and mitochondrial defects. *J Biol Chem*, 280(48), 40032–40040. doi:10.1074/jbc.M506510200

Surma, M. A., Klose, C., Peng, D., et al. 2013. A lipid E-MAP identifies Ubx2 as a critical regulator of lipid saturation and lipid bilayer stress. *Mol Cell*, 51(4), 519–530. doi:10.1016/j.molcel.2013.06.014

Travers, K. J., Patil, C. K., Wodicka, L., et al. 2000. Functional and genomic analyses reveal an essential coordination between the unfolded protein response and ER-associated degradation. *Cell*, 101(3), 249–258. doi:10.1016/s0092-8674(00)80835-1

Uemura, A., Oku, M., Mori, K., et al. 2009. Unconventional splicing of XBP1 mRNA occurs in the cytoplasm during the mammalian unfolded protein response. *J Cell Sci*, 122(Pt 16), 2877–2886. doi:10.1242/jcs.040584

Van der Veen, J. N., Kennelly, J. P., Wan, S., et al. 2017. The critical role of phosphatidylcholine and phosphatidylethanolamine metabolism in health and disease. *Biochim Biophys Acta Biomembr*, 1859(9 Pt B), 1558–1572. doi:10.1016/j.bbamem.2017.04.006

Vance, D. E. 2014. Phospholipid methylation in mammals: from biochemistry to physiological function. *Biochim Biophys Acta*, 1838(6), 1477–1487. doi:10.1016/j.bbamem.2013.10.018

Vance, J. E. 2008. Phosphatidylserine and phosphatidylethanolamine in mammalian cells: two metabolically related aminophospholipids. *J Lipid Res*, 49(7), 1377–1387. doi:10.1194/jlr.R700020-JLR200

Vance, J. E., & Tasseva, G. 2013. Formation and function of phosphatidylserine and phosphatidylethanolamine in mammalian cells. *Biochim Biophys Acta*, 1831(3), 543–554. doi:10.1016/j.bbalip.2012.08.016

Volmer, R., van der Ploeg, K., & Ron, D. 2013. Membrane lipid saturation activates endoplasmic reticulum unfolded protein response transducers through their transmembrane domains. *Proc Natl Acad Sci U S A*, 110(12), 4628–4633. doi:10.1073/pnas.1217611110

Wang, B., & Tontonoz, P. 2019. Phospholipid remodeling in physiology and disease. *Annu Rev Physiol*, 81, 165–188. doi:10.1146/annurev-physiol-020518-114444

Wang, P., Li, J., Tao, J., et al. 2018. The luminal domain of the ER stress sensor protein PERK binds misfolded proteins and thereby triggers PERK oligomerization. *J Biol Chem*, 293(11), 4110–4121. doi:10.1074/jbc.RA117.001294

Wang, Y. T., Lim, Y., McCall, M. N., et al. 2019. Cardioprotection by the mitochondrial unfolded protein response requires ATF5. *Am J Physiol Heart Circ Physiol*, 317(2), H472–H478. doi:10.1152/ajpheart.00244.2019

Yang, Y., Lee, M., & Fairn, G. D. 2018. Phospholipid subcellular localization and dynamics. *J Biol Chem*, 293(17), 6230–6240. doi:10.1074/jbc.R117.000582

Ye, J., Rawson, R. B., Komuro, R., et al. 2000. ER stress induces cleavage of membrane-bound ATF6 by the same proteases that process SREBPs. *Mol Cell*, 6(6), 1355–1364. doi:10.1016/s1097-2765(00)00133-7

Ubiquitin Ligases Involved in Progeroid Syndromes and Age-Associated Pathologies

Lisa Fechtner and Thorsten Pfirrmann

CONTENTS

7.1 INTRODUCTION

An irreversible and yet unpreventable physiological change all organisms face during lifetime is the process of aging. This process can be defined as a time-dependent decline of cellular and organismal physiological functions correlating with typical aging phenotypes that manifest on a cellular, molecular and organismal scale. Phenotypic changes promoting the aging process, established on a cellular and molecular level, are generally classified into nine hallmarks of aging. They include genomic instability, telomere attrition, epigenetic alterations, altered intercellular communication, stem cell exhaustion, cellular senescence, mitochondrial dysfunction, deregulated nutrient sensing and the loss of proteostasis (Lopez-Otin et al., 2013).

The term proteostasis combines the words protein and homeostasis, defined as the fine-tuned balance between protein synthesis, protein folding and protein degradation at the level of functional proteins. Young cells and organisms preserve this balance through regulatory protein quality control systems that guarantee the removal or refolding of misfolded, dysfunctional proteins. In aging and aged cells, these control

DOI: 10.1201/9781003048138-7

systems are dysfunctional, and cells accumulate damaged and misfolded proteins. Consequently, several age-associated diseases are caused by dysfunction in these processes. A central intracellular protein degradation system responsible for removing those proteins is the ubiquitin modification system (UMS). The ubiquitination cascade allows the targeted degradation of misfolded proteins through the recognition of hydrophobic patches. Besides that, it is also capable of degrading virtually any specific protein of the cell in a spatiotemporal manner, and consequently, not only proteostasis but also all other hallmarks of aging can be affected. The substrate specificity of the ubiquitination cascade is guaranteed through a class of enzymes called ubiquitin ligases (E3s). This chapter introduces ubiquitin ligases involved in premature aging (progeroid) syndromes, age-associated diseases or E3s with a general function in the process of aging.

7.2 THE UBIQUITIN MODIFICATION SYSTEM

Posttranslational modification of proteins with ubiquitin (ubiquitination) plays an essential role in virtually every cellular process. Ubiquitin is a stable and highly abundant 76-amino acid polypeptide that is highly evolutionarily conserved from yeast to man (Ozkaynak et al., 1984). The vast majority of proteins undergo ubiquitination within their cellular lifetime, and besides many other functions, ubiquitination targets up to 80% of proteins for degradation by the 26S proteasome (Rock et al., 1994). The nature of this regulatory diversity lays in the ubiquitination machinery's substrate specificity and the specific topology of the ubiquitin modification, e.g. the attachment of mono- or polyubiquitin residues to target proteins. Ubiquitination starts with the formation of an isopeptide bond between the C-terminal glycine residue of ubiquitin and a lysine residue within the substrate, thus forming a mono-ubiquitinated substrate. Ubiquitin itself contains seven lysine residues that can be further modified, allowing linkage-specific polyubiquitination. These variable kinds of modifications enable the regulation of biological processes like DNA repair, endocytosis, receptor trafficking, protein degradation by the 26S proteasome and others (Yau and Rape, 2016). The process of ubiquitination consists of three consecutive enzymatic steps involving a ubiquitin-activating enzyme (E1), a ubiquitin-conjugating enzyme (E2) and a ubiquitin ligase (E3) (see Figure 7.1) (Ciechanover, 2005). Substrate specificity is mostly regulated by the interaction of a substrate with its corresponding specific ubiquitin ligase. As a consequence, ubiquitin ligases form the biggest and most complex group within the hierarchical ubiquitination machinery. In mammals, there are two E1 enzymes, about 30 E2 enzymes, and nearly 800 ubiquitin ligases (Hutchins et al., 2013).

7.2.1 Ubiquitin Ligases

Ubiquitin ligases can be divided into three major groups: the most extensive group are the RING (Really Interesting New Gene/U-box)-type E3s with about 600 members (Metzger et al., 2012), followed by the HECT (homologous to E6AP C-terminus)-type E3s with about 30 members (Sluimer and Distel, 2018) and the RBR (RING between RING)-type E3s with about 14 members (Reiter and Klevit, 2018). Some ubiquitin ligases function as monomers; others are heterooligomers and form protein complexes.

7.2.1.1 RING Domain-Containing Ubiquitin Ligases

The RING domain-containing ubiquitin ligases represent the largest family of ubiquitin ligases (Li et al., 2008). RING (also known as RING finger, RING motif or RING domain) and RING-like E3s (plant homeodomain/leukemia-associated protein (PHD/LAP) and U-box) are collectively referred to as RING-type ubiquitin ligases (Metzger et al., 2014). RING ubiquitin ligases are characterized by the presence of a RING domain that binds two zinc ions coordinated by eight conserved cysteine or histidine residues (Deshaies and Joazeiro, 2009). Non-canonical RING-type E3s, like the RING/U-box-type ubiquitin ligases, have a RING-fold without zinc coordination sites. Both RING and U-box domains are responsible for binding the ubiquitin-charged E2 to enable ubiquitin transfer to the target substrate (Metzger et al., 2014).

7.2.1.2 HECT Ubiquitin Ligases

Among the ubiquitin-ligase family, the HECT domain ubiquitin ligases belong to a smaller group of roughly 30 known HECT ligases in

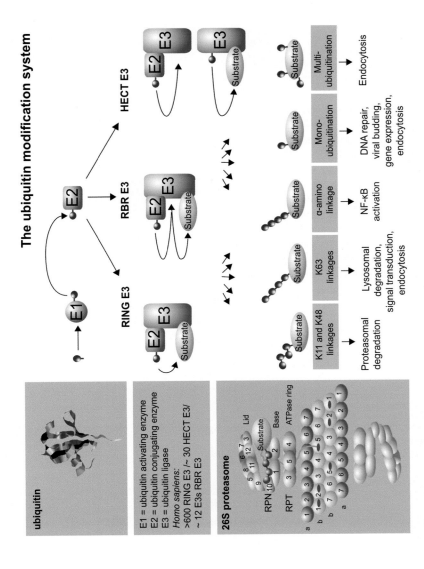

Figure 7.1 **The Ubiquitin modification system (UMS).** This overview shows the three enzymatic steps required for the ubiquitination of a substrate protein. (1) A C-terminal glycine carboxyl group of ubiquitin forms a thioester linkage with a ubiquitin-activating enzyme (E1). (2) Ubiquitin is transferred to a sulfhydryl group of a ubiquitin-conjugating enzyme (E2). (3) The ubiquitin ligase (E3) featuring the so-called RING-, RBR- or HECT domains mediates the isopeptide formation between the C-terminus ubiquitin and a lysine ε-amino group of the substrate or ubiquitin, resulting in mono-, multi- or poly-ubiquitinated substrates. *Left corner:* Structure of the 26S proteasome with its three subcomplexes: (1) the 20S core with four stacked rings combined of 1–7 α-subunits and two rings composed of 1–7 β-subunits (pink), β-subunits with proteolytic activity (dark gray); (2) the 19S base (with RPN1-2 and the ATPase subunits RPT1–RPT6 (light gray) and (3) the lid with Rpn3, RPN5-9, RPN12 (gray) and RPN10, RPT5 (blue). *Abbreviation:* RPN, regulatory particle non-ATPase; RPT, regulatory particle ATPase.

mammals (Foote et al., 2017). The human HECTs can be divided into three subfamilies: the NEDD4 family that contains tryptophan-tryptophan (WW) motifs (Sluimer and Distel, 2018), the HERC (HECT and RCC1-like domain) family that possesses one or more regulators of chromosome condensation 1 (RCC1)-like domains (RLDs), and HECTs that contain various domains (Sala-Gaston et al., 2020). The HECT domain has a molecular weight of approximately 40 kDa and consists of two flexibly tethered lobes (the N- and C-lobes). The N-lobe includes the docking surface for the E2 (Huang et al., 1999) and the shorter C-lobes contain the active-site cysteine (Verdecia et al., 2003). In contrast to RING E3s, HECT E3s form a thioester intermediate with the loaded ubiquitin. Initially, the HECT domain binds a specific E2, and a ubiquitin-thioester intermediate with the active-site cysteine is formed. After that, the activated ubiquitin is transferred to either the ε-amino group of a substrate lysine residue or to a lysine residue of ubiquitin to form the elongated polyubiquitin chain (Dye and Schulman, 2007). HECT ubiquitin ligases are known to generate polyubiquitin chains such as K48-linked (e.g. E6AP), and K29-linked (e.g. KIAA10), and also mono-ubiquitination of a specific substrate (e.g. NEDD4) (Weber et al., 2019).

7.2.1.3 RBR Ubiquitin Ligases

The RBR ubiquitin ligases are a group of multidomain enzymes that share a combination of RING and HECT ubiquitination mechanisms. These enzymes form a ubiquitin-thioester intermediate with an active-site cysteine like observed in HECT E3s but recruit thioester-activated E2s via a RING domain (Wenzel et al., 2011). The RBR ubiquitin ligase has two predicted RING domains separated by an in-between RING (IBR) domain. RING1 recruits the ubiquitin-charged E2, and the RING2 domain, also called Rcat (required-for-catalysis) domain, possesses a catalytic cysteine that forms the ubiquitin-thioester intermediate. The IBR domain has a similar fold like the RING2 (Rcat) domain but lacks the catalytic cysteine residue and is, therefore, also called BRcat (benign-catalytic) domain (Spratt et al., 2014). An essential RBR ubiquitin ligase is parkin that plays a role in the early onset of Parkinson's disease (PD) (PRKN; Table 7.1).

7.2.1.4 Heterooligomeric Ubiquitin Ligases

Some RING finger/U-box E3s form homooligomeric- and/or heterooligomeric complexes with another RING finger/U-box E3. They can be divided into two subclasses: (1) one subclass

TABLE 7.1

Ubiquitin ligases involved in progeroid syndromes and age-associated disease

Protein (#UniProt)	Disease (#OMIM)	Function	Affected hallmarks of aging
PRKN (#O60260)	Parkinson disease, juvenile, type 2 (#600116)	**RING domain ubiquitin ligase involved in:** • Mono-ubiquitination (K-6, K-11, K-48 and K-63) • Polyubiquitination • Autophagic degradation of depolarized mitochondria (mitophagy)	Mitochondrial dysfunction, loss of proteostasis
MDM2 (#Q00987)	Werner-like syndrome (# 277700)	**RING domain ubiquitin ligase involved in:** • Mono- and polyubiquitination of p53	Altered intercellular communication, genomic instability, cellular senescence, loss of proteostasis
CSA (#Q13216)	Cockayne syndrome type A (#216400)	**Substrate-recognition component (DCAF) of the CRL4^{CSA} ubiquitin ligase complex involved in:** • DNA repair • Ubiquitination of RNA polymerase II (RNAPII)	Genomic instability, loss of proteostasis

PROTEOSTASIS AND PROTEOLYSIS

FBXO7 (#Q9Y3I1)	Parkinson disease 15, early-onset (#260300)	**Substrate-recognition component (F-BOX) of the SCF^{FBXO7} ubiquitin ligase complex involved in:** • Autophagic degradation of depolarized mitochondria (mitophagy)	Mitochondrial dysfunction, loss of proteostasis
VHL (#Q19213)	Von Hippel-Lindau syndrome (#193300)	**Substrate-recognition component of the CRL2^{VHL} ubiquitin ligase complex involved in:** • Oxygen-dependent ubiquitination and subsequent proteasomal degradation of HIF1A	Cellular senescence, altered intercellular communication, genomic instability, loss of proteostasis
NEDD4 (#P46934)	Cockayne syndrome (#602278)	**HECT ubiquitin ligase involved in:** • Ubiquitination and subsequent endocytosis of membrane proteins • Ubiquitination of RNA polymerase II (RNAPII)	Altered cellular communication, genomic instability, loss of proteostasis
BRCA1 (#P38398)	Cockayne syndrome (#113705/601593)	**RING domain ubiquitin ligase involved in:** • K6-linked polyubiquitination • BRCA1/BARD1 dimer regulates cell cycle, DNA damage repair	Altered cellular communication, genomic instability, loss of proteostasis
FBXO32 (#Q969P5)	Muscle atrophy (#606604)	**Substrate recognition component of a SCF (SKP1-CUL1-F-box protein) ubiquitin-protein ligase complex involved in:** • Ubiquitination and subsequent proteasomal degradation • Recognizes and binds to phosphorylated target proteins during skeletal muscle atrophy • Recognizes TERF1	Loss of proteostasis, dysregulated nutrient sensing
TRIM63 (#Q969Q1)	Muscle atrophy (#606131)	**RING domain ubiquitin ligase involved in:** • Proteasomal degradation of muscle proteins under amino acid starvation • Inhibition of skeletal muscle protein synthesis under amino acid starvation • Organization of myofibrils in muscle cells	Loss of proteostasis, dysregulated nutrient sensing
GID/CTLH complex	–	**RING domain containing ubiquitin ligase involved in:** • AMPK regulation	Loss of proteostasis, dysregulated nutrient sensing
KLHL22 (#Q53GT1)	–	**Adapter protein of a BCR (BTB-CUL3-RBX1) ubiquitin ligase involved in:** • Chromosome alignment, silencing of the spindle assembly checkpoint (SAC) and chromosome segregation • Amino-acid-stimulated K-48 poly-ubiquitination and proteasomal degradation of DEPDC5	Loss of proteostasis, dysregulated nutrient sensing

affects the activity of the respective E3s, involving physical contacts between another RING finger or U-box domain-containing second E3, for example, in the case of the RING finger/U-box E3s BRCA1-BARD1 and Mdm2-MdmX (Kosztyu et al., 2019). Here, RING- and U-box domains serve as interfaces for the interaction with E2s and other E3s. (2) Another subclass consists of multi-component E3s, in which the RING finger and the substrate-binding units are assembled as different individual proteins (Zhao and Sun, 2013). The cullin-RING E3s (CRLs) are composed of a cullin scaffold protein (e.g. CUL1-5), a conserved RING-box protein (RBX) (e.g. RBX1 or RBX2), a substrate receptor (SR) (e.g. F-box protein or SKP2) and an adaptor protein (e.g. SKP1) (Petroski and Deshaies, 2005). Six closely related cullin proteins, CUL1, CUL2, CUL3, CUL4A, CUL4B and CUL5, are encoded by the human genome (Zimmerman et al., 2010). CRLs have a modular design to achieve high specificity toward a large number of substrates. Each cullin-RING complex can assemble with its subset of SR proteins, individually linked to the cullin scaffold by an adaptor protein. SR proteins have various protein–protein interaction domains to bind specific substrates. Six known cullin-RING complexes, combined with numerous adaptor-SR partners, create a yield of approximately 500 CRL family members in total (Zhao and Sun, 2013). The S-phase kinase-associated protein 1 (SKP1)–cullin 1 (CUL1)–F-box protein (SCF) complex, as a prominent example, is composed of the scaffold protein CUL1, the RING protein RBX1, SKP1 and one out of several variable F-box proteins responsible for substrate specificity (Skaar et al., 2013).

7.2.2 Ubiquitin Ligases Involved in Progeroid Syndromes

Progeroid syndromes constitute a group of rare genetic disorders characterized by an early onset of aging phenotypes. Many progeroid diseases are linked to defects in genome maintenance (Kyng and Bohr, 2005), a classical hallmark of aging (Figure 7.2), and several gene products involved in progeroid syndromes are ubiquitin ligases.

7.2.2.1 Cockayne Syndrome

Cockayne syndrome (CS) is an autosomal recessive disorder characterized by cachectic dwarfism, mental deficiency, microcephaly, intracranial calcifications, neurological degeneration, eye disorders, photosensitivity and premature aging (Bertola et al., 2006). The disease is divided into two primary forms, namely CS complementation group A (CSA; OMIM #216400) and CS complementation group B (CSB; OMIM #133540) (Lehmann et al., 1994). Cells derived from CS patients are deficient in RNA polymerase II (RNAPII)-dependent transcription-coupled DNA repair mechanisms and show defects in repairing UV-induced DNA lesions. Around 80% of CS patients carry mutations in the CS group B gene (CSB, also called ERCC6), with the remaining cases caused by mutations in the CS group A gene (CSA, also called ERCC8) (Licht et al., 2003). Both functional CSA and CSB are required for UV-radiation-induced ubiquitination of the large subunit of RNAPII, a modification absent in fibroblasts derived from persons with either CS A or B (Lee et al., 2002). Recent studies show that RNAPII ubiquitination is essential for transcription recovery and DNA repair (Nakazawa et al., 2020). A lack of transcription recovery in CS is caused by the dysregulation of global RNAPII levels (Tufegdzic Vidakovic et al., 2020). Several ubiquitin ligases were described to be involved in this process, among them are BRCA1/BARD1 (Kleiman et al., 2005), elongin–cullin complexes (Yasukawa et al., 2008) and the HECT ubiquitin ligase NEDD4 (Anindya et al., 2007), suggesting a complex interplay and redundancy of different ubiquitination processes to control and regulate transcription-coupled DNA repair. BRCA1, as a classical susceptibility breast-cancer gene, contributes to transcription-coupled DNA repair and polyubiquitinates CSB upon UV irradiation for its subsequent proteasomal degradation (Wei et al., 2011). NEDD4 was shown to ubiquitinate RNAPII as a response to UV-induced DNA damage in vivo and ubiquitination can be reproduced also in vitro (Anindya et al., 2007). Elongin A has a dual function: on the one hand, it is part of the heterotrimeric RNAPII elongation factor Elongin; on the other hand, it is part of a ubiquitin ligase complex assembled with the Cullin family member CUL5 and the RING protein RBX. This ubiquitin ligase targets the RPB1 subunit of RNAPII for ubiquitination and proteasomal degradation (Yasukawa et al., 2008). Most strikingly, the CS A (CSA) protein itself is a ubiquitin ligase substrate receptor from the family of DDB1 and CUL4-associated factors

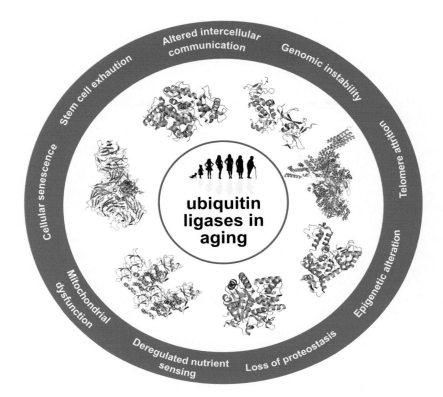

Figure 7.2 **Ubiquitin-ligases and the nine hallmarks of aging.** The cartoon displays the nine hallmarks of aging (genomic instability, telomere attrition, epigenetic alterations, loss of proteostasis, deregulated nutrient sensing, mitochondrial dysfunction, cellular senescence, stem cell exhaustion and altered intercellular communication) in the outer circle. The inner circle displays structures of selected ubiquitin ligases involved in the process of aging (e.g. PRKN (RBR E3), CSA (CRL4^{CSA}), MDM2 (RING E3), GID/CTLH complex (RING E3), NEDD4 (HECT E3), BRCA1/BARD1 (RING E3) and VHL (CRL2^{VHL}). *Abbreviation:* PRKN, parkin; CSA, Cockayne syndrome 1; MDM2, mouse double minute 2; GID/CTLH, glucose-induced degradation deficient/C-terminal to LisH complex; BRCA1/BARD1, breast cancer 1/BRCA1-associated RING domain 1; VHL, von Hippel-Lindau disease tumor suppressor. PDB accession numbers: PRKN (#5C9V); CSA (#6FCV); MDM2 (#5TRF); GID/CTLH complex (#6SXY); NEDD4 (#4BE8); BRCA1/BARD1 (#2NTE); VHL (#1VCB).

(DCAFs) that specifically assembles with cullin 4 (CUL4) through the DDB1 adaptor to form the CRL4^{CSA} ubiquitin ligase complex (Groisman et al., 2003). A reported CRL4^{CSA} ubiquitination substrate is the CS B protein (CSB) (Groisman et al., 2006). Several lines of evidence even support that CRL4^{CSA} directly targets RNAPII for ubiquitination (Nakazawa et al., 2020; Tornaletti, 2009; Tufegdzic Vidakovic et al., 2020).

7.2.2.2 Werner Syndrome

Werner syndrome (WRN, OMIM # 277700) is a rare autosomal recessive progeroid syndrome. Patients exhibit premature aging phenotypes like graying and thinning of hair and skin atrophy. Additionally, they develop early onset of diseases associated with aging, including bilateral cataract, diabetes mellitus, osteoporosis and premature arteriosclerosis and are predisposed to sarcomas (Oshima et al., 1996). WRN is caused by a homozygous or compound heterozygous mutation in the RECQL2 gene, which gene product is a member of the RecQ helicase family (Monnat, 2010). Recently, a patient with the putative diagnosis of a Werner syndrome-like progeroid disorder was discovered to bear a homozygous antiterminating mutation that removes the stop codon of MDM2 (Lessel et al., 2017). The concentration of the tumor suppressor p53 in the nucleus is tightly regulated by the ubiquitin ligase MDM2 by several mechanisms. The underlying mechanisms include (1) MDM2 binding and blockage of its transactivation, (2) mono-ubiquitining that induces nuclear export and (3) polyubiquitination with subsequent proteasomal degradation (Lee and Gu, 2010; Sasaki et al., 2011). Defects in MDM2 or p53 correlate with increased risk of cancer (Malkin et al., 1990) but there is also evidence that the MDM2/p53 axis influences the process of aging. Several mouse models that directly interfere with MDM2/p53 display accelerated aging phenotypes (Maier et al., 2004; Tyner et al., 2002), a p53 ARG72Pro polymorphism is even associated with increased longevity (Orsted et al., 2007) and several human progeroid syndromes and phenotypes were linked to p53 signaling (Scaffidi and Misteli, 2006; Varela et al., 2005). Concluding, the MDM2/p53 axis is associated with both premature aging and also with increased longevity. Hallmarks of aging that are affected by the MDM2/p53 axis are genome instability, mitochondrial dysfunction and cellular senescence (Wu and Prives, 2018), but the underlying mechanisms are currently unclear.

7.2.3 Ubiquitin Ligases Involved in Age-Associated Pathologies

This paragraph discusses the involvement of ubiquitin ligases in a selected class of age-associated pathologies.

7.2.3.1 Parkinson's Disease

PD affects over 1% of the population over the age of 60 and reaches 5% over the age of 85, illustrating that aging is likely the most significant risk factor for its manifestation (Reeve et al., 2014; Rodriguez et al., 2015). The disease is caused by the selective loss of neurons in the *substantia nigra* and other brain regions accompanied by the pathological accumulation of the protein α-synuclein, so-called Lewy bodies (Wakabayashi et al., 2013). A small percentage of patients (around 5%) develop juvenile PD, an autosomal recessive form of parkinsonism with typical symptoms like bradikynesia, rigidity and tremor that manifest before the age of 40, often without the formation of Lewy bodies in the brain (Takahashi et al., 1994). Several Parkinson risk genes have E3 activity or are subunits of multisubunit ubiquitin ligase complexes, suggesting that defects in removing protein aggregates via the UMS are a potential pathomechanism of PD (Walden and Muqit, 2017). More than 15% of PD patients in a

UK-based study show mutations in either *PINK1* or *PRKN* (Kilarski et al., 2012). The PRKN gene (Parkinson Disease 2; OMIM #600116) encodes for parkin, a RING domain-containing ubiquitin ligase involved in proteasome-dependent degradation of proteins like the mitochondrial Rho GTPases RHOT1 (Miro) (Wang et al., 2011). Another primary function of parkin is autophagic clearance of damaged, depolarized mitochondria, a process called mitophagy (Palikaras et al., 2018). Mitophagy is induced by the phosphatase and tensin homolog (PTEN)-induced putative kinase 1 (PINK1) that phosphorylates parkin and induces a conformational change that allows binding to the mitochondrial surface of damaged mitochondria (Aguirre et al., 2017). Parkin association with the mitochondrial surface results in polyubiquitination of several outer mitochondrial membrane proteins that induce the uptake of damaged mitochondria by mitophagy. Defects in this quality control system are a crucial pathomechanism of PD (Truban et al., 2017).

The Parkinson Disease 15 gene (OMIM #260300) also plays a role in mitophagy. The gene encodes for the F-box protein FBXO7. Autosomal-recessive mutations in FBXO7 cause rare Parkinsonian-pyramidal syndrome (Di Fonzo et al., 2009). FBXO7 is a substrate recognition factor of the multisubunit SCF-ubiquitin ligase complex and functions downstream of PINK1 in the clearance of damaged mitochondria. FBXO7 binds to both Parkin and PINK1 and is required to recruit parkin to damaged mitochondria and for mitophagy (Burchell et al., 2013).

7.2.3.2 Sarcopenia

The progressive loss of muscle mass and muscle function in aged organisms is referred to as sarcopenia and can be observed in different organisms ranging from *Caenorhabditis elegans* to *Homo sapiens*. In humans, the loss of muscle mass after the age of 30 years adds up to 0.5–1% per year with an even more dramatic rate of muscle degeneration after 65 years of age (Nair, 2005). Sarcopenia can thus be considered as an age-associated pathology. Several mechanisms are responsible for sarcopenia and include extrinsic factors like a reduction of nutrient intake and physical activity in the aged organism, reduced response to anabolic and catabolic stimuli, a decrease in the regenerative capacity of muscle cells and of mitochondrial function as well as declined ability to maintain proteostasis (Demontis et al., 2013). The UMS, an essential mechanism to control proteostasis, plays a central role in regulating muscle protein degradation to control muscle size. Several components of the UMS are altered in muscles of aged rats when compared to their adult controls, e.g. a two to threefold higher level of 26S proteasomes, increased activity of several deubiquitinating enzymes and increased levels of the ubiquitin ligase CHIP (Altun et al., 2010). Other important ubiquitin ligases with a crucial role in sarcopenia are the F-box protein Atrogin-1 (FBXO32) and the ubiquitin ligase MuRF-1 (TRIM63) (Gumucio and Mendias, 2013). FBXO32 and TRIM63 are highly expressed in skeletal muscle during muscle atrophy and direct polyubiquitination of muscle proteins to induce their proteasomal degradation (Bodine et al., 2001; Gomes et al., 2001). Furthermore, the neural precursor cell-expressed developmentally downregulated protein 4 (NEDD4-1) is another ubiquitin ligase described as necessary to activate muscle atrophy. Some observations suggest a function of NEDD4-1 in denervation-induced sarcopenia, i.e. NEDD4-1 is upregulated as a consequence of denervation and a muscle-specific NEDD4-1 knockout mouse shows increased weights and cross-sectional area of denervated gastrocnemius muscle when compared to wild-type mice indicating that sarcopenia is slowed down (Nagpal et al., 2012). Other studies show that the overexpression of NEDD4-1 in muscles did not induce myofiber atrophy (Koncarevic et al., 2007). The reason for this discrepancy is currently unclear.

7.2.4 Aging-Associated Ubiquitin Ligase Genes

Genes or corresponding gene products involved in lifespan extension are termed as gerontogenes (Guarente and Kenyon, 2000). In this section, we discuss some gerontogenes that encode for ubiquitin ligases.

7.2.4.1 VHL (Von Hippel–Lindau Factor)

The *C. elegans* VHL-1 is homologous to the mammalian von Hippel–Lindau tumor suppressor protein. It constitutes a part of a conserved ubiquitin ligase complex that inhibits hypoxic signaling in the presence of oxygen via ubiquitination and subsequent proteasomal degradation of the alpha subunit of the hypoxic response transcription factor HIF-1 (Kaelin, 2003). In *C. elegans*, the loss of VHL-1 increases the average life span by approximately 30% and enhances resistance to toxic protein aggregates. This effect is independent of dietary restriction- and insulin-like signaling (Mehta et al., 2009; Muller et al., 2009). VHL belongs to a small group of significantly upregulated genes during aging in several organisms (de Magalhaes et al., 2009).

7.2.4.2 GID-Complex

The glucose-induced degradation deficient (GID) complex is an evolutionarily conserved ubiquitin ligase complex involved in regulating glucose metabolism (Liu and Pfirrmann, 2019). In response to glucose supplementation, the *Saccharomyces cerevisiae* GID complex targets critical enzymes of gluconeogenesis for polyubiquitination and subsequent proteasomal degradation in a process called catabolite degradation (Santt et al., 2008). Similarly, in higher vertebrates, the GID complex regulates metabolism by targeting AMPK for ubiquitination. Lack of this AMPK ubiquitination sustains AMPK activity resulting in increased autophagic flux, reduced mTOR activity, increased branched-chain amino acids and fatty acid degradation (Liu et al., 2019). In *C. elegans*, knockdown of several GID-complex subunits results in an approximately 30% increase in lifespan (Liu et al., 2019).

7.2.4.3 CUL3-KLHL22

The mechanistic target of rapamycin complex 1 (mTORC1) is a well-known regulator of energy homeostasis and lifespan (Zoncu et al., 2011). KLHL22 is a CRL ubiquitin ligase adaptor protein that forms a functional complex with the scaffold protein CUL3 and the RING finger protein RBX1. The ubiquitin ligase complex promotes amino-acid-stimulated K48-linked polyubiquitination and degradation of DEPDC5, a subunit of the GATOR1 complex and an inhibitor of mTORC1. KLHL22-dependent degradation of DEPDC5 abolishes mTORC1 inhibition and plays an evolutionarily conserved role in the activation of mTORC1. The depletion of the KLHL22 orthologous *mel-26* was shown to extend the lifespan in *C. elegans* (Chen et al., 2018).

7.3 OUTLOOK AND CONCLUSIONS

Protein ubiquitination by the ubiquitination cascade regulates processes far beyond protein half-life by hundreds of different specific ubiquitin ligases. This specificity allows both interference and control of virtually every cellular process including the process of aging. In this chapter, we focus on defects in ubiquitin ligases that can cause syndromes of premature aging but also on loss-of-function mutations in specific ubiquitin ligases that prolong organismal lifespan. We predict that more ubiquitin ligases with specific function in the aging process will be discovered and described in the coming years.

Due to their specificity, ubiquitin ligases are interesting drug targets with potentially low side effects. This turns ubiquitin ligases in the focus for drug development (Huang and Dixit, 2016). As an example, the so-called PROteolysis TArgeting Chimeras (PROTACs) have become a promising technology to target E3 substrates for ubiquitination and subsequent degradation. These chimeric molecules bind a specific substrate and the corresponding ubiquitin ligase to pharmacologically induce the formation of a ternary complex that allows substrate ubiquitination and often subsequent substrate degradation (Sun et al., 2019). Given the fast progress in the field, it seems only a matter of short time until ubiquitin ligases will be in the focus of drug development with the aim to pharmacologically treat age-associated pathologies or even to extend lifespan.

ACKNOWLEDGMENT

This work is supported by the Deutsche Forschungsgemeinschaft, Funder ID: 10.13039/501100001659 (ProMoAge GRK 2155).

REFERENCES

Aguirre, J. D., Dunkerley, K. M., Mercier, P., et al. 2017. Structure of phosphorylated UBL domain and insights into PINK1-orchestrated Parkin activation. *Proc Natl Acad Sci U S A*, 114(2), 298–303. doi:10.1073/pnas.1613040114

Altun, M., Besche, H. C., Overkleeft, H. S., et al. 2010. Muscle wasting in aged, sarcopenic rats is associated with enhanced activity of the ubiquitin proteasome pathway. *J Biol Chem*, 285(51), 39597–39608. doi:10.1074/jbc.M110.129718

Anindya, R., Aygun, O., & Svejstrup, J. Q. 2007. Damage-induced ubiquitylation of human RNA polymerase II by the ubiquitin ligase Nedd4, but not Cockayne syndrome proteins or BRCA1. *Mol Cell*, 28(3), 386–397. doi:10.1016/j.molcel.2007.10.008

Bertola, D. R., Cao, H., Albano, L. M. J., et al. 2006. Cockayne syndrome type A: novel mutations in eight typical patients. *J Hum Genet*, 51(8), 701–705. doi:10.1007/s10038-006-0011-7

Bodine, S. C., Latres, E., Baumhueter, S., et al. 2001. Identification of ubiquitin ligases required for skeletal muscle atrophy. *Science*, 294(5547), 1704–1708. doi:10.1126/science.1065874

Burchell, V. S., Nelson, D. E., Sanchez-Martinez, A., et al. 2013. The Parkinson's disease-linked proteins Fbxo7 and Parkin interact to mediate mitophagy. *Nat Neurosci*, 16(9), 1257–1265. doi:10.1038/nn.3489

Chen, J., Ou, Y., Yang, Y., et al. 2018. KLHL22 activates amino-acid-dependent mTORC1 signalling to promote tumorigenesis and ageing. *Nature*, 557(7706), 585–589. doi:10.1038/s41586-018-0128-9

Ciechanover, A. 2005. Proteolysis: from the lysosome to ubiquitin and the proteasome. *Nat Rev Mol Cell Biol*, 6(1), 79–87. doi:10.1038/nrm1552

de Magalhaes, J. P., Curado, J., & Church, G. M. 2009. Meta-analysis of age-related gene expression profiles identifies common signatures of aging. *Bioinformatics*, 25(7), 875–881. doi:10.1093/bioinformatics/btp073

Demontis, F., Piccirillo, R., Goldberg, A. L., et al. 2013. Mechanisms of skeletal muscle aging: insights from Drosophila and mammalian models. *Dis Model Mech*, 6(6), 1339–1352. doi:10.1242/dmm.012559

Deshaies, R. J., & Joazeiro, C. A. 2009. RING domain E3 ubiquitin ligases. *Annu Rev Biochem*, 78, 399–434. doi:10.1146/annurev.biochem.78.101807.093809

Di Fonzo, A., Dekker, M. C., Montagna, P., et al. 2009. FBXO7 mutations cause autosomal recessive, early-onset Parkinsonian-pyramidal syndrome. *Neurology*, 72(3), 240–245. doi:10.1212/01.wnl.0000338144.10967.2b

Dye, B. T., & Schulman, B. A. 2007. Structural mechanisms underlying posttranslational modification by ubiquitin-like proteins. *Annu Rev Biophys Biomol Struct*, 36, 131–150. doi:10.1146/annurev.biophys.36.040306.132820

Foote, P. K., Krist, D. T., & Statsyuk, A. V. 2017. High-throughput screening of HECT E3 ubiquitin ligases using UbFluor. *Curr Protocols Chem Biol*, 9(3), 174–195. doi:10.1002/cpch.24

Gomes, M. D., Lecker, S. H., Jagoe, R. T., et al. 2001. Atrogin-1, a muscle-specific F-box protein highly expressed during muscle atrophy. *Proc Natl Acad Sci U S A*, 98(25), 14440–14445. doi:10.1073/pnas.251541198

Groisman, R., Polanowska, J., Kuraoka, I., et al. 2003. The ubiquitin ligase activity in the DDB2 and CSA complexes is differentially regulated by the COP9 signalosome in response to DNA damage. *Cell*, 113(3), 357–367. doi:10.1016/s0092-8674(03)00316-7

Groisman, R., Kuraoka, I., Chevallier, O., et al. 2006. CSA-dependent degradation of CSB by the ubiquitin-proteasome pathway establishes a link between complementation factors of the Cockayne syndrome. *Genes Dev*, 20(11), 1429–1434. doi:10.1101/gad.378206

Guarente, L., & Kenyon, C. 2000. Genetic pathways that regulate ageing in model organisms. *Nature*, 408(6809), 255–262. doi:10.1038/35041700

Gumucio, J. P., & Mendias, C. L. 2013. Atrogin-1, MuRF-1, and sarcopenia. *Endocrine*, 43(1), 12–21. doi:10.1007/s12020-012-9751-7

Huang, X., & Dixit, V. M. 2016. Drugging the undruggables: exploring the ubiquitin system for drug development. *Cell Res*, 26(4), 484–498. doi:10.1038/cr.2016.31

Huang, L., Kinnucan, E., Wang, G., et al. 1999. Structure of an E6AP-UbcH7 complex: insights into ubiquitination by the E2-E3 enzyme cascade. *Science*, 286(5443), 1321–1326. doi:10.1126/science.286.5443.1321

Hutchins, A. P., Liu, S., Diez, D., et al. 2013. The repertoires of ubiquitinating and deubiquitinating enzymes in eukaryotic genomes. *Mol Biol Evol*, 30(5), 1172–1187. doi:10.1093/molbev/mst022

Kaelin, W. G., Jr. 2003. The von Hippel-Lindau gene, kidney cancer, and oxygen sensing. *J Am Soc Nephrol*, 14(11), 2703–2711. doi:10.1097/01.asn.0000092803.69761.41

Kilarski, L. L., Pearson, J. P., Newsway, V., et al. 2012. Systematic review and UK-based study of PARK2 (Parkin), PINK1, PARK7 (DJ-1) and LRRK2 in early-onset Parkinson's disease. *Mov Disord*, 27(12), 1522–1529. doi:10.1002/mds.25132

Kleiman, F. E., Wu-Baer, F., Fonseca, D., et al. 2005. BRCA1/BARD1 inhibition of mRNA 3' processing involves targeted degradation of RNA polymerase II. *Genes Dev*, 19(10), 1227–1237. doi:10.1101/gad.1309505

Koncarevic, A., Jackman, R. W., & Kandarian, S. C. 2007. The ubiquitin-protein ligase Nedd4 targets Notch1 in skeletal muscle and distinguishes the subset of atrophies caused by reduced muscle tension. *FASEB J*, 21(2), 427–437. doi:10.1096/fj.06-6665com

Kosztyu, P., Slaninová, I., Valčíková, B., et al. 2019. A single conserved amino acid residue as a critical context-specific determinant of the differential ability of Mdm2 and MdmX RING domains to dimerize. *Front Physiol*, 10, 390–390. doi:10.3389/fphys.2019.00390

Kyng, K. J., & Bohr, V. A. 2005. Gene expression and DNA repair in progeroid syndromes and human aging. *Ageing Res Rev*, 4(4), 579–602. doi:10.1016/j.arr.2005.06.008

Lee, J. T., & Gu, W. 2010. The multiple levels of regulation by p53 ubiquitination. *Cell Death Differ*, 17(1), 86–92. doi:10.1038/cdd.2009.77

Lee, K. B., Wang, D., Lippard, S. J., et al. 2002. Transcription-coupled and DNA damage-dependent ubiquitination of RNA polymerase II in vitro. *Proc Natl Acad Sci U S A*, 99(7), 4239–4244. doi:10.1073/pnas.072068399

Lehmann, A. R., Bootsma, D., Clarkson, S. G., et al. 1994. Nomenclature of human DNA repair genes. *Mutat Res*, 315(1), 41–42. doi:10.1016/0921-8777(94)90026-4

Lessel, D., Wu, D., Trujillo, C., et al. 2017. Dysfunction of the MDM2/p53 axis is linked to premature aging. *J Clin Invest*, 127(10), 3598–3608. doi:10.1172/JCI92171

Li, W., Bengtson, M. H., Ulbrich, A., et al. 2008. Genome-wide and functional annotation of human E3 ubiquitin ligases identifies MULAN, a mitochondrial E3 that regulates the organelle's dynamics and signaling. *PLoS One*, 3(1), e1487. doi:10.1371/journal.pone.0001487

Licht, C. L., Stevnsner, T., & Bohr, V. A. 2003. Cockayne syndrome group B cellular and biochemical functions. *Am J Hum Genet*, 73(6), 1217–1239. doi:10.1086/380399

Liu, H., & Pfirrmann, T. 2019. The Gid-complex: an emerging player in the ubiquitin ligase league. *Biol Chem*, 400(11), 1429–1441. doi:10.1515/hsz-2019-0139

Liu, H., Ding, J., Kohnlein, K., et al. 2019. The GID ubiquitin ligase complex is a regulator of AMPK activity and organismal lifespan. *Autophagy*, 1–17. doi:10.1080/15548627.2019.1695399

Lopez-Otin, C., Blasco, M. A., Partridge, L., et al. 2013. The hallmarks of aging. *Cell*, 153(6), 1194–1217. doi:10.1016/j.cell.2013.05.039

Maier, B., Gluba, W., Bernier, B., et al. 2004. Modulation of mammalian life span by the short isoform of p53. *Genes Dev*, 18(3), 306–319. doi:10.1101/gad.1162404

Malkin, D., Li, F. P., Strong, L. C., et al. 1990. Germ line p53 mutations in a familial syndrome of breast cancer, sarcomas, and other neoplasms. *Science*, 250(4985), 1233–1238. doi:10.1126/science.1978757

Mehta, R., Steinkraus, K. A., Sutphin, G. L., et al. 2009. Proteasomal regulation of the hypoxic response modulates aging in C. elegans. *Science*, 324(5931), 1196–1198. doi:10.1126/science.1173507

Metzger, M. B., Hristova, V. A., & Weissman, A. M. 2012. HECT and RING finger families of E3 ubiquitin ligases at a glance. *J Cell Sci*, 125(Pt 3), 531–537. doi:10.1242/jcs.091777

Metzger, M. B., Pruneda, J. N., Klevit, R. E., et al. 2014. RING-type E3 ligases: master manipulators of E2 ubiquitin-conjugating enzymes and ubiquitination. *Biochim Biophys Acta – Mol Cell Res*, 1843(1), 47–60. doi:https://doi.org/10.1016/j.bbamcr.2013.05.026

Monnat, R. J., Jr. 2010. Human RECQ helicases: roles in DNA metabolism, mutagenesis and cancer biology. *Semin Cancer Biol*, 20(5), 329–339. doi:10.1016/j.semcancer.2010.10.002

Muller, R. U., Fabretti, F., Zank, S., et al. 2009. The von Hippel Lindau tumor suppressor limits longevity. *J Am Soc Nephrol*, 20(12), 2513–2517. doi:10.1681/ASN.2009050497

Nagpal, P., Plant, P. J., Correa, J., et al. 2012. The ubiquitin ligase Nedd4-1 participates in denervation-induced skeletal muscle atrophy in mice. *PLoS One*, 7(10), e46427. doi:10.1371/journal.pone.0046427

Nair, K. S. 2005. Aging muscle. *Am J Clin Nutr*, 81(5), 953–963. doi:10.1093/ajcn/81.5.953

Nakazawa, Y., Hara, Y., Oka, Y., et al. 2020. Ubiquitination of DNA damage-stalled RNAPII promotes transcription-coupled repair. *Cell*, 180(6), 1228–1244 e1224. doi:10.1016/j.cell.2020.02.010

Orsted, D. D., Bojesen, S. E., Tybjaerg-Hansen, A., et al. 2007. Tumor suppressor p53 Arg72Pro polymorphism and longevity, cancer survival, and risk of cancer in the general population. *J Exp Med*, 204(6), 1295–1301. doi:10.1084/jem.20062476

Oshima, J., Yu, C. E., Piussan, C., et al. 1996. Homozygous and compound heterozygous mutations at the Werner syndrome locus. *Hum Mol Genet*, 5(12), 1909–1913. doi:10.1093/hmg/5.12.1909

Ozkaynak, E., Finley, D., & Varshavsky, A. 1984. The yeast ubiquitin gene: head-to-tail repeats encoding a polyubiquitin precursor protein. *Nature*, 312(5995), 663–666. doi:10.1038/312663a0

Palikaras, K., Lionaki, E., & Tavernarakis, N. 2018. Mechanisms of mitophagy in cellular homeostasis, physiology and pathology. *Nat Cell Biol*, 20(9), 1013–1022. doi:10.1038/s41556-018-0176-2

Petroski, M. D., & Deshaies, R. J. 2005. Function and regulation of cullin-RING ubiquitin ligases. *Nat Rev Mol Cell Biol*, 6(1), 9–20. doi:10.1038/nrm1547

Reeve, A., Simcox, E., & Turnbull, D. 2014. Ageing and Parkinson's disease: why is advancing age the biggest risk factor? *Ageing Res Rev*, 14, 19–30. doi:10.1016/j.arr.2014.01.004

Reiter, K. H., & Klevit, R. E. 2018. Characterization of RING-between-RING E3 ubiquitin transfer mechanisms. *Methods Mol Biol*, 1844, 3–17. doi:10.1007/978-1-4939-8706-1_1

Rock, K. L., Gramm, C., Rothstein, L., et al. 1994. Inhibitors of the proteasome block the degradation of most cell proteins and the generation of peptides presented on MHC class I molecules. *Cell*, 78(5), 761–771. doi:10.1016/s0092-8674(94)90462-6

Rodriguez, M., Rodriguez-Sabate, C., Morales, I., et al. 2015. Parkinson's disease as a result of aging. *Aging Cell*, 14(3), 293–308. doi:10.1111/acel.12312

Sala-Gaston, J., Martinez-Martinez, A., Pedrazza, L., et al. 2020. HERC ubiquitin ligases in cancer. *Cancers*, 12(6), 1653. doi:10.3390/cancers12061653

Santt, O., Pfirrmann, T., Braun, B., et al. 2008. The yeast GID complex, a novel ubiquitin ligase (E3) involved in the regulation of carbohydrate metabolism. *Mol Biol Cell*, 19(8), 3323–3333. doi:10.1091/mbc.E08-03-0328

Sasaki, M., Kawahara, K., Nishio, M., et al. 2011. Regulation of the MDM2-P53 pathway and tumor growth by PICT1 via nucleolar RPL11. *Nat Med*, 17(8), 944–951. doi:10.1038/nm.2392

Scaffidi, P., & Misteli, T. 2006. Lamin A-dependent nuclear defects in human aging. *Science*, 312(5776), 1059–1063. doi:10.1126/science.1127168

Skaar, J. R., Pagan, J. K., & Pagano, M. 2013. Mechanisms and function of substrate recruitment by F-box proteins. *Nat Rev Mol Cell Biol*, 14(6), 369–381. doi:10.1038/nrm3582

Sluimer, J., & Distel, B. 2018. Regulating the human HECT E3 ligases. *Cell Mol Life Sci*, 75(17), 3121–3141. doi:10.1007/s00018-018-2848-2

Spratt, D. E., Walden, H., & Shaw, G. S. 2014. RBR E3 ubiquitin ligases: new structures, new insights, new questions. *Biochem J*, 458(3), 421–437. doi:10.1042/BJ20140006

Sun, X., Gao, H., Yang, Y., et al. 2019. PROTACs: great opportunities for academia and industry. *Signal Transduct Target Ther*, 4, 64. doi:10.1038/s41392-019-0101-6

Takahashi, H., Ohama, E., Suzuki, S., et al. 1994. Familial juvenile parkinsonism: clinical and pathologic study in a family. *Neurology*, 44(3 Pt 1), 437–441. doi:10.1212/wnl.44.3_part_1.437

Tornaletti, S. 2009. DNA repair in mammalian cells: transcription-coupled DNA repair: directing your effort where it's most needed. *Cell Mol Life Sci*, 66(6), 1010–1020. doi:10.1007/s00018-009-8738-x

Truban, D., Hou, X., Caulfield, T. R., et al. 2017. PINK1, Parkin, and mitochondrial quality control: what can we learn about Parkinson's disease pathobiology? *J Parkinsons Dis*, 7(1), 13–29. doi:10.3233/JPD-160989

Tufegdzic Vidakovic, A., Mitter, R., Kelly, G. P., et al. 2020. Regulation of the RNAPII pool is integral to the DNA damage response. *Cell*, 180(6), 1245–1261 e1221. doi:10.1016/j.cell.2020.02.009

Tyner, S. D., Venkatachalam, S., Choi, J., et al. 2002. p53 mutant mice that display early ageing-associated phenotypes. *Nature*, 415(6867), 45–53. doi:10.1038/415045a

Varela, I., Cadinanos, J., Pendas, A. M., et al. 2005. Accelerated ageing in mice deficient in Zmpste24 protease is linked to p53 signalling activation. *Nature*, 437(7058), 564–568. doi:10.1038/nature04019

Verdecia, M. A., Joazeiro, C. A., Wells, N. J., et al. 2003. Conformational flexibility underlies ubiquitin ligation mediated by the WWP1 HECT domain E3 ligase. *Mol Cell*, 11(1), 249–259. doi:10.1016/s1097-2765(02)00774-8

Wakabayashi, K., Tanji, K., Odagiri, S., et al. 2013. The Lewy body in Parkinson's disease and related neurodegenerative disorders. *Mol Neurobiol*, 47(2), 495–508. doi:10.1007/s12035-012-8280-y

Walden, H., & Muqit, M. M. 2017. Ubiquitin and Parkinson's disease through the looking glass of genetics. *Biochem J*, 474(9), 1439–1451. doi:10.1042/BCJ20160498

Wang, X., Winter, D., Ashrafi, G., et al. 2011. PINK1 and Parkin target Miro for phosphorylation and degradation to arrest mitochondrial motility. *Cell*, 147(4), 893–906. doi:10.1016/j.cell.2011.10.018

Weber, J., Polo, S., & Maspero, E. 2019. HECT E3 ligases: a tale with multiple facets. *Front Physiol*, 10(370). doi:10.3389/fphys.2019.00370

Wei, L., Lan, L., Yasui, A., et al. 2011. BRCA1 contributes to transcription-coupled repair of DNA damage through polyubiquitination and degradation of Cockayne syndrome B protein. *Cancer Sci*, 102(10), 1840–1847. doi:10.1111/j.1349-7006.2011.02037.x

Wenzel, D. M., Lissounov, A., Brzovic, P. S., et al. 2011. UBCH7 reactivity profile reveals Parkin and HHARI to be RING/HECT hybrids. *Nature*, 474(7349), 105–108. doi:10.1038/nature09966

Wu, D., & Prives, C. 2018. Relevance of the p53-MDM2 axis to aging. *Cell Death Differ*, 25(1), 169–179. doi:10.1038/cdd.2017.187

Yasukawa, T., Kamura, T., Kitajima, S., et al. 2008. Mammalian elongin A complex mediates DNA-damage-induced ubiquitylation and degradation of Rpb1. *EMBO J*, 27(24), 3256–3266. doi:10.1038/emboj.2008.249

Yau, R., & Rape, M. 2016. The increasing complexity of the ubiquitin code. *Nat Cell Biol*, 18(6), 579–586. doi:10.1038/ncb3358

Zhao, Y., & Sun, Y. 2013. Cullin-RING Ligases as attractive anti-cancer targets. *Curr Pharm Des*, 19(18), 3215–3225. doi:10.2174/13816128113199990300

Zimmerman, E. S., Schulman, B. A., & Zheng, N. 2010. Structural assembly of cullin-RING ubiquitin ligase complexes. *Curr Opin Struct Biol*, 20(6), 714–721. doi:10.1016/j.sbi.2010.08.010

Zoncu, R., Efeyan, A., & Sabatini, D. M. 2011. mTOR: from growth signal integration to cancer, diabetes and ageing. *Nat Rev Mol Cell Biol*, 12(1), 21–35. doi:10.1038/nrm3025

CHAPTER EIGHT

Role of SUMOylation in Neurodegenerative Diseases and Inflammation

Esmeralda Parra-Peralbo, Veronica Muratore, Orhi Barroso-Gomila,
Ana Talamillo, James D. Sutherland and Rosa Barrio

CONTENTS

8.1 INTRODUCTION

To understand how SUMOylation may contribute to both physiological and pathological conditions, it is worth to consider the complexity underlying the "writing, reading, and editing" of this important post-translational modification. Small ubiquitin-like modifiers (SUMOs) belong to the ubiquitin-like (UbL) family of proteins that attach covalently to target substrates in a transient and reversible way. There are at least three major mammalian SUMO paralogues (SUMO1, -2, -3). Human SUMO2 and SUMO3 share 97% sequence identity, whereas they share 47% of sequence identity with SUMO1 (Flotho and Melchior, 2013). Protein SUMOylation is a rigorously regulated cycle involving an enzymatic machinery that acts in a stepwise manner (Figure 8.1). First, maturation of newly synthesized SUMO by SUMO isopeptidases (SENPs) exposes the C-terminal di-glycine motif. Then, SUMO is activated by forming a thioester bond between its C-terminal glycine and the catalytic cysteine of the heterodimeric E1 SUMO-activating enzyme SAE1/SAE2. SUMO is then passed via a thioester bond to the E2 conjugating enzyme UBC9. Substrate lysines, typically found within the consensus sequence ψKXE (where ψ represents hydrophobic amino acids and X any amino acid), can then be directly modified by SUMO and UBC9. SUMO E3 ligases can enhance conjugation rates by associating with both UBC9 and the substrate and confer substrate specificity (Pichler et al., 2017). The best characterized SUMO E3 ligases concern the protein inhibitor of activated STAT family (PIAS1, PIAS2, PIAS3 and PIAS4), MMS21, Ran binding protein 2 (RanBP2) and the ZNF451 family. If required, SUMO as well as the substrate can be recycled by the action of SENPs that cleave the isopeptide bond. There are six members of the SENP family of proteins in humans: SENP1, SENP2, SENP3, SENP5, SENP6 and SENP7. SENP1 has a preference for SUMO1 deconjugation, and SENP6 and SENP7 preferentially cleave SUMO2/3 chains (Jansen and Vertegaal, 2020). Substrates

DOI: 10.1201/9781003048138-8

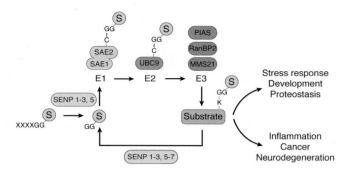

Figure 8.1 **A schematic of SUMO conjugation.** The different steps in SUMO conjugation are represented. SUMO modification of substrates is essential for many physiological processes and disruption of SUMO signaling regulation can favor the development of diseases. *Abbreviations:* S: SUMO; ~: thioester bond; -: isopeptide bond.

can be preferentially modified by SUMO1 or SUMO2/3, or by either (Vertegaal et al., 2006). One of the main differences is that SUMO2/3 can form chains and might induce protein degradation through SUMO-targeted ubiquitin ligases or STUbLs (Ring Finger Protein 4, RNF4, Arkadia). SENP6-mediated cleavage of such chains is a good example to counteract this phenomenon. Arkadia, interestingly, has preferences for SUMO1-capped SUMO2/3 chains (Sriramachandran et al., 2019). Mono-versus poly-SUMOylation should be thus considered as different signals. SUMO paralogue specificity of SUMO-interacting motifs (SIMs) in proteins has been observed, but how this specificity is achieved is not well understood.

According to recent studies, more than 40,700 SUMO sites within more than 6,700 substrates have been identified in the human proteome (Hendriks et al., 2017). Targets are monoSUMOylated, multi-monoSUMOylated or even polySUMOylated through SUMO2/3 chains (Hendriks et al., 2017; Hendriks and Vertegaal, 2016). Proteins can be preferentially modified by SUMO1 or by SUMO2/3 or can be conjugated by either homologue (Vertegaal et al., 2006). In addition, SUMO itself can be further modified by Ub and NEDD8 chains, or by smaller moieties such as acetyl or phosphoryl groups (El Motiam et al., 2019; Matic et al., 2008; Ullmann et al., 2012). This complexity of the SUMO signaling is translated in diverse fates for particular substrates, including protein complex assembly, protein-protein interactions, subcellular localization or degradation.

Much of knowledge about how SUMO can influence targets stems from major substrates such as PML (promyelocytic leukemia). The protein PML forms nuclear bodies (NBs) considered as hubs for SUMO-dependent signaling where

many components of the SUMO machinery localize (Lallemand-Breitenbach and de The, 2018).

SUMO plays crucial roles in nuclear processes such as DNA replication, DNA damage response, cell cycle regulation, transcription and proteostasis (Flotho and Melchior, 2013). SUMO also controls vital biological processes including development and lipid metabolism (Oishi et al., 2008; Sapir, 2020; Talamillo, Ajuria, et al., 2020; Talamillo, Barroso-Gomila, et al., 2020; Yau et al., 2020). It is thus not surprising that alterations in such a versatile functional modifier are linked to many diseases, including cancer and neurodegeneration.

Recent reviews highlighting the role of SUMO in various diseases, including neurodegenerative diseases, can be found in the literature (Celen and Sahin, 2020; Chang and Yeh, 2020; Princz and Tavernarakis, 2020; Vijayakumaran and Pountney, 2018; Yau et al., 2020). In this chapter, we will focus on implications of SUMOylation for neurodegeneration and the relation of neurodegenerative diseases to inflammation in the context of regulation by SUMOylation.

8.2 SUMO IN NEURODEGENERATIVE DISEASES

A role for SUMOylation in neurodegenerative diseases is highly likely given the importance of SUMO for physiological neuronal roles, such as controlling synaptic activity and dendritic spine density, whereas its overexpression affects learning and memory. Loss of SUMO causes reduced capacity of recovery after neuronal injury (Matsuzaki et al., 2015; L. Zhang et al., 2017). In addition, numerous proteins involved in neuronal development and degeneration have been found to be SUMOylated (Hendriks et al., 2017; Hendriks and Vertegaal, 2016). The role of SUMOylation in neuronal activity has been reviewed elsewhere (Anderson et al., 2017; Henley et al., 2018; Princz and Tavernarakis, 2020).

8.2.1 Alzheimer's Disease

Alzheimer's disease (AD) is the most common form of dementia in the elderly and occurs with rising incidence (Prince et al., 2013). It is characterized by presenting cerebral amyloid plaques formed by aggregation of β-amyloid peptide (Aβ) and prominent neurofibrillary tangles, resulting from hyperphosphorylation of the microtubule associated protein tau. Phosphorylated tau losses its affinity to microtubules and, as a consequence, it aggregates in neuronal bodies to form neurofibrillary tangles, which disrupts axonal transport and leads to synaptic dysfunction (Serrano-Pozo et al., 2011).

Amyloid precursor protein (APP) is normally cleaved by α and γ-secretases, with the peptide resulting from a normal cleavage having roles in neuroprotection and neuroplasticity. In pathological conditions, APP is first cleaved by a distinct β-secretase (BACE-1), leading to Aβ peptide generation. This accumulates and forms fibrils resulting in cell dysfunction and death (Hardy, 2017). Aβ aggregation together with synaptic dysfunction, oxidative stress, loss of calcium regulation and inflammation contribute to the development and progression of the disease.

A role for SUMOylation in AD is supported by various lines of evidence, although some of these findings might seem partial or even contradictory. In some human populations, polymorphisms in SUMO1 and 2 have been associated to AD risk and increased plasma levels of SUMO1 have been found in AD patients (Cho, Yun, Lee, et al., 2015; Mun et al., 2016).

Well-established AD-associated proteins are substrates for SUMOylation. In the case of tau, SUMO1ylation at lysine K340 stimulates its hyperphosphorylation, which in turn reduces its ubiquitination and consequent degradation (Dorval and Fraser, 2006; Luo et al., 2014). Furthermore, SUMO1 colocalizes with phosphorylated tau aggregates (Takahashi et al., 2008). In this case, SUMOylation seems to have a detrimental effect on the disease.

APP can be modified by SUMO1 and SUMO2 at lysines K587 and K595 adjacent to the β-secretase cleavage site, decreasing Aβ aggregation levels in HeLa cells (Zhang and Sarge, 2008). In agreement, SUMO3 overexpression in HEK293T cells reduces Aβ production by regulating APP processing (Li et al., 2003). In contrast, other studies show that exogenous SUMO3 in HEK293T cells increases Aβ peptides by increasing the expression of BACE (Dorval et al., 2007). Likewise, SUMO1 promotes Aβ generation, affecting BACE-1 accumulation by autophagy activation in neuroglioma H4 cell (Cho, Yun, Jo, et al., 2015; Yun et al., 2013).

Work in animal models supports a protective action of SUMO against AD. In an APP-PS1 mouse model of AD, high levels of Aβ increase the expression of the E3 ligase PIAS1, which in turn enhances SUMOylation of histone deacetylase 1 (HDAC1) in the hippocampus (Tao et al., 2017). SUMOylated HDAC1 reduces amyloid plaques and apoptosis in the brain, representing a defense mechanism against Aβ toxicity. PIAS1 also stimulates the SUMOylation of Elk-1, which reduces the number of apoptotic neurons in the hippocampus of model mice (Liu et al., 2019). PIAS1 also promotes the SUMOylation of the AICD (APP intracellular domain) at K43 in the hippocampus (Y. C. Liu et al., 2020). Interestingly, SUMOylation of AICD increases its association with cyclic AMP-responsive element-binding protein (CREB) and p65, a complex which stimulates the transcriptional activation of two major Aβ-degrading enzymes. As a result, SUMOylated AICD decreases Aβ oligomerization and amyloid plaques, rescuing spatial memory in model mice (Liu et al., 2020).

SENP1 and SENP2 cause deSUMOylation of APP. In agreement with the protective role of SUMOylation, SENP1 expression increases in an age-dependent manner in female mice (Maruyama et al., 2018). Higher SUMO2/3 levels in the mouse hippocampus correlate to increased performance in memory tasks, although the affected substrates are unknown (Yang et al., 2012).

In the 5xFAD AD mouse model, age-related region-specific changes in SUMO1 modification levels were observed, but no specific changes in SUMO1 levels or targets due to AD were detected (Stankova et al., 2018). Specific roles for SUMO2/3 conjugation, more typical in stress response, are still possible and warrant further study. In contrast, overexpression of SUMO1 in TgCRND8, a mouse model with transgenic human APP, increased insoluble Aβ and plaque density at later ages (Knock et al., 2018). In a distinct human APP-expressing mouse model (Tg2576), SUMO1 conjugation increased in cortex and hippocampus during aging, while SUMO2 levels decreased (Nistico et al., 2014). While contradictory, these results might depend on specifics of the different transgenic models and SUMO detection methods, genetic backgrounds, and the age of the analyzed animals, highlighting the necessity for more studies to reproduce and augment these findings.

8.2.2 Parkinson's Disease

Parkinson's disease (PD) is the most common movement disorder and is the second most common neurodegenerative disease, after AD (Dorsey et al., 2018). Only a minority of PD cases are familiar, i.e. have a known heritable genetic origin. In those cases, mutations in a number of genes can cause an increase of intracellular toxicity (e.g. α-synuclein) or lead to a loss of intrinsic protective function (e.g. LRRK2, PRKN, PINK1, PARK7/DJ-1). Most of the PD cases, however, are sporadic and influenced by environmental factors (Gasser, 2009).

Similarly to AD, post-mortem PD histology is defined by the accumulation and aggregation of misfolded α-synuclein, known as Lewy bodies (LB) and Lewy neurites, and by the loss of dopaminergic neurons in the substantia nigra (Glass et al., 2010). α-Synuclein is located in presynaptic terminals of neurons in normal physiological conditions but, in pathological conditions, α-synuclein changes conformation and forms intermediate oligomers that can assemble forming large protein aggregates.

α-Synuclein is SUMOylated in lysines K96 and K102 and mutations in those sites increase its aggregation (Krumova et al., 2011). Interestingly, α-synuclein is regulated by SUMO1 and SUMO2 in different ways. While SUMO2 increases its solubility, SUMO1 inhibits its degradation and promotes its aggregation, with SUMO1 colocalizing with LBs (Kim et al., 2011; Krumova et al., 2011; Rott et al., 2017; Zhu et al., 2018). Trafficking of α-synuclein by means of extracellular vesicles may contribute to interneuronal spread, and a role for SUMOylation for regulating ESCRT (endosomal-sorting complex required for transport) has been proposed (Kunadt et al., 2015).

Parkin, one of the main players in PD, is a ubiquitin E3 ligase encoded by PARK3. Mutations in PARK3 are implicated in autosomal recessive juvenile Parkinsonism. Parkin locates to dysfunctional mitochondria that have lost membrane potential and promotes their degradation through selective autophagy (mitophagy) by increasing the ubiquitination of mitochondrial proteins (Narendra et al., 2008; Vives-Bauza et al., 2010). Parkin also ubiquitinates RanBP2, a SUMO E3 ligase that SUMOylates nuclear pore components, which are then degraded by the proteasome (Um et al., 2006). In cellular studies, Parkin binds SUMO1, but not SUMO2, in a non-covalent way, but was not detected to be SUMOylated (Um and Chung, 2006). The interaction with SUMO1 influences Parkin self-ubiquitination and its localization in the nucleus. Parkin regulates the degradation of PARIS (Parkin-Interacting Substrate; ZNF746), which is involved in dopamine cell loss through repression of PGC-1α (Proliferator-Activated Receptor Gamma Coactivator-1-alpha) (Lee et al., 2017). PARIS SUMOylation regulates its ubiquitination and proteasomal degradation, alleviating PGC-1α repression (Nishida and Yamada, 2016, 2020).

DJ-1, encoded by PARK7, is a mitochondrial peroxiredoxin-like peroxidase, which act as an oxidative stress sensor to maintain mitochondrial function. Upon UV irradiation, DJ-1 is SUMOylated in lysine K130, leading to its nuclear translocation (Shinbo et al., 2006). Mutant DJ-1 L166P found in PD patients is improperly SUMOylated and becomes insoluble (Bonifati et al., 2003). Mutations in the SUMOylation site of DJ-1 are also associated with a greater sensitivity to UV (Shinbo et al., 2006). The relation between SUMO and mitochondrial regulation, as well as the regulation of α-synuclein by SUMOylation, Parkin and DJ-1, might represent a link between SUMO and PD (Guerra de Souza et al., 2016).

8.2.3 Polyglutamine Diseases

Huntington's disease (HD) is a neurodegenerative disorder caused by abnormal expansion of a polyglutamine (polyQ) stretch at the N-terminal part of huntingtin (HTT), which leads to its misfolding (Ross and Tabrizi, 2011). HTT is modified by ubiquitin and SUMO2 in the N-terminal part at lysines K6 and K9, with PIAS1 being the likely SUMO E3 ligase involved in its modification (O'Rourke et al., 2013). SUMO2-HTT is insoluble in HeLa cells and accumulates in HD post-mortem brains. In agreement with a detrimental function of SUMO2 in HD, reduction of PIAS1 in HTT *Drosophila* and mouse models is protective for the disease (O'Rourke et al., 2013; Ochaba et al., 2016). HTT is also modified by SUMO1 on the same residues, and the effect of this modification exacerbates neurodegeneration (Ohkuni et al., 2018; Steffan et al., 2004).

Another polyQ-expansion disease with connections to SUMO is the autosomal dominant disorder known as spinocerebellar ataxia-3 or Machado–Joseph disease (MJD) (Dantuma and Herzog, 2020). MJD is caused by the polyQ expansion in the ataxin-3 protein (ATXN3). This is a member of the Josephin family of Deubiquitinating enzymes, which contains a SIM. ATXN3 binds SUMO1 in a non-covalent way, which does not alter its aggregation capacity (Jung and Lee, 2013; Pfeiffer et al., 2017). ATXN3 might have a deep impact on cellular processes regulated by SUMO, including DNA double-strand breaks response, as it counteracts the activity of the SUMO-dependent E3 ubiquitin ligase RNF4 (Pfeiffer et al., 2017). Furthermore, ATXN3 is itself covalently modified by SUMO1 and SUMO2 in lysine K356, with SUMOylation modifying its capacity to interact with other cellular agents (Almeida et al., 2015).

SUMOylation appears to be involved in other neuropathies. For instance, PML-mediated SUMOylation regulates the stability of ATXN1 through its SUMO-dependent ubiquitination by RNF4 in a mouse model of spinocerebellar ataxia (SCA1) (Guo et al., 2014). A similar situation is true for soluble mutant ATXN7 in SCA7, which accumulates in PML NBs and is degraded by the proteasome (Janer et al., 2006). ATXN7 is modified by SUMO in K257 and SUMOylation reduces its aggregation and toxicity (Janer et al., 2010). Interestingly, interferon (IFN) beta stimulates the degradation of ATXN7 by inducing the expression of PML, improving the SCA7 symptoms in a mouse model (Chort et al., 2013).

Beyond specific disease-related proteins, other factors involved in neuronal functions, like ion channels, mitochondria etc., are also modified by SUMOylation, placing this modification at the core of nervous system function (Henley et al., 2018).

8.3 SUMO IN INFLAMMATION

Chronic inflammation is characterized by the activation of inflammatory molecules (IFNs, interleukins [ILs] and tumor necrosis factor [TNF]-α). This in turn can lead to autoimmunity, insulin resistance and tissue damage (Germolec et al., 2018).

Inflammasomes are macromolecular complexes accounting for caspase 1 activation in response to microbial, but also non-microbial signals (Tweedell and Kanneganti, 2020). Dysfunctional regulation of inflammasomes leads to an excessive activation of caspase 1, which is linked to neurodegenerative diseases (Heneka et al., 2013; Tan et al., 2013). NLRP3 (NLR Family Pyrin Domain Containing 3) induces the assembly of inflammasome macrocomplexes upon infection and tissue damage. NLRP3 is negatively regulated by SUMOylation via modification of lysine K689 by MUL1 (Mitochondrial E3 Ubiquitin Protein Ligase 1; also known as MALP) (Barry et al., 2018).

IL-17 is a proinflammatory cytokine produced by Th17 lymphocytes, which targets the production of IL-6, TNF-α and GCSF (granulocyte-colony-stimulating factor) (Patel and Kuchroo, 2015). IL-17 transcription is induced by RORγt (retinoic acid-related orphan receptor γt). RORγt is SUMOylated at lysine K187 and SUMO-RORγt represses IL-17 gene expression through HDAC2 interaction (Singh et al., 2018). Albeit indirect, SUMOylation is a means to control IL-17 expression and chronic inflammation.

Inflammation also depends on ROS (reactive oxygen species) and the oxidative stress response, where SUMO also has an important function (Nathan and Cunningham-Bussel, 2013; Stankovic-Valentin and Melchior, 2018). Redox signaling depends on the oxidative modification of amino acids like methionine, tyrosine and cysteine by low or moderate ROS levels (Berlett and Stadtman, 1997; Wall et al., 2012). Enzymes driving SUMO, Ub and other UbL modifications (E1s, E2s, some E3s and isopeptidases) contain cysteines in their catalytic domains, which makes them susceptible to oxidation. Interestingly, oxidative stress increases SUMOylation, while ROS inhibits it (Bossis and Melchior, 2006). This is caused by the formation of disulfide bonds involving the catalytic cysteines of E1 and E2s, which inhibits their enzymatic activity and can be reverted in presence of reducing agents. This reversible, transient inhibition is important for cell survival under acute oxidative stresses, via regulation of the ATM–Chk2 DNA damage response pathway (Stankovic-Valentin et al., 2016). Indirectly, ROS can also regulate SUMOylation through the regulation of ubiquitination, phosphorylation and acetylation (Stankovic-Valentin and Melchior, 2018).

SENP deSUMOylases are sensitive to changes in ROS (Xu et al., 2008; Yan et al., 2010). SENP3 presents a redox-sensing domain that interacts with the chaperone HSP90 in the presence of H_2O_2, protecting SENP3 from ubiquitination and degradation (Yan et al., 2010). Stabilized SENP3 translocates to the nucleoplasm and interacts with PML and the histone acetyltransferase p300, which is involved in the regulation of the hypoxia response (Huang et al., 2009; Kunz et al., 2016). As neurons are particularly sensitive to hypoxic stress, SUMOylation can be considered as a protective mechanism.

SUMOylation also has an important role in the regulation of the NF-κB pathway, which plays an important role in inflammation and immunity (Morgan and Liu, 2011). Although ROS restricts the transcriptional activity of P50 (NFKB1), it can also induce the SUMO1 modification of the IKKγ kinase regulatory subunit NEMO via PIAS4, which triggers the activation of the pathway through a phosphorylation cascade (Mabb et al., 2006).

8.4 INFLAMMATION IN NEURODEGENERATIVE DISEASES

Peripheral chronic inflammation can lead to neuroinflammation, which is mediated by astrocytes and microglia and plays an important role in the progression of neurodegenerative diseases (Glass et al., 2010), so regulation of inflammation might be an effective therapy.

IFN1-stimulated astrocytes express proinflammatory cytokines, such as TNF-α, IL-6 and IRF1 (interferon regulatory factor 1), the expression of which is induced by p-STAT1 (phosphorylated-STAT1). The induction of proinflammatory factors can be blocked by the SUMOylation of LXR-α and LXR-β (ligand-activated liver X receptors α and β), which prevent p-STAT1 interaction with the promoters of inflammatory genes (Lee et al., 2009, 2016).

Inflammation, mitochondrial dynamics, synaptic transmission and plasticity, as well as protective responses to cell stress, are all processes underlying AD (Martins et al., 2016). SUMOylation target substrates include Drp1, potassium channels, glutamate receptors and transporter, iNOS (induced nitric oxide synthases), GSK-3β (glycogen synthase kinase 3β) and JNK (c-Jun N-terminal Kinase). Importantly, Aβ induces NO (nitric oxide) production, a well-established actor in inflammation, by promoting NOS2 expression in microglia and astrocytes. More specifically, Aβ stimulates the activation of NFκB in rat astrocytes; therefore, Aβ induces NO synthase expression and NO production occurs through an NFκB-dependent mechanism. This finding provided a key mechanistic link between Aβ and the generation of oxidative damage (Akama et al., 1998). In addition, Aβ promotes the release of IL-1β and TNF-α, which induces NO production as well as peroxynitrite synthesis, leading to protein and lipid modification, mitochondrial damage, apoptosis and increased formation of Aβ (Akama et al., 1998; Goodwin et al., 1997). NOS2 promoter activity is downregulated by overexpression of SUMO1, UBC9 and SENP1. These factors decrease in astrocytes during inflammation and aging, which is in agreement with higher NO production (Akar and Feinstein, 2009).

Inflammation is also an important component in PD etiology. Astrocyte dysfunction has been shown to be at the core of PD pathogenesis, linked to dopaminergic neuronal death (Booth et al., 2017). α-Synuclein intermediate-state oligomers are released by dying neurons. The oligomers activate microglia, leading to NF-κB activation and ROS production, which in turn leads to proinflammatory mediators including NO, TNF-α and IL-1β (Glass et al., 2010). In parallel to dopamine synthesis, an increase in intracellular oxidative processes occurs in dopaminergic neurons, which makes these neurons vulnerable to oxidative stress. Furthermore, DJ-1 downregulation causes ROS increases, leading to mitochondrial defects and hypersensitivity to oxidative stress in animal models (Heo et al., 2012; Kim et al., 2005; Park et al., 2005; Yang et al., 2007).

Finally, inflammation has also been associated with polyQ diseases. Activated microglia and reactive astrocytes have been observed in MJD (Evert et al., 2001, 2003, 2006). Interestingly, a polymorphism in the IL-6 locus, IL6 174G>C, is associated with lower IL-6 mRNA levels and an early onset of MJD, suggesting an anti-inflammatory role for IL-6 in this disease (Raposo et al., 2017). Inflammation of microglia has been also described in HD (Illarioshkin et al., 2018).

8.5 CONCLUDING REMARKS

SUMOylation can affect neurodegeneration in a dual way (Figure 8.2). On one side, SUMOylation regulates diverse proteins involved in the pathogenesis of AD, PD and polyQ diseases (Anderson

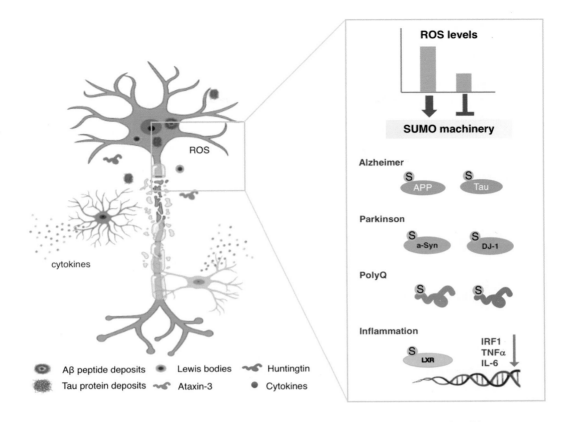

Figure 8.2 **Protective role of SUMOylation in neurodegenerative diseases.** Degenerating neuron (purple), astrocyte (violet) and microglia (yellow), Aβ deposits, tau deposits, Lewis bodies, huntingtin protein, ATXN3 and cytokines are indicated. Inset: Regulation of SUMO machinery by ROS levels. Pathogenic proteins related to Alzheimer's, Parkinson's and polyglutamine diseases are indicated. SUMOylation of LXR regulates inflammatory factors.

et al., 2017; Princz and Tavernarakis, 2020). On the other side, SUMO also regulates inflammation, an inherent factor to these neurodegenerative diseases affecting their early onset, progression and aggravation (Glass et al., 2010; Raposo et al., 2017).

Interestingly, in the last years, a growing number of pharmacological agents have been developed to target the posttranslational modification machineries by members of the ubiquitin family (Cox and Huber, 2019; Gatel et al., 2020; Krajnak and Dahl, 2018; Nalepa et al., 2006). Most are used to target the E1 and E2 enzymes, which, although effective, lack specificity due to the many affected substrates. Emerging targets are the "removal" enzymes (DUBs, SENPs), which may offer better specificity (Schauer et al., 2020). Most inhibitors are designed with cancer in mind, but in light of the role that SUMOylation plays in neurodegenerative and age-associated diseases, these therapies deserve a closer look.

ACKNOWLEDGMENTS

We acknowledge funding by grants BFU2017-84653-P (MINECO/FEDER, EU), SEV-2016-0644 (Severo Ochoa Excellence Program), 765445-EU (UbiCODE Program), SAF2017-90900-REDT (UBIRed Program) and IT1165-19 (Basque Country Government).

REFERENCES

Akama, K. T., Albanese, C., Pestell, R. G., et al. 1998. Amyloid beta-peptide stimulates nitric oxide production in astrocytes through an NFkappaB-dependent mechanism. *Proc Natl Acad Sci U S A*, 95(10), 5795–5800. doi:10.1073/pnas.95.10.5795

Akar, C. A., & Feinstein, D. L. 2009. Modulation of inducible nitric oxide synthase expression by sumoylation. *J Neuroinflammation*, 6, 12. doi:10.1186/1742-2094-6-12

Almeida, B., Abreu, I. A., Matos, C. A., et al. 2015. SUMOylation of the brain-predominant Ataxin-3 isoform modulates its interaction with p97. *Biochim Biophys Acta*, 1852(9), 1950–1959. doi:10.1016/j.bbadis.2015.06.010

Anderson, D. B., Zanella, C. A., Henley, J. M., et al. 2017. Sumoylation: implications for neurodegenerative diseases. *Adv Exp Med Biol*, 963, 261–281. doi:10.1007/978-3-319-50044-7_16

Barry, R., John, S. W., Liccardi, G., et al. 2018. SUMO-mediated regulation of NLRP3 modulates inflammasome activity. *Nat Commun*, 9(1), 3001. doi:10.1038/s41467-018-05321-2

Berlett, B. S., & Stadtman, E. R. 1997. Protein oxidation in aging, disease, and oxidative stress. *J Biol Chem*, 272(33), 20313–20316. doi:10.1074/jbc.272.33.20313

Bonifati, V., Rizzu, P., van Baren, M. J., et al. 2003. Mutations in the DJ-1 gene associated with autosomal recessive early-onset parkinsonism. *Science*, 299(5604), 256–259. doi:10.1126/science.1077209

Booth, H. D. E., Hirst, W. D., & Wade-Martins, R. 2017. The role of astrocyte dysfunction in Parkinson's disease pathogenesis. *Trends Neurosci*, 40(6), 358–370. doi:10.1016/j.tins.2017.04.001

Bossis, G., & Melchior, F. 2006. Regulation of SUMOylation by reversible oxidation of SUMO conjugating enzymes. *Mol Cell*, 21(3), 349–357. doi:10.1016/j.molcel.2005.12.019

Celen, A. B., & Sahin, U. 2020. Sumoylation on its 25th anniversary: mechanisms, pathology, and emerging concepts. *FEBS J*, 287(15), 3110–3140. doi:10.1111/febs.15319

Chang, H. M., & Yeh, E. T. H. 2020. SUMO: from bench to bedside. *Physiol Rev*, 100(4), 1599–1619. doi:10.1152/physrev.00025.2019

Cho, S. J., Yun, S. M., Jo, C., et al. 2015. SUMO1 promotes Abeta production via the modulation of autophagy. *Autophagy*, 11(1), 100–112. doi:10.4161/15548627.2014.984283

Cho, S. J., Yun, S. M., Lee, D. H., et al. 2015. Plasma SUMO1 protein is elevated in Alzheimer's disease. *J Alzheimers Dis*, 47(3), 639–643. doi:10.3233/JAD-150103

Chort, A., Alves, S., Marinello, M., et al. 2013. Interferon beta induces clearance of mutant ataxin 7 and improves locomotion in SCA7 knock-in mice. *Brain*, 136(Pt 6), 1732–1745. doi:10.1093/brain/awt061

Cox, O. F., & Huber, P. W. 2019. Developing practical therapeutic strategies that target protein SUMOylation. *Curr Drug Targets*, 20(9), 960–969. doi:10.2174/1389450119666181026151802

Dantuma, N. P., & Herzog, L. K. 2020. Machado-Joseph disease: a stress combating deubiquitylating enzyme changing sides. *Adv Exp Med Biol*, 1233, 237–260. doi:10.1007/978-3-030-38266-7_10

Dorsey, E. R., Sherer, T., Okun, M. S., et al. 2018. The emerging evidence of the Parkinson pandemic. *J Parkinsons Dis*, 8(s1), S3–S8. doi:10.3233/JPD-181474

Dorval, V., & Fraser, P. E. 2006. Small ubiquitin-like modifier (SUMO) modification of natively unfolded proteins tau and alpha-synuclein. *J Biol Chem*, 281(15), 9919–9924. doi:10.1074/jbc.M510127200

Dorval, V., Mazzella, M. J., Mathews, P. M., et al. 2007. Modulation of Abeta generation by small ubiquitin-like modifiers does not require conjugation to target proteins. *Biochem J*, 404(2), 309–316. doi:10.1042/BJ20061451

El Motiam, A., Vidal, S., de la Cruz-Herrera, C. F., et al. 2019. Interplay between SUMOylation and NEDDylation regulates RPL11 localization and function. *FASEB J*, 33(1), 643–651. doi:10.1096/fj.201800341RR

Evert, B. O., Vogt, I. R., Kindermann, C., et al. 2001. Inflammatory genes are upregulated in expanded ataxin-3-expressing cell lines and spinocerebellar ataxia type 3 brains. *J Neurosci*, 21(15), 5389–5396.

Evert, B. O., Vogt, I. R., Vieira-Saecker, A. M., et al. 2003. Gene expression profiling in ataxin-3 expressing cell lines reveals distinct effects of normal and mutant ataxin-3. *J Neuropathol Exp Neurol*, 62(10), 1006–1018. doi:10.1093/jnen/62.10.1006

Evert, B. O., Schelhaas, J., Fleischer, H., et al. 2006. Neuronal intranuclear inclusions, dysregulation of cytokine expression and cell death in spinocerebellar ataxia type 3. *Clin Neuropathol*, 25(6), 272–281.

Flotho, A., & Melchior, F. 2013. Sumoylation: a regulatory protein modification in health and disease. *Annu Rev Biochem*, 82, 357–385. doi:10.1146/annurev-biochem-061909-093311

Gasser, T. 2009. Molecular pathogenesis of Parkinson disease: insights from genetic studies. *Expert Rev Mol Med*, 11, e22. doi:10.1017/S1462399409001148

Gatel, P., Piechaczyk, M., & Bossis, G. 2020. Ubiquitin, SUMO, and Nedd8 as therapeutic targets in cancer. *Adv Exp Med Biol*, 1233, 29–54. doi:10.1007/978-3-030-38266-7_2

Germolec, D. R., Shipkowski, K. A., Frawley, R. P., et al. 2018. Markers of inflammation. *Methods Mol Biol*, 1803, 57–79. doi:10.1007/978-1-4939-8549-4_5

Glass, C. K., Saijo, K., Winner, B., et al. 2010. Mechanisms underlying inflammation in neurodegeneration. *Cell*, 140(6), 918–934. doi:10.1016/j.cell.2010.02.016

Goodwin, J. L., Kehrli, M. E., Jr., & Uemura, E. 1997. Integrin Mac-1 and beta-amyloid in microglial release of nitric oxide. *Brain Res*, 768(1-2), 279–286. doi:10.1016/s0006-8993(97)00653-7

Guerra de Souza, A. C., Prediger, R. D., & Cimarosti, H. 2016. SUMO-regulated mitochondrial function in Parkinson's disease. *J Neurochem*, 137(5), 673–686. doi:10.1111/jnc.13599

Guo, L., Giasson, B. I., Glavis-Bloom, A., et al. 2014. A cellular system that degrades misfolded proteins and protects against neurodegeneration. *Mol Cell*, 55(1), 15–30. doi:10.1016/j.molcel.2014.04.030

Hardy, J. 2017. The discovery of Alzheimer-causing mutations in the APP gene and the formulation of the "amyloid cascade hypothesis". *FEBS J*, 284(7), 1040–1044. doi:10.1111/febs.14004

Hendriks, I. A., Lyon, D., Young, C., et al. 2017. Site-specific mapping of the human SUMO proteome reveals co-modification with phosphorylation. *Nat Struct Mol Biol*, 24(3), 325–336. doi:10.1038/nsmb.3366

Hendriks, I. A., & Vertegaal, A. C. 2016. A comprehensive compilation of SUMO proteomics. *Nat Rev Mol Cell Biol*, 17(9), 581–595. doi:10.1038/nrm.2016.81

Heneka, M. T., Kummer, M. P., Stutz, A., et al. 2013. NLRP3 is activated in Alzheimer's disease and contributes to pathology in APP/PS1 mice. *Nature*, 493(7434), 674–678. doi:10.1038/nature11729

Henley, J. M., Carmichael, R. E., & Wilkinson, K. A. 2018. Extranuclear SUMOylation in neurons. *Trends Neurosci*, 41(4), 198–210. doi:10.1016/j.tins.2018.02.004

Heo, J. Y., Park, J. H., Kim, S. J., et al. 2012. DJ-1 null dopaminergic neuronal cells exhibit defects in mitochondrial function and structure: involvement of mitochondrial complex I assembly. *PLoS One*, 7(3), e32629. doi:10.1371/journal.pone.0032629

Huang, C., Han, Y., Wang, Y., et al. 2009. SENP3 is responsible for HIF-1 transactivation under mild oxidative stress via p300 de-SUMOylation. *EMBO J*, 28(18), 2748–2762. doi:10.1038/emboj.2009.210

Illarioshkin, S. N., Klyushnikov, S. A., Vigont, V. A., et al. 2018. Molecular pathogenesis in Huntington's disease. *Biochemistry (Mosc)*, 83(9), 1030–1039. doi:10.1134/S0006297918090043

Janer, A., Martin, E., Muriel, M. P., et al. 2006. PML clastosomes prevent nuclear accumulation of mutant ataxin-7 and other polyglutamine proteins. *J Cell Biol*, 174(1), 65–76. doi:10.1083/jcb.200511045

Janer, A., Werner, A., Takahashi-Fujigasaki, J., et al. 2010. SUMOylation attenuates the aggregation propensity and cellular toxicity of the polyglutamine expanded ataxin-7. *Hum Mol Genet*, 19(1), 181–195. doi:10.1093/hmg/ddp478

Jansen, N. S., & Vertegaal, A. C. O. 2020. A chain of events: regulating target proteins by SUMO polymers. *Trends Biochem Sci*. doi:10.1016/j.tibs.2020.09.002

Jung, N. R., & Lee, D. H. 2013. SUMO-1 promotes degradation of the polyglutamine disease protein ataxin-3. *Anim Cells Syst*, 17(1), 7–14.

Kim, R. H., Smith, P. D., Aleyasin, H., et al. 2005. Hypersensitivity of DJ-1-deficient mice to 1-methyl-4-phenyl-1,2,3,6-tetrahydropyridine (MPTP) and oxidative stress. *Proc Natl Acad Sci U S A*, 102(14), 5215–5220. doi:10.1073/pnas.0501282102

Kim, Y. M., Jang, W. H., Quezado, M. M., et al. 2011. Proteasome inhibition induces alpha-synuclein SUMOylation and aggregate formation. *J Neurol Sci*, 307(1-2), 157–161. doi:10.1016/j.jns.2011.04.015

Knock, E., Matsuzaki, S., Takamura, H., et al. 2018. SUMO1 impact on Alzheimer disease pathology in an amyloid-depositing mouse model. *Neurobiol Dis*, 110, 154–165. doi:10.1016/j.nbd.2017.11.015

Krajnak, K., & Dahl, R. 2018. Small molecule SUMOylation activators are novel neuroprotective agents. *Bioorg Med Chem Lett*, 28(3), 405–409. doi:10.1016/j.bmcl.2017.12.028

Krumova, P., Meulmeester, E., Garrido, M., et al. 2011. Sumoylation inhibits alpha-synuclein aggregation and toxicity. *J Cell Biol*, 194(1), 49–60. doi:10.1083/jcb.201010117

Kunadt, M., Eckermann, K., Stuendl, A., et al. 2015. Extracellular vesicle sorting of alpha-Synuclein is regulated by sumoylation. *Acta Neuropathol*, 129(5), 695–713. doi:10.1007/s00401-015-1408-1

Kunz, K., Wagner, K., Mendler, L., et al. 2016. SUMO signaling by hypoxic inactivation of SUMO-specific isopeptidases. *Cell Rep*, 16(11), 3075–3086. doi:10.1016/j.celrep.2016.08.031

Lallemand-Breitenbach, V., & de The, H. 2018. PML nuclear bodies: from architecture to function. *Curr Opin Cell Biol*, 52, 154–161. doi:10.1016/j.ceb.2018.03.011

Lee, J. H., Park, S. M., Kim, O. S., et al. 2009. Differential SUMOylation of LXRalpha and LXRbeta mediates transrepression of STAT1 inflammatory signaling in IFN-gamma-stimulated brain astrocytes. *Mol Cell*, 35(6), 806–817. doi:10.1016/j.molcel.2009.07.021

Lee, J. H., Kim, H., Park, S. J., et al. 2016. Small heterodimer partner SHP mediates liver X receptor (LXR)-dependent suppression of inflammatory signaling by promoting LXR SUMOylation specifically in astrocytes. *Sci Signal*, 9(439), ra78. doi:10.1126/scisignal.aaf4850

Lee, Y., Stevens, D. A., Kang, S. U., et al. 2017. PINK1 primes Parkin-mediated ubiquitination of PARIS in dopaminergic neuronal survival. *Cell Rep*, 18(4), 918–932. doi:10.1016/j.celrep.2016.12.090

Li, Y., Wang, H., Wang, S., et al. 2003. Positive and negative regulation of APP amyloidogenesis by sumoylation. *Proc Natl Acad Sci U S A*, 100(1), 259–264. doi:10.1073/pnas.0235361100

Liu, S. Y., Ma, Y. L., Hsu, W. L., et al. 2019. Protein inhibitor of activated STAT1 Ser(503) phosphorylation-mediated Elk-1 SUMOylation promotes neuronal survival in APP/PS1 mice. *Br J Pharmacol*, 176(11), 1793–1810. doi:10.1111/bph.14656

Liu, Y. C., Hsu, W. L., Ma, Y. L., et al. 2020. Melatonin induction of APP intracellular domain 50 SUMOylation alleviates AD through enhanced transcriptional activation and Abeta degradation. *Mol Ther*. doi:10.1016/j.ymthe.2020.09.003

Luo, H. B., Xia, Y. Y., Shu, X. J., et al. 2014. SUMOylation at K340 inhibits tau degradation through deregulating its phosphorylation and ubiquitination. *Proc Natl Acad Sci U S A*, 111(46), 16586–16591. doi:10.1073/pnas.1417548111

Mabb, A. M., Wuerzberger-Davis, S. M., & Miyamoto, S. 2006. PIASy mediates NEMO sumoylation and NF-kappaB activation in response to genotoxic stress. *Nat Cell Biol*, 8(9), 986–993. doi:10.1038/ncb1458

Martins, W. C., Tasca, C. I., & Cimarosti, H. 2016. Battling Alzheimer's disease: targeting SUMOylation-mediated pathways. *Neurochem Res*, 41(3), 568–578. doi:10.1007/s11064-015-1681-3

Maruyama, T., Abe, Y., & Niikura, T. 2018. SENP1 and SENP2 regulate SUMOylation of amyloid precursor protein. *Heliyon*, 4(4), e00601. doi:10.1016/j.heliyon.2018.e00601

Matic, I., Macek, B., Hilger, M., et al. 2008. Phosphorylation of SUMO-1 occurs in vivo and is conserved through evolution. *J Proteome Res*, 7(9), 4050–4057. doi:10.1021/pr800368m

Matsuzaki, S., Lee, L., Knock, E., et al. 2015. SUMO1 affects synaptic function, spine density and memory. *Sci Rep*, 5, 10730. doi:10.1038/srep10730

Morgan, M. J., & Liu, Z. G. 2011. Crosstalk of reactive oxygen species and NF-kappaB signaling. *Cell Res*, 21(1), 103–115. doi:10.1038/cr.2010.178

Mun, M. J., Kim, J. H., Choi, J. Y., et al. 2016. Polymorphisms of small ubiquitin-related modifier genes are associated with risk of Alzheimer's disease in Korean: a case-control study. *J Neurol Sci*, 364, 122–127. doi:10.1016/j.jns.2016.03.023

Nalepa, G., Rolfe, M., & Harper, J. W. 2006. Drug discovery in the ubiquitin-proteasome system. *Nat Rev Drug Discov*, 5(7), 596–613. doi:10.1038/nrd2056

Narendra, D., Tanaka, A., Suen, D. F., et al. 2008. Parkin is recruited selectively to impaired mitochondria and promotes their autophagy. *J Cell Biol*, 183(5), 795–803. doi:10.1083/jcb.200809125

Nathan, C., & Cunningham-Bussel, A. 2013. Beyond oxidative stress: an immunologist's guide to reactive oxygen species. *Nat Rev Immunol*, 13(5), 349–361. doi:10.1038/nri3423

Nishida, T., & Yamada, Y. 2016. SUMOylation of the KRAB zinc-finger transcription factor PARIS/ZNF746 regulates its transcriptional activity. *Biochem Biophys Res Commun*, 473(4), 1261–1267. doi:10.1016/j.bbrc.2016.04.051

Nishida, T., & Yamada, Y. 2020. RNF4-mediated SUMO-targeted ubiquitination relieves PARIS/ZNF746-mediated transcriptional repression. *Biochem Biophys Res Commun*, 526(1), 110–116. doi:10.1016/j.bbrc.2020.03.063

Nistico, R., Ferraina, C., Marconi, V., et al. 2014. Age-related changes of protein SUMOylation balance in the AbetaPP Tg2576 mouse model of Alzheimer's disease. *Front Pharmacol*, 5, 63. doi:10.3389/fphar.2014.00063

O'Rourke, J. G., Gareau, J. R., Ochaba, J., et al. 2013. SUMO-2 and PIAS1 modulate insoluble mutant huntingtin protein accumulation. *Cell Rep*, 4(2), 362–375. doi:10.1016/j.celrep.2013.06.034

Ochaba, J., Monteys, A. M., O'Rourke, J. G., et al. 2016. PIAS1 regulates mutant Huntingtin accumulation and Huntington's disease-associated phenotypes in vivo. *Neuron*, 90(3), 507–520. doi:10.1016/j.neuron.2016.03.016

Ohkuni, K., Pasupala, N., Peek, J., et al. 2018. SUMO-targeted ubiquitin ligases (STUbLs) reduce the toxicity and abnormal transcriptional activity associated with a mutant, aggregation-prone fragment of Huntingtin. *Front Genet*, 9, 379. doi:10.3389/fgene.2018.00379

Oishi, Y., Manabe, I., Tobe, K., et al. 2008. SUMOylation of Kruppel-like transcription factor 5 acts as a molecular switch in transcriptional programs of lipid metabolism involving PPAR-delta. *Nat Med*, 14(6), 656–666. doi:10.1038/nm1756

Park, J., Kim, S. Y., Cha, G. H., et al. 2005. Drosophila DJ-1 mutants show oxidative stress-sensitive locomotive dysfunction. *Gene*, 361, 133–139. doi:10.1016/j.gene.2005.06.040

Patel, D. D., & Kuchroo, V. K. 2015. Th17 cell pathway in human immunity: lessons from genetics and therapeutic interventions. *Immunity*, 43(6), 1040–1051. doi:10.1016/j.immuni.2015.12.003

Pfeiffer, A., Luijsterburg, M. S., Acs, K., et al. 2017. Ataxin-3 consolidates the MDC1-dependent DNA double-strand break response by counteracting the SUMO-targeted ubiquitin ligase RNF4. EMBO J, 36(8), 1066–1083. doi:10.15252/embj.201695151

Pichler, A., Fatouros, C., Lee, H., et al. 2017. SUMO conjugation – a mechanistic view. Biomol Concepts, 8(1), 13–36. doi:10.1515/bmc-2016-0030

Prince, M., Bryce, R., Albanese, E., et al. 2013. The global prevalence of dementia: a systematic review and metaanalysis. Alzheimers Dement, 9(1), 63–75.e62. doi:10.1016/j.jalz.2012.11.007

Princz, A., & Tavernarakis, N. 2020. SUMOylation in neurodegenerative diseases. Gerontology, 66(2), 122–130. doi:10.1159/000502142

Raposo, M., Bettencourt, C., Ramos, A., et al. 2017. Promoter variation and expression levels of inflammatory genes IL1A, IL1B, IL6 and TNF in blood of spinocerebellar ataxia type 3 (SCA3) patients. Neuromolecular Med, 19(1), 41–45. doi:10.1007/s12017-016-8416-8

Ross, C. A., & Tabrizi, S. J. 2011. Huntington's disease: from molecular pathogenesis to clinical treatment. Lancet Neurol, 10(1), 83–98. doi:10.1016/S1474-4422(10)70245-3

Rott, R., Szargel, R., Shani, V., et al. 2017. SUMOylation and ubiquitination reciprocally regulate alpha-synuclein degradation and pathological aggregation. Proc Natl Acad Sci U S A, 114(50), 13176–13181. doi:10.1073/pnas.1704351114

Sapir, A. 2020. Not so slim anymore-evidence for the role of SUMO in the regulation of lipid metabolism. Biomolecules, 10(8). doi:10.3390/biom10081154

Schauer, N. J., Magin, R. S., Liu, X., et al. 2020. Advances in discovering deubiquitinating enzyme (DUB) inhibitors. J Med Chem, 63(6), 2731–2750. doi:10.1021/acs.jmedchem.9b01138

Serrano-Pozo, A., Frosch, M. P., Masliah, E., et al. 2011. Neuropathological alterations in Alzheimer disease. Cold Spring Harb Perspect Med, 1(1), a006189. doi:10.1101/cshperspect.a006189

Shinbo, Y., Niki, T., Taira, T., et al. 2006. Proper SUMO-1 conjugation is essential to DJ-1 to exert its full activities. Cell Death Differ, 13(1), 96–108. doi:10.1038/sj.cdd.4401704

Singh, A. K., Khare, P., Obaid, A., et al. 2018. SUMOylation of ROR-gammat inhibits IL-17 expression and inflammation via HDAC2. Nat Commun, 9(1), 4515. doi:10.1038/s41467-018-06924-5

Sriramachandran, A. M., Meyer-Teschendorf, K., Pabst, S., et al. 2019. Arkadia/RNF111 is a SUMO-targeted ubiquitin ligase with preference for substrates marked with SUMO1-capped SUMO2/3 chain. Nat Commun, 10(1), 3678. doi:10.1038/s41467-019-11549-3

Stankova, T., Piepkorn, L., Bayer, T. A., et al. 2018. SUMO1-conjugation is altered during normal aging but not by increased amyloid burden. Aging Cell, 17(4), e12760. doi:10.1111/acel.12760

Stankovic-Valentin, N., & Melchior, F. 2018. Control of SUMO and ubiquitin by ROS: signaling and disease implications. Mol Aspects Med, 63, 3–17. doi:10.1016/j.mam.2018.07.002

Stankovic-Valentin, N., Drzewicka, K., Konig, C., et al. 2016. Redox regulation of SUMO enzymes is required for ATM activity and survival in oxidative stress. EMBO J, 35(12), 1312–1329. doi:10.15252/embj.201593404

Steffan, J. S., Agrawal, N., Pallos, J., et al. 2004. SUMO modification of huntingtin and Huntington's disease pathology. Science, 304(5667), 100–104. doi:10.1126/science.1092194

Takahashi, K., Ishida, M., Komano, H., et al. 2008. SUMO-1 immunoreactivity co-localizes with phospho-tau in APP transgenic mice but not in mutant tau transgenic mice. Neurosci Lett, 441(1), 90–93. doi:10.1016/j.neulet.2008.06.012

Talamillo, A., Ajuria, L., Grillo, M., et al. 2020. SUMOylation in the control of cholesterol homeostasis. Open Biol, 10(5), 200054. doi:10.1098/rsob.200054

Talamillo, A., Barroso-Gomila, O., Giordano, I., et al. 2020. The role of SUMOylation during development. Biochem Soc Trans, 48(2), 463–478. doi:10.1042/BST20190390

Tan, M. S., Yu, J. T., Jiang, T., et al. 2013. The NLRP3 inflammasome in Alzheimer's disease. Mol Neurobiol, 48(3), 875–882. doi:10.1007/s12035-013-8475-x

Tao, C. C., Hsu, W. L., Ma, Y. L., et al. 2017. Epigenetic regulation of HDAC1 SUMOylation as an endogenous neuroprotection against Abeta toxicity in a mouse model of Alzheimer's disease. Cell Death Differ, 24(4), 597–614. doi:10.1038/cdd.2016.161

Tweedell, R. E., & Kanneganti, T. D. 2020. Advances in inflammasome research: recent breakthroughs and future hurdles. Trends Mol Med. doi:10.1016/j.molmed.2020.07.010

Ullmann, R., Chien, C. D., Avantaggiati, M. L., et al. 2012. An acetylation switch regulates SUMO-dependent protein interaction networks. Mol Cell, 46(6), 759–770. doi:10.1016/j.molcel.2012.04.006

Um, J. W., & Chung, K. C. 2006. Functional modulation of Parkin through physical interaction with SUMO-1. J Neurosci Res, 84(7), 1543–1554. doi:10.1002/jnr.21041

Um, J. W., Min, D. S., Rhim, H., et al. 2006. Parkin ubiquitinates and promotes the degradation of RanBP2. *J Biol Chem*, 281(6), 3595–3603. doi:10.1074/jbc.M504994200

Vertegaal, A. C., Andersen, J. S., Ogg, S. C., et al. 2006. Distinct and overlapping sets of SUMO-1 and SUMO-2 target proteins revealed by quantitative proteomics. *Mol Cell Proteomics*, 5(12), 2298–2310. doi:10.1074/mcp.M600212-MCP200

Vijayakumaran, S., & Pountney, D. L. 2018. SUMOylation, aging and autophagy in neurodegeneration. *Neurotoxicology*, 66, 53–57. doi:10.1016/j.neuro.2018.02.015

Vives-Bauza, C., Zhou, C., Huang, Y., et al. 2010. PINK1-dependent recruitment of Parkin to mitochondria in mitophagy. *Proc Natl Acad Sci U S A*, 107(1), 378–383. doi:10.1073/pnas.0911187107

Wall, S. B., Oh, J. Y., Diers, A. R., et al. 2012. Oxidative modification of proteins: an emerging mechanism of cell signaling. *Front Physiol*, 3, 369. doi:10.3389/fphys.2012.00369

Xu, Z., Lam, L. S., Lam, L. H., et al. 2008. Molecular basis of the redox regulation of SUMO proteases: a protective mechanism of intermolecular disulfide linkage against irreversible sulfhydryl oxidation. *FASEB J*, 22(1), 127–137. doi:10.1096/fj.06-7871com

Yan, S., Sun, X., Xiang, B., et al. 2010. Redox regulation of the stability of the SUMO protease SENP3 via interactions with CHIP and Hsp90. *EMBO J*, 29(22), 3773–3786. doi:10.1038/emboj.2010.245

Yang, W., Chen, L., Ding, Y., et al. 2007. Paraquat induces dopaminergic dysfunction and proteasome impairment in DJ-1-deficient mice. *Hum Mol Genet*, 16(23), 2900–2910. doi:10.1093/hmg/ddm249

Yang, Q. G., Wang, F., Zhang, Q., et al. 2012. Correlation of increased hippocampal Sumo3 with spatial learning ability in old C57BL/6 mice. *Neurosci Lett*, 518(2), 75–79. doi:10.1016/j.neulet.2012.04.051

Yau, T. Y., Molina, O., & Courey, A. J. 2020. SUMOylation in development and neurodegeneration. *Development*, 147(6). doi:10.1242/dev.175703

Yun, S. M., Cho, S. J., Song, J. C., et al. 2013. SUMO1 modulates Abeta generation via BACE1 accumulation. *Neurobiol Aging*, 34(3), 650–662. doi:10.1016/j.neurobiolaging.2012.08.005

Zhang, L., Liu, X., Sheng, H., et al. 2017. Neuron-specific SUMO knockdown suppresses global gene expression response and worsens functional outcome after transient forebrain ischemia in mice. *Neuroscience*, 343, 190–212. doi:10.1016/j.neuroscience.2016.11.036

Zhang, Y. Q., & Sarge, K. D. 2008. Sumoylation of amyloid precursor protein negatively regulates Abeta aggregate levels. *Biochem Biophys Res Commun*, 374(4), 673–678. doi:10.1016/j.bbrc.2008.07.109

Zhu, L. N., Qiao, H. H., Chen, L., et al. 2018. SUMOylation of alpha-synuclein influences on alpha-synuclein aggregation induced by methamphetamine. *Front Cell Neurosci*, 12, 262. doi:10.3389/fncel.2018.00262

CHAPTER NINE

NEDD8 and Oxidative Stress

Elah Pick and Giovanna Serino

CONTENTS

9.1 INTRODUCTION

All living organisms are exposed to oxidative stress produced by reactive oxygen species (ROS) (McCord, 2000; Pajares et al., 2015). ROS, which are highly unstable and reactive, play an important role in various cellular reactions and signals. The protein post-translation modifications (PTM) induced by ROS trigger progressive cell and tissue damage, eventually leading to pathophysiological defects (Devine et al., 2011; Sedelnikova et al., 2010). However, the presence of oxidized molecules does not necessarily imply stress, because many signaling pathways are triggered by ROS ("redox signaling") and act primarily to restore cellular homeostasis (Finkel, 2000). To maintain the cellular redox homeostasis, cells regulate the level of these free radicals through strictly controlled mechanisms, consisting of highly specific cellular antioxidant defense systems, including free radical neutralizing enzymes and antioxidant molecules (Antelmann and Helmann, 2011; Reichmann et al., 2018).

Oxidation of amino acids can lead to reversible or irreversible PTM, of which the first can play a significant role in redox signaling. Amino acids with a sulfur atom in their "R" group, such as cysteine or methionine, are known for their high susceptibility to oxidative modifications (Fra et al., 2017). Indeed, the cysteine thiol (Cys-thiol) group is one of the most reactive amino acid residues, and it is able to function as a reversible sensor of oxidation, translating redox status into specific cellular responses (Barford, 2004). Switches of Cys-thiols between their reduced and oxidized forms are involved in various cellular pathways, including signaling kinases, phosphatases, transcriptional regulation and the control of DNA damage repair (Stankovic-Valentin and Melchior, 2018). An abundant cellular mechanism that is driven by Cys-thiol-based enzymes is the PTM of proteins by ubiquitin (Ub) and Ub-like proteins (Ubls) (Hochstrasser, 2009; Tanaka et al., 1998). The attachment of Ub, SUMO/Smt3, NEDD8/Rub1 or other Ubls to target proteins

DOI: 10.1201/9781003048138-9

requires analogous cascades of E1–E3 enzymes. Within these cascades, Ubls are activated by an ATP-consuming E1 enzyme, which forms a Ubl–E1 thioester bond with a C-terminal Gly–Gly (GG) motif of the Ubl through an acceptor Cys-thiol (Cappadocia and Lima, 2017). In a second step, the Ubl is *trans*-thiolized from the E1 enzyme to a catalytic Cys-thiol of an E2-conjugating enzyme, forming once again an E2–Ubl thioester bond. Finally, the activated Ubl is transferred to a Lys residue of target proteins in a step mediated by a substrate-specific E3 ligase (Cappadocia and Lima, 2017). The presence of Cys-thiol residues in the enzymes' active site raises the possibility that oxidation might interfere with Ub and Ubl attachment cascades. However, it has been proposed that Ubl E2 enzymes are not targeted by oxidation, because of the relatively high pKa of their active sites. This high pKa has in fact been proposed as a regulatory mechanism to protect free E2s from exposure to oxidation (Tolbert et al., 2005). Still, few oxidation-driven alterations of Ubl cascade enzymes have been reported (Jahngen-Hodge et al., 1997; Stankovic-Valentin and Melchior, 2018; Yao et al., 2004), suggesting that the transient thiolate anions (RS$^-$) in Ubl E2 enzymes could be potentially exposed to oxidation signals during the *trans*-thiolation step. This chapter discusses and summarizes the current knowledge on the sensitivity of Ubl enzymatic cascades to oxidation and their involvement in the anti-oxidation response. Particular emphasis will be given to the emerging data on NEDD8/Rub1 conjugation and deconjugation enzymes and on the involvement of their activities in the antioxidant signaling pathways.

9.2 THE CELLULAR REDOX STATE AND ITS EFFECT ON UBL ENZYME ACTIVITY

Although Ubl enzymatic cascades harbor thiol-based enzymes, there is limited evidence that they are targets of reversible oxidation (Brandes et al., 2009; Stankovic-Valentin and Melchior, 2018). Still, several examples for negative or positive regulation of Ubls function by oxidation exist, as described below.

Comprehensive evidence for manipulation of a Ubl pathway by oxidation came from studies on SUMO. SUMO is a Ubl that modifies proteins, altering their function or localization (Cappadocia and Lima, 2017). The induction of ROS production by hydrogen peroxide (H_2O_2) inhibits human SUMOylation in a reversible manner. The exposure to low ROS concentrations of H_2O_2 leads to the formation of a reversible disulfide bridge between the catalytic cysteines of the human SUMO-E1 and SUMO-E2 enzymes, thus interfering with their activity (Bossis and Melchior, 2006). Reversibility of this thiol bridge was confirmed in a cell-free assay, through the use of reducing chemicals such as dithiothreitol, as well as *in vitro*, by adding reduced glutathione, which plays a key role in the detoxification of various electrophilic compounds in living cells. This reversible thiol-switch is required for cell survival, especially when high ROS levels induce genotoxic modifications of DNA (Stankovic-Valentin et al., 2016). Indeed, the SUMO E1–E2 thiol bridge that is formed during oxidation has been suggested as a redox-based signal mechanism for cell survival, essential to sustain activity of the ATM-Chk2 DNA damage repair pathway during oxidation (Stankovic-Valentin et al., 2016).

In the case of Ub, it has been found that exposure to ROS leads to oxidation of active cysteines of the UBA1 and Ub E2 enzymes of the eye retina cells until their recovery by the antioxidant glutathione, resulting in Ub E1 (UBA1) hyperactivation and increased ubiquitination (Jahngen-Hodge et al., 1997; Obin et al., 1998). An interesting observation was made on some specific Ub E2 enzymes, which display differential sensitivities to oxidizing chemicals. For example, Cdc34, the yeast Ub E2 for Cullin RING E3 ligases (CRLs), is highly sensitive to ROS, while Ubc4 E2 is not. The differential sensitivity of these enzymes suggests that the redox state of a cell is part of a mechanism that directs Ub toward different Ub E2s for accurate and selective ubiquitination of specific substrates (Doris et al., 2012). In certain cases, Ub E2s can function as game-changers of the oxidative stress response, as it was shown in the case of UbE2E3, a Ub E2 enzyme that regulates the nuclear factor (erythroid-derived 2)-like 2 (NRF2) activity. NRF2 is a transcription factor that induces the expression of antioxidant genes to neutralize ROS and restore redox homeostasis (will be further discussed below). Alkylation of the non-catalytic Cys 136 residue of UbE2E3 (to mimic its oxidation) results in constitutive UbE2E3-NRF2 binding, thus decreasing NRF2 turnover and stabilizing its transcriptional activity (Plafker et al., 2010).

Unlike the abovementioned inhibitory effect of ROS on SUMO and Ub cascade enzymes, oxidation can sometimes also lead to activation of Ubl cascades. A unique example was shown by Zhao et al. in 2018, who described an allosteric hyperactivation of UbE2N, a genuine Ub E2 enzyme, through thiol oxidation of its Ub E2 variant-binding partner Ube2V2 (Zhao et al., 2018). In contrast to the conventional Cys-thiol-modifications of E2 enzymes that regulate ubiquitination, this modification of Ube2V2 allosterically hyperactivates the enzymatic function of its binding partner UbE2N. This in turn promotes K63-linked ubiquitination of H2AX and stimulates the H2AX-dependent DNA damage response. Hence, similar to the SUMO E1–E2 thiol bridge during oxidation, also in this case, oxidative stress regulates the DNA damage response, but it does it by activating the Ubl E2 rather than blocking its activity (Bossis and Melchior, 2006; Zhao et al., 2018). Another example for oxidation as a signal for activation of Ubls has been reported for Urm1, which interestingly functions as a sulfur carrier. Unlike other Ubls, the E1 of Urm1 is uniquely activated by oxidation, in a mechanism occurring *via* the formation of a thiocarboxilate intermediate at its C-terminus (Van der Veen et al., 2011).

9.3 THE NEDD8/RUB1 CATALYTIC CASCADE OF ENZYMES IS SENSITIVE TO ROS

Of all Ubls, NEDD8 is the closest orthologue of Ub in sequence and in structure (Whitby et al., 1998). However, Ub and NEDD8 have distinct functions. Ub modifies a large number of substrates, whereas most of our knowledge on NEDD8 substrates is related to one family of proteins: cullins. Cullins are the scaffold component of CRLs, which are responsible for up to 20% of all ubiquitinated substrates, as was reported for human HCT116 colorectal carcinoma cells (Soucy et al., 2009). In CRL complexes, the cullin subunit (Cul1-3, 4a/b, 5 and 7 in human; Cdc53/yCul1, yCul3 and Rtt101/yCul4 in *Saccharomyces cerevisiae*) functions as a modular scaffold which is attached through its N-terminus to an adaptor and substrate receptor (SR) module and through its C-terminus to a RING box E3 protein (Rbx1/Rbx2) (Mahon et al., 2014). Assembled CRL complexes allow the conjugation of Ub from Ub-E2 to the protein target.

NEDD8 is translated as a precursor and it requires a specific hydrolytic cysteine protease (Yuh1/UCHL3 in *S. cerevisiae*; UCHL3 and NEDP1 in humans) to expose the C-terminal GG motif (Table 9.1). Conjugation of mature NEDD8 (NEDD8-GG) to a specific lysine on cullins is mediated by a NEDD8-activating enzyme (NAE1), a heterodimeric E1 enzymatic complex and an E2 conjugating enzyme (Ubc12/UbE2M, UbE2F) (Table 9.1; Figure 9.1) (Finley et al., 2012; Laplaza et al., 2004; Liakopoulos et al., 1998). It has been shown that among the genes encoding the two mammalian E2 enzymes, the expression of Ubc12/UbE2M, but not that of UBE2F, is induced by various stressors. Indeed, UbE2M regulates the ubiquitination and degradation of UBE2F under stress conditions (Zhou et al., 2018) and some lines of evidence suggest that accumulated ROS can inhibit UbE2M function in the NEDDylation pathway, although the molecular mechanism has not been detailed (Downs et al., 2013; Ly et al., 2013). Indeed, butyric acid secreted by bacteria in the human colon causes ROS increase, which leads to oxidative catalytic inactivation of UbE2M, resulting in the loss of Cul1 NEDDylation (Collier-Hyams et al., 2005; A. Kumar et al., 2009). The sensitivity of Cul1 NEDDylation cascade to oxidation is conserved from yeast to mammals. Bramasole et al. showed that in *S. cerevisiae* cells, yCul1 NEDDylation status is dramatically decreased during the diauxic shift, a molecular transition from anaerobic glycolysis to mitochondrial respiration accompanied by high ROS production. Indeed, Ubc12–NEDD8/Rub1 thioester forms are rare and hard to find at the diauxic shift and eventually correlate with the low NEDDylation status of yCul1 (Bramasole et al., 2019).

9.4 THE COP9 SIGNALOSOME CULLIN DENEDDYLASE AND ROS

Hydrolysis of NEDD8 from cullins is performed exclusively by the CSN, an evolutionarily conserved cullin-de-NEDDylating complex (Table 9.1) (Figure 9.1) (Cope et al., 2002; D. Dubiel et al., 2015; Lingaraju et al., 2014; Maytal-Kivity et al., 2002; Pick et al., 2009; Schmaler and Dubiel, 2010; Schwechheimer, 2018; Wei et al., 2008). The core CSN subunits (CSN1–8 in most eukaryotes) are broadly expressed and essential for viability among all studied multicellular organisms, and their loss-of-function mutants display critical pleiotropic defects such as abnormal response to DNA damage, defects in cell-cycle progression

TABLE 9.1
Comparison of the NEDDylation and deNEDDylation enzymes in S. cerevisiae, Arabidopsis thaliana *and human*

Function	Other names	*Saccharomyces cerevisiae*	*A. thaliana*	Human	References
NEDD8		RUB1	RUB1, 2, 3	NEDD8	Laplaza et al. (2004), Rao-Naik et al. (1998), Kumar et al. (1993)
NEDDylation enzymes					
E1	NAE		AXR1/AXL1	APP-BP1	Pozo et al. (1998), Gong et al. (1999)
		Ula/Uba3	ECR1	NAEβ/UBA3	Dharmasiri (2007), Gong et al. (1999)
E2		Ubc12	RCE1	UBE2M/UBC12	Lyakopoulos (1998), Huang (2009), Zhou et al. (2018), Dharmasiri et al. (2003), Gong et al. (1999)
		N/A	N/A	UBE2F	Huang et al. (2009)
co-E3		Dcn1		DCN1	Kurz et al. (2005, 2008)
E3		Rbx1	RBX1	RBX1,2	Scott et al. (2014)
deNEDDylation enzymes					
C-terminal hydrolase		Yuh1	N/A	UCHL3	Johnston et al. (1999)
DeNEDDylase		N/A	DEN1	NEDP1/SENP8	Christmann et al. (2013), Mergner et al. (2015), Gan-Erdene et al. (2003)
Cullin deNEDDylase	CSN	CSN (Rri1/ Csn5)	CSN (CSN5)	CSN (CSN5)	Wee et al. (2002), Maytal-Kivity et al. (2002), Schwechheimer et al. (2001), Lyapina et al. (2001), Cope et al. (2002)

Abbreviations: AXL1, AXR1-like; AXR1, auxin-resistant 1; CSN, COP9 signalosome; CSN5, COP9 signalosome subunit 5; DCN1, defective in cullin NEDDylation 1; DEN1, de-NEDDylase 1; ECR, E1 C-terminal related 1; NAE, NEDD8 E1-activating enzyme; NEDD8, neural precursor cell-expressed developmentally downregulated 8; RCE1, RUB1-conjugating enzyme 1; RUB1–3, related to ubiquitin 1–3; SENP8, Sentrin-specific protease 8; UBE2F, ubiquitin-conjugating enzyme 2 F; UBE2M, ubiquitin-conjugating enzyme 2 M; UCHL3, ubiquitin C-terminal hydrolase 3; YUH1, yeast ubiquitin hydrolase 1

and development (Nezames and Deng, 2012; Oron et al., 2002; Schwechheimer and Isono, 2010).

The de-NEDDylase activity center is located within a conserved metalloprotease enzymatic motif in the catalytic subunit Csn5/Jab1, known as MPN+ (**M**pr1, **P**ad1 **N**-terminal)/JAMM (**JAB**1/**MPN**/**M**ov34). Cullin deNEDDylation is carried out only when Csn5 is integrated into the CSN and upon attachment to a CRL (Ambroggio et al., 2004; Cope et al., 2002; Scheel and Hofmann, 2005; Schmaler and Dubiel, 2010). In human, CSN-CRL interactions lead to a series of conformational changes within the CSN complex, starting with Csn4 and Csn7, triggering rearrangements in Csn6 and finally priming the 104 Glu residue in Csn5 for deNEDDylation (Cavadini et al., 2016; Lingaraju et al., 2014). Interestingly, although structural studies suggest a relatively conserved disulfide bridge within the MPN+/JAMM motif under oxidizing conditions (Cao et al., 2017), a

link between this oxidation and CSN activity has not yet been identified.

CSN activity is highly conserved from yeast to mammals. Remarkably, *S. cerevisiae* and human CSN complexes demonstrate cross-species activity by reciprocal hydrolysis of NEDD8 from CRLs (Pick et al., 2012; Wee et al., 2002). Additionally, the *S. cerevisiae* NEDD8 and Cdc34 (Ub E2 for CRLs) could be functionally swapped with their human orthologues (Toro et al., 2013). In line with the above, CSN5i-3, a specific inhibitor designed for the human Csn5 (Schlierf et al., 2016), targets the *S. cerevisiae* orthologue, as well. While studying the mechanism of action of CSN5i-3 in *S. cerevisiae*, it was shown that the CSN stays active even upon metabolic oxidization during mitochondrial respiration at the post-diauxic phase (Bramasole et al., 2019). This suggests that, unlike the ROS-sensitive NEDDylation pathway, CRL deNEDDylation by the CSN complex is largely

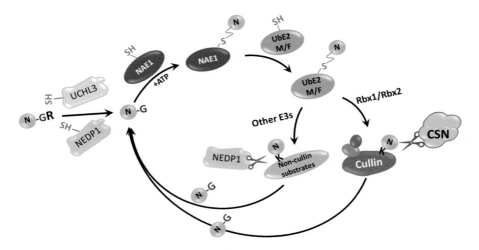

Figure 9.1 **NEDD8 modification cycle.** The precursor form of NEDD8/Rub1 (-GR) is processed by two hydrolases: NEDP1 and UCHL3, that edit NEDD8 to the mature form (-G). Subsequently, NEDD8 in an ATP-dependent reaction forms an intermediate thioester with a cysteine of NAE1 that *trans*-thiolation to UBE2M or UBE2F. Next, NEDD8 is transferred to a *specific lysine (K) on the substrate through an E3 ligase. De-NEDDylases such as COP9 signalosome (CSN) (for cullins) and* NEDP1 (for other substrates) deconjugate NEDD8 from the substrates. The terminology in this figure is mostly based on human.

ROS-insensitive, and it is in line with the findings that the CSN activity is implicated in the oxidative stress response in different organisms (Deng et al., 2011; Nahlik et al., 2010). However, some interesting observations made in the model plant *Arabidopsis thaliana* add a level of complexity to this story. In fact, in *Arabidopsis*, CRL NEDDylation increases upon light exposure (Christians et al., 2018) and during seed maturation, with a peak in dry seeds (Franciosini et al., 2015). Although not reported in the context of these experiments, both conditions could be related to oxidative stress: (i) Upon high light exposure, plants absorb more light than they can utilize in photosynthesis, resulting in overexcitation of the photosynthetic apparatus and formation of ROS such as the singlet oxygen, which cannot be completely quenched by carotenoids. (ii) The severe dehydration that occurs during seed maturation might limit carbon fixation and trigger ROS formation (Tripathy and Oelmuller, 2012). Accordingly, it is possible to speculate that unusual high ROS levels might hinder CSN function and lead to accumulation of NEDDylated CRLs, at least in some contexts. Nevertheless, *Arabidopsis csn* mutants accumulate NEDDylated CRLs and molecules with antioxidant properties, such as anthocyanins and ascorbic acid (AsA) (Wang et al., 2013). This suggests that CSN might have a specific function in redox perception and/or signaling through some of its known targets. Promising targets include the transcription factor LONG HYPOCOTYL5 (HY5), a key regulator of anthocyanin biosynthesis, and the AsA biosynthetic enzyme GDP-Man pyrophosphorylase (VTC1), which are both stabilized in *csn* mutants (Osterlund et al., 2000; Wang et al., 2013).

9.5 CRL SUBSTRATES THAT PLAY A ROLE IN THE ANTI-OXIDANT RESPONSE

Sensitivity of the NEDDylation cascade to ROS may also alter the turnover of CRL substrates. Indeed, under oxidizing conditions, Ubc12/UbE2M is not charged with NEDD8 and CRL1 remains inactive (see part 3 in this chapter). Inactivity of CRL1 leads to stabilization of the lung epithelial sodium channel (ENaC), a key player in the maintenance of osmotic balance in alveolar cells (Downs et al., 2013). Similarly, epithelia contaminated by bacteria rapidly generate ROS that inactivates UbE2M and consequently Cul1 NEDDylation and CRL1 activity. The loss of CRL1 activity results in stabilization of both β-catenin and the NF-κB inhibitor IκB-α, which cannot be recruited to the SR β-transducing repeat-containing protein (βTrCP) (Collier-Hyams et al., 2005; A. Kumar et al., 2009) (Figure 9.2A).

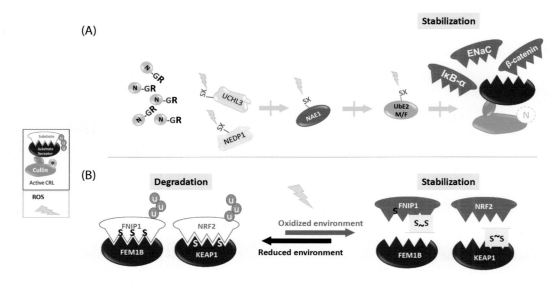

Figure 9.2 **Various mechanisms for CRLs inactivation during oxidation.** Two mechanistic models for CRLs inactivity during oxidative stress. (A) Many enzymes of the NEDDylation cascade such as NEDP1, UCHL3, NAE1 and UBE2M/UBE2F include a catalytic cysteine that could potentially be prone to oxidation (yellow bolt of lightning). Indeed, few CRLs, such as CRL1, become unNEDDylated upon the exposure to ROS leading to the accumulation of substrates such as IκB-α, β-catenin and ENaC. (B) Oxidative modification of cysteines can interfere with substrate recognition by SRs of cullin RING ligases (CRLs), as in the case of the substrates FNIP1 and NRF2 and their cognate SR Fem1 homologue beta protein (FEM1b) and KEAP1 (see text for details). ([A] from Downs et al., 2013; A. Kumar et al., 2009; [B] from Jaramillo and Zhang, 2013; Manford et al., 2020.)

In addition to the ROS-mediated blockage of the NEDDylation cascade, redox signaling directly regulates the turnover of central redox modulators that are also CRL substrates by hindering the interaction between them and their cognate CRL during hypoxic (HIF-1α), oxidative (NRF2) or reducing (Folliculin-interacting protein 1 (FNIP1)) environments in mammals (Lefaki et al., 2017; Manford et al., 2020; Pajares et al., 2015) (Figure 9.2B). These three redox-signaling players also control central metabolic pathways: HIF-1α increases glycolysis, NRF2 is involved in mitochondria biogenesis and FNIP1 fine-tunes mitochondrial respiration (Manford et al., 2020). These and other CRL substrates are commonly degraded in non-oxidized environments and protected from degradation during oxidative stress as described in Figure 9.2.

9.5.1 HIF-1α

Hypoxia is intimately related to oxidative stress. HIF-1α is an essential transcription factor in hypoxic response. HIF-1α is stable under hypoxic conditions and activates the transcription of genes that drive developmental programs such as angiogenesis and erythropoiesis (Ivan et al., 2001). Yet, in normal oxygen levels (normoxia), HIF-1α is ubiquitinated for proteasomal degradation by a CRL E3 Ub ligase containing Cul2, the adaptor subunits Elongin B (EloB) and C (EloC), and the SR subunit Von Hippel–Lindau tumor suppressor protein (VHL). VHL recognizes HIF-1α only when it is modified by prolyl hydroxylation in its degradation domain (degron) by specific prolyl hydroxylases (PHD) (Bruick and McKnight, 2001). Mutations in VHL that disrupt its interaction with the HIF-1α degron lead to the VHL disease and renal cell carcinomas (Kaelin, 2002; Nordstrom-O'Brien et al., 2010).

9.5.2 NRF2

NRF2 is a transcription factor that activates the cellular antioxidant response and hence is commonly considered a major regulator of cell survival (Antelmann and Helmann, 2011). Indeed, NRF2 protects against various diseases, including cancer, diabetes, pulmonary and cardiovascular diseases (Motohashi and Yamamoto, 2004).

PROTEOSTASIS AND PROTEOLYSIS

Normally, the levels of NRF2 are low since it is constitutively ubiquitinated by the Kelch-like erythroid cell-derived protein with CNC homology [ECH]-Associated Protein 1 (KEAP1), a SR of CRL3, and then degraded by the proteasome. During oxidative stress, two of the KEAP1 Cys-thiol residues become oxidized, leading to conformational changes that disrupt the KEAP1-NRF2 interaction and stabilize NRF2, which enters the nucleus for transcriptional activity (Jaramillo and Zhang, 2013). It is known that CRL SRs are prone to self-ubiquitination, which could lead to their proteasomal degradation and accordingly to substrate stabilization. Yet, as long as the degradation of NRF2 is needed, KEAP1 levels are protected from self-ubiquitination by deubiquitination (DUB) activity of Ub-specific protease 15 (USP15), which stabilizes the CRL3-KEAP1 complex (Villeneuve et al., 2013). Interestingly, USP15 associates with the CSN complex (W. Dubiel et al., 2020). Altogether, the above suggest that NRF2 is regulated by the CSN at two levels: (i) the CSN terminates CRL3-KEAP1 activity through Cul3 deNEDDylation; and (ii) the CSN associated DUB, USP15, shields KEAP1 from degradation by its DUB. In addition, in several circumstances, NRF2 is recognized for ubiquitination also through the tryptophan (W) aspartic (D) acid Repeat protein 23 (WDR-23) SR of CRL4. This alternative ubiquitination pathway has been suggested as a parallel mechanism required to finetune NRF2 levels in cases of loss of its KEAP1-dependent regulation (Lo et al., 2017).

9.5.3 FNIP1

FNIP1 was recently suggested as a mitochondrial gatekeeper, required to counteract oxidative stress. Stabilization of FNIP1 induces changes in mitochondrial morphology and leads to decrease in mitochondria-derived ROS (Manford et al., 2020). Similar to NRF2 and HIF-1α, the recognition of FNIP1 by a CRL SR depends on a redox sensitive degron. The studies by Manford et al. (2020) showed that in muscle cells, upon cellular oxidation, three Cys-thiol residues of FNIP1 become oxidized, hence leading to substrate stabilization. On the other hand, in a reduced environment achieved through a thiol-based redox switch, the three cysteine thiolate residues are recognized by the FEM1b SR of CRL2 for ubiquitination followed by proteasomal degradation. FNIP1 degradation leads to excessive entrance of metabolites into

mitochondria, increased mitochondria activity, which leads to accelerated oxidation (Manford et al., 2020).

9.5.4 NPR1

Redox signaling regulates CRL activity also in plants, as shown in the case of NPR1 (nonexpressor of pathogenesis-related genes 1), a transcription factor that senses the cellular redox state and helps fight pathogens in *Arabidopsis*. Under normal conditions, NPR1 is sequestered in the cytoplasm as a large redox-sensitive oligomer that is formed by intermolecular disulfide linkages between conserved cysteine residues (Mou et al., 2003). The redox changes triggered by salicylic acid (SA), a phytohormone produced in response to pathogen attack, induce the release of NPR1 from the oligomers to an active, monomeric form by reduction of the intermolecular disulfide bridges. Once NPR1 monomers enter the nucleus, they can promote transcription of the NPR-1 gene and simultaneously get ubiquitinated by an abscisic acid (ABA)-induced CRL3, resulting in their degradation by the proteasome. This indicates that two different plant hormones – ABA and SA – might play opposite roles in NPR1 activity (Spoel et al., 2009).

9.6 NON-CULLIN NEDD8 SUBSTRATES AND THEIR RESPONSE TO STRESS

Recent studies have documented CRL-independent roles for NEDD8, which can be found covalently attached to non-cullin substrates during various natural cellular stress responses. These studies led to the postulation of a secondary role for NEDP1 as the deNEDDylase that detaches NEDD8 from non-cullin substrates (Table 9.1) (Coleman et al., 2017; Gan-Erdene et al., 2003; Schwechheimer, 2018). A decade ago, VHL was identified as directly modified and stabilized by NEDD8. Considering that VHL is the SR of CRL2 for HIF-1α degradation (see Section 9.1 in this chapter), these data suggest NEDD8 impact on HIF-1α function via two regulatory mechanisms: (i) enhancement of HIF-1α ubiquitination and turnover by the NEDDylation of Cul2 in the CRL2 VHL complex and (ii) reduction of HIF-1α ubiquitination and turnover upon the self-NEDDylation of VHL. Both mechanisms impact the function of HIF-1α on cell survival during hypoxia and oxidative stress (Ryu et al., 2011).

NEDD8 can also regulate non-cullin substrates during nucleolar stress caused by abnormal metabolic conditions or cytotoxic compounds. Upon nucleolar stress, ribosomal proteins (RPs), including RPL11 and RPL5, become increasingly bound to the p53 E3 ligase MDM2. This interaction blocks the ligase activity of MDM2 and causes stabilization and activation of p53, which mediates the nucleolar stress response (Holmberg Olausson et al., 2012). It was shown that NEDDylation protects RPL11 from degradation, since non-NEDDylated RPL11 accumulates in the nucleoplasm (Sundqvist et al., 2009). Complementary studies showed that myeloma overexpressed 2 (Myeov2) associates with RPL11 to hinder its NEDDylation and leading to its translocation to the nucleus to modulate nucleolar integrity (Ebina et al., 2013).

NEDDylation of non-cullin substrates is also a critical anti-viral stress response and it triggers initiation of immunity in the Zebrafish *Danio rerio*. The latter stress response is initiated through the activation of IRF3, IRF7 and NF-κB transcription factors (Tan et al., 2018), and it has been suggested that IRF3 and IRF7 NEDDylation facilitate the antiviral response (Yu et al., 2019). However, NEDDylation *per se* does not always promote stress suppression, as it was shown for NEDDylation of the Serine-rich splicing factor 3 (SRSF3) during progressive liver disease. SRSF3 plays a critical role in liver function and regeneration, as its deletion leads to liver cirrhosis and hepatocellular carcinoma (Jayabalan et al., 2016; D. Kumar et al., 2019). Oxidative stress triggers SRSF3 poly-NEDDylation at Lys 11, following proteasomal degradation. Preventing SRSF3 NEDDylation *in vivo* leads to its stabilization and protection of mice from hepatic steatosis, fibrosis and inflammation (D. Kumar et al., 2019). This is in line with the findings by Bailly et al. (2019), who showed that decreased NEDP1 levels in hepatocellular carcinoma lead to accumulation of NEDD8 conjugates, while enhanced NEDP1 activity stabilizes SRSF3 by restricting its poly-NEDDylation at Lys 11, ultimately suppressing hepatic tumorigenesis.

An interesting finding is that NEDDylation of non-cullin substrates could be achieved not only by NAE1 but also through UBA1 and the Ub pathway. Indeed, mass spectrometry of stressed proteomes revealed the existence of mixed chains between NEDD8 and Ub, mediated mainly by UBA1 (Leidecker et al., 2012). This mixed chain formation has been termed "atypical

NEDDylation" (Enchev et al., 2015). The functional consequence of atypical NEDDylation is currently unknown. Yet, activation of atypical NEDDylation by overexpression of NEDD8 or treatment with H_2O_2 impairs proteasome function (Hjerpe et al., 2012; Li et al., 2015). Moreover, suppression of atypical NEDDylation by the negative regulator of NEDDylation, NEDD8 Ultimate Buster 1 Long (NUB1L), enhances proteasome function under stress conditions, suggesting that NEDDylation of ubiquitinated proteins impairs their proteasomal degradation (Li et al., 2015).

There is growing evidence for a link between cellular stress and the accumulation of NEDD8 molecules attached to non-cullin substrates through a canonical NEDDylation cascade. Indeed, oxidation has been suggested as a strong stress inducer of the formation of non-cullin NEDD8 conjugates (Leidecker et al., 2012; Li et al., 2015). Oxidative stress through the generation of ROS could be induced by exposure to arsenite, which leads to NEDDylation of few RPs, translation factors and RNA-binding proteins (RBPs). These NEDDylation events promote the assembly of stress granules (SGs), cytoplasmic aggregates that harbor translationally stalled messenger ribonucleoprotein (mRNP) (Jayabalan et al., 2016). Inhibition or depletion of key components of the NEDDylation machinery inhibits both stress-induced polysome disassembly and SG assembly.

It was recently shown that the electron transport chain flavoproteins ETFA and ETFB are NEDDylated, and that NEDDylation protects them from proteasomal degradation as well as from pathologies related to fatty acid oxidation disorders (Zhang et al., 2020). Oxidation by H_2O_2 leads to accumulation of NEDD8-modified proliferating cell nuclear antigen (PCNA) (Guan et al., 2018). PCNA NEDDylation promotes recruitment of polymerase η (polη) to bypass DNA lesions. The attachment of NEDD8 to PCNA is promoted by the E3 ligase RAD18 and antagonized by NEDP1. In response to H_2O_2 stimulation, NEDP1 disassociates from PCNA and RAD18-dependent PCNA NEDDylation increases markedly (Guan et al., 2018). In correlation with these findings, Keuss et al. (2019) showed sensitivity of NEDP1 to H_2O_2-induced oxidative stress, resulting in accumulation of unanchored NEDD8 trimers. These NEDD8 trimers bind and inhibit Poly ADP-Ribose Polymerase 1 (PARP1) to prevent the activation of apoptosis.

Consistently, inactivity of the *A. thaliana* and human orthologues of NEDP1 lead to the autoNED-Dylation of NEDD8 cascade enzymes, which has an inhibitory effect on the NEDDylation process (Coleman et al., 2017; Mergner et al., 2017).

9.7 CONCLUDING REMARKS

The interplay between NEDDylation of cullin and non-cullin substrates and the regulation of cellular redox status appears to be linked: cullin NEDDylation is decreased with ROS, and non-cullin NEDDylation increases. Indeed, NEDD8-mediated degradation of the two main regulators of ROS, NRF2 and FNIP1, is avoided during oxidation due to steric clashes with their associated CRL-SRs (Figure 9.2B). Furthermore, CRL activity is inhibited by ROS in a more general mechanism, based on a decrease in cullin NEDDylation status (Figure 9.2A). This inhibition leads to stabilization and activation of the NF-κB signal in order to transcribe inflammatory proteins. On the other side of the equilibrium, non-cullin substrates of NEDD8 are more NEDDylated during oxidative stress due to NEDP1 inhibition, which is required for cell survival. Overall, NEDD8 appears to play a critical and extensive role in maintaining the health of living cells during ROS accumulation.

ACKNOWLEDGMENTS

We would like to thank the Israel Ministry of Science and Technology (MOST) – Italy Ministry of Foreign Affairs (MAE) [grant no. 3-9022 to EP and GS] and the Israel Science Foundation [grant no. 162/17 for EP] for supporting our studies.

REFERENCES

Ambroggio, X. I., Rees, D. C., & Deshaies, R. J. 2004. JAMM: a metalloprotease-like zinc site in the proteasome and signalosome. *PLoS Biol*, 2(1), e2. Retrieved from http://dx.doi.org/10.1371/journal.pbio.0020002

Antelmann, H., & Helmann, J. D. 2011. Thiol-based redox switches and gene regulation. *Antioxid Redox Signal*, 14(6), 1049–1063. doi:10.1089/ars.2010.3400

Bailly, A. P., Perrin, A., Serrano-Macia, M., et al. 2019. The balance between mono- and NEDD8-chains controlled by NEDP1 upon DNA damage is a regulatory module of the HSP70 ATPase activity. *Cell Rep*, 29(1), 212–224.e218. doi:10.1016/j.celrep.2019.08.070

Barford, D. 2004. The role of cysteine residues as redox-sensitive regulatory switches. *Curr Opin Struct Biol*, 14(6), 679–686. doi:10.1016/j.sbi.2004.09.012

Bossis, G., & Melchior, F. 2006. Regulation of SUMOylation by reversible oxidation of SUMO conjugating enzymes. *Mol Cell*, 21(3), 349–357. doi:10.1016/j.molcel.2005.12.019

Bramasole, L., Sinha, A., Gurevich, S., et al. 2019. Proteasome lid bridges mitochondrial stress with Cdc53/Cullin1 NEDDylation status. *Redox Biol*, 20, 533–543. doi:10.1016/j.redox.2018.11.010

Brandes, N., Schmitt, S., & Jakob, U. 2009. Thiol-based redox switches in eukaryotic proteins. *Antioxid Redox Signal*, 11(5), 997–1014. doi:10.1089/ARS.2008.2285

Bruick, R. K., & McKnight, S. L. 2001. A conserved family of prolyl-4-hydroxylases that modify HIF. *Science*, 294(5545), 1337–1340. doi:10.1126/science.1066373

Cao, S., Engilberge, S., Girard, E., et al. 2017. Structural insight into ubiquitin-like protein recognition and oligomeric states of JAMM/MPN(+) proteases. *Structure*, 25(6), 823–833.e826. doi:10.1016/j.str.2017.04.002

Cappadocia, L., & Lima, C. D. 2017. Ubiquitin-like protein conjugation: structures, chemistry, and mechanism. *Chem Rev*. doi:10.1021/acs.chemrev.6b00737

Cavadini, S., Fischer, E. S., Bunker, R. D., et al. 2016. Cullin-RING ubiquitin E3 ligase regulation by the COP9 signalosome. *Nature*, 531(7596), 598–603. doi:10.1038/nature17416

Christians, M. J., Rottier, A., & Wiersma, C. 2018. Light regulates the RUBylation levels of individual cullin proteins in Arabidopsis thaliana. *Plant Mol Biol Report*, 36(1), 123–134. doi:10.1007/s11105-017-1064-9

Christmann, M., Schmaler, T., Gordon, C., et al. 2013. Control of multicellular development by the physically interacting deneddylases DEN1/DenA and COP9 signalosome. *PLoS Genet*, 9(2), e1003275. doi:10.1371/journal.pgen.1003275

Coleman, K. E., Bekes, M., Chapman, J. R., et al. 2017. SENP8 limits aberrant NEDDylation of NEDD8 pathway components to promote cullin-RING ubiquitin ligase function. *Elife*, 6. doi:10.7554/eLife.24325

Collier-Hyams, L. S., Sloane, V., Batten, B. C., et al. 2005. Cutting edge: bacterial modulation of epithelial signaling via changes in NEDDylation of cullin-1. *J Immunol*, 175(7), 4194–4198.

Cope, G. A., Suh, G. S., Aravind, L., et al. 2002. Role of predicted metalloprotease motif of Jab1/Csn5 in cleavage of NEDD8 from CUL1. *Science*, 298(5593), 608–611.

Deng, Z., Pardi, R., Cheadle, W., et al. 2011. Plant homologue constitutive photomorphogenesis 9 (COP9) signalosome subunit CSN5 regulates innate immune responses in macrophages. *Blood*, 117(18), 4796–4804. doi:10.1182/blood-2010-10-314526

Devine, M. J., Plun-Favreau, H., & Wood, N. W. 2011. Parkinson's disease and cancer: two wars, one front. *Nat Rev Cancer*, 11(11), 812–823. doi:10.1038/nrc3150

Dharmasiri, S., Dharmasiri, N., Hellmann, H., Estelle, M. The RUB/Nedd8 conjugation pathway is required for early development in Arabidopsis. 2003 *EMBO J*, 22(8):1762–1770. doi: 10.1093/emboj/cdg190.

Dharmasiri, N., Dharmasiri S., Weijers, D., Karunarathna, N., Jurgens, G., Estelle, M. 2007. AXL and AXR1 have redundant functions in RUB conjugation and growth and development in Arabidopsis. *Plant J*, 52(1), 114–123. doi: 10.1111/j.1365-313X.2007.03211.x.

Doris, K. S., Rumsby, E. L., & Morgan, B. A. 2012. Oxidative stress responses involve oxidation of a conserved ubiquitin pathway enzyme. *Mol Cell Biol*, 32(21), 4472–4481. doi:10.1128/MCB.00559-12

Downs, C. A., Kumar, A., Kreiner, L. H., et al. 2013. H_2O_2 regulates lung epithelial sodium channel (ENaC) via ubiquitin-like protein NEDD8. *J Biol Chem*, 288(12), 8136–8145. doi:10.1074/jbc.M112.389536

Dubiel, D., Rockel, B., Naumann, M., et al. 2015. Diversity of COP9 signalosome structures and functional consequences. *FEBS Lett*, 589(19 Pt A), 2507–2513. doi:10.1016/j.febslet.2015.06.007

Dubiel, W., Chaithongyot, S., Dubiel, D., et al. 2020. The COP9 signalosome: a multi-DUB complex. *Biomolecules*, 10(7). doi:10.3390/biom10071082

Ebina, M., Tsuruta, F., Katoh, M. C., et al. 2013. Myeloma overexpressed 2 (Myeov2) regulates L11 subnuclear localization through NEDD8 modification. *PLoS One*, 8(6), e65285. doi:10.1371/journal.pone.0065285

Enchev, R. I., Schulman, B. A., & Peter, M. 2015. Protein NEDDylation: beyond cullin-RING ligases. *Nat Rev Mol Cell Biol*, 16(1), 30–44. doi:10.1038/nrm3919

Finkel, T. 2000. Redox-dependent signal transduction. *FEBS Lett*, 476(1–2), 52–54. doi:10.1016/s0014-5793(00)01669-0

Finley, D., Ulrich, H. D., Sommer, T., et al. 2012. The ubiquitin-proteasome system of Saccharomyces cerevisiae. *Genetics*, 192(2), 319–360. doi:10.1534/genetics.112.140467

Fra, A., Yoboue, E. D., & Sitia, R. 2017. Cysteines as redox molecular switches and targets of disease. *Front Mol Neurosci*, 10, 167. doi:10.3389/fnmol.2017.00167

Franciosini, A., Moubayidin, L., Du, K., et al. 2015. The COP9 SIGNALOSOME is required for postembryonic meristem maintenance in Arabidopsis thaliana. *Mol Plant*, 8:1623-1634. doi:10.1016/j.molp.2015.08.003

Gan-Erdene, T., Nagamalleswari, K., Yin, L., et al. 2003. Identification and characterization of DEN1, a deneddylase of the ULP family. *J Biol Chem*, 278(31), 28892–28900. Retrieved from http://www.jbc.org/cgi/content/abstract/278/31/28892

Gong, L., Yeh, E.T. 1999. Identification of the activating and conjugating enzymes of the NEDD8conjugation pathway. *J Biol Chem*, 274(17):12036–12042. doi: 10.1074/jbc.274.17.12036.

Guan, J., Yu, S., & Zheng, X. 2018. NEDDylation antagonizes ubiquitination of proliferating cell nuclear antigen and regulates the recruitment of polymerase eta in response to oxidative DNA damage. *Protein Cell*, 9(4), 365–379. doi:10.1007/s13238-017-0455-x

Hjerpe, R., Thomas, Y., Chen, J., et al. 2012. Changes in the ratio of free NEDD8 to ubiquitin triggers NEDDylation by ubiquitin enzymes. *Biochem J*, 441(3), 927–936. doi:10.1042/BJ20111671

Hochstrasser, M. 2009. Origin and function of ubiquitin-like proteins. *Nature*, 458(7237), 422–429. doi:10.1038/nature07958

Holmberg Olausson, K., Nister, M., & Lindstrom, M. S. 2012. p53-dependent and -independent nucleolar stress responses. *Cells*, 1(4), 774–798. doi:10.3390/cells1040774

Huang, D.T., Ayrault, O., Hunt H.W. et al. 2009. E2-RING expansion of the NEDD8 cascade confers specificity to cullin modification. *Mol Cell*, 33(4),483–495. doi: 10.1016/j.molcel.2009.01.011.

Ivan, M., Kondo, K., Yang, H., et al. 2001. HIFa targeted for VHL-mediated destruction by proline hydroxylation: implications for O_2 sensing. *Science*, 292, 464–468.

Jahngen-Hodge, J., Obin, M. S., Gong, X., et al. 1997. Regulation of ubiquitin-conjugating enzymes by glutathione following oxidative stress. *J Biol Chem*, 272(45), 28218–28226. doi:10.1074/jbc.272.45.28218

Jaramillo, M. C., & Zhang, D. D. 2013. The emerging role of the Nrf2-Keap1 signaling pathway in cancer. *Genes Dev*, 27(20), 2179–2191. doi:10.1101/gad.225680.113

Jayabalan, A. K., Sanchez, A., Park, R. Y., et al. 2016. NEDDylation promotes stress granule assembly. *Nat Commun*, 7, 12125. doi:10.1038/ncomms12125

Johnston, S.C., Riddle, S.M., Cohen, R.E., Hill, C. 1999. Structural basis for the specificity of ubiquitin C-terminal hydrolases. *EMBO J*, 18, 3877–3887. doi: 10.1093/emboj/18.14.3877.

Kaelin, W. G., Jr. 2002. Molecular basis of the VHL hereditary cancer syndrome. *Nat Rev Cancer*, 2(9), 673–682. doi:10.1038/nrc885

Keuss, M. J., Hjerpe, R., Hsia, O., et al. 2019. Unanchored tri-NEDD8 inhibits PARP-1 to protect from oxidative stress-induced cell death. *EMBO J*, 38(6). doi:10.15252/embj.2018100024

Kumar, A., Wu, H., Collier-Hyams, L. S., et al. 2009. The bacterial fermentation product butyrate influences epithelial signaling via reactive oxygen species-mediated changes in cullin-1 NEDDylation. *J Immunol*, 182(1), 538–546. Retrieved from http://www.ncbi.nlm.nih.gov/pubmed/19109186

Kumar, D., Das, M., Sauceda, C., et al. 2019. Degradation of splicing factor SRSF3 contributes to progressive liver disease. *J Clin Invest*, 129(10), 4477–4491. doi:10.1172/JCI127374

Kurz T., Chou Y.C., Willems A.R., et al. 2008. Control of multicellular development by the physically interacting deneddylases DEN1/DenA and COP9 signalosome. *Mol Cell*, 29(1), 23–35. doi: 10.1016/j.molcel.2007.12.012.

Liakopoulos, D., Doenges, G., Matuschewski, K., Jentsch, S. 1998. A novel protein modification pathway related to the ubiquitin system. *EMBO J*, 17(8), 2208–2214. doi: 10.1093/emboj/17.8.2208. PMID: 9545234

Ly, Y. K., McIntosh, C. J., Biasio, W., et al. 2013. Regulation of the delta and alpha epithelial sodium channel (ENaC) by ubiquitination and NEDD8. *J Cell Physiol*, 228(11), 2190–2201. doi:10.1002/jcp.24390

Laplaza, J. M., Bostick, M., Scholes, D. T., et al. 2004. Saccharomyces cerevisiae ubiquitin-like protein Rub1 conjugates to cullin proteins Rtt101 and Cul3 in vivo. *Biochem J*, 377, 459–467.

Lefaki, M., Papaevgeniou, N., & Chondrogianni, N. 2017. Redox regulation of proteasome function. *Redox Biol*, 13, 452–458. doi:10.1016/j.redox.2017.07.005

Leidecker, O., Matic, I., Mahata, B., et al. 2012. The ubiquitin E1 enzyme Ube1 mediates NEDD8 activation under diverse stress conditions. *Cell Cycle*, 11(6), 1142–1150. doi:10.4161/cc.11.6.19559

Li, J., Ma, W., Li, H., et al. 2015. NEDD8 ultimate buster 1 long (NUB1L) protein suppresses atypical NEDDylation and promotes the proteasomal degradation of misfolded proteins. *J Biol Chem*, 290(39), 23850–23862. doi:10.1074/jbc.M115.664375

Liakopoulos, D., Doenges, G., Matuschewski, K., et al. 1998. A novel protein modification pathway related to the ubiquitin system. *EMBO J*, 17, 2208–2214.

Lingaraju, G. M., Bunker, R. D., Cavadini, S., et al. 2014. Crystal structure of the human COP9 signalosome. *Nature*, 512(7513), 161–165. doi:10.1038/nature13566

Lo, J. Y., Spatola, B. N., & Curran, S. P. 2017. WDR23 regulates NRF2 independently of KEAP1. *PLoS Genet*, 13(4), e1006762. doi:10.1371/journal.pgen.1006762

Lyapina, S., Cope, G., Shevchenko, A., et al. 2001. Promotion of NEDD-CUL1 conjugate cleavage by COP9 signalosome. *Science*, 18, 292(5520), 1382–1385. doi: 10.1126/science.1059780.

Mahon, C., Krogan, N. J., Craik, C. S., et al. 2014. Cullin E3 ligases and their rewiring by viral factors. *Biomolecules*, 4(4), 897–930. doi:10.3390/biom4040897

Manford, A. G., Rodriguez-Perez, F., Shih, K. Y., et al. 2020. A cellular mechanism to detect and alleviate reductive stress. *Cell*. doi:10.1016/j.cell.2020.08.034

Maytal-Kivity, V., Piran, R., Pick, E., et al. 2002. COP9 signalosome components play a role in the mating pheromone response of S. cerevisiae. *EMBO Rep*, 12(3), 1215–1221.

McCord, J. M. 2000. The evolution of free radicals and oxidative stress. *Am J Med*, 108(8), 652–659. Retrieved from http://www.ncbi.nlm.nih.gov/pubmed/10856414

Mergner, J., Heinzlmeir, S., Kuster, B., Schwechheimer, C. 2015 DENEDDYLASE1 deconjugates NEDD8 from non-cullin protein substrates in *Arabidopsis thaliana*. *Plant Cell*, 27(3), 741–753. doi: 10.1105/tpc.114.135996

Mergner, J., Kuster, B., & Schwechheimer, C. 2017. DENEDDYLASE1 protein counters automodification of NEDDylating enzymes to maintain NEDD8 protein homeostasis in Arabidopsis. *J Biol Chem*, 292(9), 3854–3865. doi:10.1074/jbc.M116.767103

Motohashi, H., & Yamamoto, M. 2004. Nrf2-Keap1 defines a physiologically important stress response mechanism. *Trends Mol Med*, 10(11), 549–557. doi:10.1016/j.molmed.2004.09.003

Mou, Z., Fan, W., & Dong, X. 2003. Inducers of plant systemic acquired resistance regulate NPR1 function through redox changes. *Cell*, 113(7), 935–944. doi:10.1016/s0092-8674(03)00429-x

Nahlik, K., Dumkow, M., Bayram, O., et al. 2010. The COP9 signalosome mediates transcriptional and metabolic response to hormones, oxidative stress protection and cell wall rearrangement during fungal development. *Mol Microbiol*, 78(4), 964–979. doi:10.1111/j.1365-2958.2010.07384.x

Nezames, C. D., & Deng, X. W. 2012. The COP9 signalosome: its regulation of cullin-based E3 ubiquitin ligases and role in photomorphogenesis. *Plant Physiol.* doi:10.1104/pp.112.198879

Nordstrom-O'Brien, M., van der Luijt, R. B., van Rooijen, E., et al. 2010. Genetic analysis of von Hippel-Lindau disease. *Hum Mutat*, 31(5), 521–537. doi:10.1002/humu.21219

Obin, M., Shang, F., Gong, X., et al. 1998. Redox regulation of ubiquitin-conjugating enzymes: mechanistic insights using the thiol-specific oxidant diamide. *FASEB J*, 12(7), 561–569. Retrieved from https://www.ncbi.nlm.nih.gov/pubmed/9576483

Oron, E., Mannervik, M., Rencus, S., et al. 2002. COP9 signalosome subunits 4 and 5 regulate multiple pleiotropic pathways in *Drosophila melanogaster*. *Development*, 129(19), 4399–4409.

Osterlund, M. T., Hardtke, C. S., Wei, N., et al. 2000. Targeted destabilization of HY5 during light-regulated development of Arabidopsis. *Nature*, 405, 462–466.

Pajares, M., Jimenez-Moreno, N., Dias, I. H. K., et al. 2015. Redox control of protein degradation. *Redox Biol*, 6, 409–420. doi:10.1016/j.redox.2015.07.003

Pick, E., Hofmann, K., & Glickman, M. H. 2009. PCI complexes: beyond the proteasome, CSN, and eIF3 Troika. *Mol Cell*, 35(3), 260–264. Retrieved from http://www.ncbi.nlm.nih.gov/entrez/query.fcgi?cmd=Retrieve&db=PubMed&dopt=Citation&list_uids=19683491

Pick, E., Golan, A., Zimbler, J. Z., et al. 2012. The minimal deneddylase core of the COP9 signalosome excludes the Csn6 MPN(−) domain. *PLoS One*, 7(8), e43980. doi:10.1371/journal.pone.0043980

Plafker, K. S., Nguyen, L., Barneche, M., et al. 2010. The ubiquitin-conjugating enzyme UbcM2 can regulate the stability and activity of the antioxidant transcription factor Nrf2. *J Biol Chem*, 285(30), 23064–23074. doi:10.1074/jbc.M110.121913

Pozo, J.C., Timpte, C., Tan, S., et al. 1998. The ubiquitin-related protein RUB1 and auxin response in Arabidopsis. *Science*. 280 (5370), 1760–1763. doi:10.1126/science.280.5370.1760.PMID: 9624055

Rao-Naik C., delaCruz W., Laplaza J.M., et al AJ.J Biol Chem. 1. 998. The rub family of ubiquitin-like proteins. Crystal structure of *Arabidopsis* rub1and expression of multiple rubs in Arabidopsis. *AJ.J Biol Chem*, 273(52),34976–34982. doi: 10.1074/jbc.273.52.34976.

Kumar, S., Yoshida, Y., Noda, M. 1993. Cloning of a cDNA which encodes a novel ubiquitin-like protein. *Biochem*,195(1), 393–399. doi: 10.1006/bbrc.1993.2056.

Reichmann, D., Voth, W., & Jakob, U. 2018. Maintaining a healthy proteome during oxidative stress. *Mol Cell*, 69(2), 203–213. doi:10.1016/j.molcel.2017.12.021

Ryu, J. H., Li, S. H., Park, H. S., et al. 2011. Hypoxia-inducible factor alpha subunit stabilization by NEDD8 conjugation is reactive oxygen species-dependent. *J Biol Chem*, 286(9), 6963–6970. doi:10.1074/jbc.M110.188706

Scheel, H., & Hofmann, K. 2005. Prediction of a common structural scaffold for proteasome lid, COP9-signalosome and eIF3 complexes. *BMC Bioinform*, 6, 71. Retrieved from http://www.ncbi.nlm.nih.gov/entrez/query.fcgi?cmd=Retrieve&db=PubMed&dopt=Citation&list_uids=15790418

Schlierf, A., Altmann, E., Quancard, J., et al. 2016. Targeted inhibition of the COP9 signalosome for treatment of cancer. *Nat Commun*, 7, 13166. doi:10.1038/ncomms13166

Schmaler, T., & Dubiel, W. 2010. Control of De-NEDDylation by the COP9 Signalosome. *Subcell Biochem*, 54, 57–68. doi:10.1007/978-1-4419-6676-6_5

Schwechheimer, C. 2018. NEDD8-its role in the regulation of Cullin-RING ligases. *Curr Opin Plant Biol*, 45(Pt A), 112–119. doi:10.1016/j.pbi.2018.05.017

Schwechheimer, C., & Isono, E. 2010. The COP9 signalosome and its role in plant development. *Eur J Cell Biol*, 89(2-3), 157–162. doi:10.1016/j.ejcb.2009.11.021

Schwechheimer, C., Serino, G., Callis J, et al. 2001. Interactions of the COP9 signalosome with the E3 ubiquitin ligase SCFTIRI in mediating auxin response. *Science*, 292(5520), 1379–1382. doi: 10.1126/science.1059776.

Scott DC, Sviderskiy VO, Monda JK, Lydeard JR, Cho SE, Harper JW, Schulman BA.*Cell*. 2014. Structure of a RING E3 trapped in action reveals ligation mechanism for the ubiquitin-like protein NEDD8. Jun 19;157(7):1671–1684. doi: 10.1016/j.cell.2014.04.037.

Sedelnikova, O. A., Redon, C. E., Dickey, J. S., et al. 2010. Role of oxidatively induced DNA lesions in human pathogenesis. *Mutat Res*, 704(1-3), 152–159. doi:10.1016/j.mrrev.2009.12.005

Soucy, T. A., Smith, P. G., & Rolfe, M. 2009. Targeting NEDD8-activated cullin-RING ligases for the treatment of cancer. *Clin Cancer Res*, 15(12), 3912–3916. doi:10.1158/1078-0432.CCR-09-0343

Spoel, S. H., Mou, Z., Tada, Y., et al. 2009. Proteasome-mediated turnover of the transcription coactivator NPR1 plays dual roles in regulating plant immunity. *Cell*, 137(5), 860–872. doi:10.1016/j.cell.2009.03.038

Stankovic-Valentin, N., & Melchior, F. 2018. Control of SUMO and ubiquitin by ROS: signaling and disease implications. *Mol Aspects Med.* doi:10.1016/j.mam.2018.07.002

Stankovic-Valentin, N., Drzewicka, K., Konig, C., et al. 2016. Redox regulation of SUMO enzymes is required for ATM activity and survival in oxidative stress. *EMBO J,* 35(12), 1312–1329. doi:10.15252/embj.201593404

Sundqvist, A., Liu, G., Mirsaliotis, A., et al. 2009. Regulation of nucleolar signalling to p53 through NEDDylation of L11. *EMBO Rep,* 10(10), 1132–1139. Retrieved from http://www.ncbi.nlm.nih.gov/entrez/query.fcgi?cmd=Retrieve&db=PubMed&dopt=Citation&list_uids=19713960

Tan, X., Sun, L., Chen, J., et al. 2018. Detection of microbial infections through innate immune sensing of nucleic acids. *Annu Rev Microbiol,* 72, 447–478. doi:10.1146/annurev-micro-102215-095605

Tanaka, K., Suzuki, T., & Chiba, T. 1998. The ligation systems for ubiquitin and ubiquitin-like proteins. *Mol Cells,* 8(5), 503–512. Retrieved from http://www.ncbi.nlm.nih.gov/pubmed/9856335

Tolbert, B. S., Tajc, S. G., Webb, H., et al. 2005. The active site cysteine of ubiquitin-conjugating enzymes has a significantly elevated pKa: functional implications. *Biochemistry,* 44(50), 16385–16391. doi:10.1021/bi0514459

Toro, T. B., Toth, J. I., & Petroski, M. D. 2013. The cyclomodulin cycle inhibiting factor (CIF) alters cullin NEDDylation dynamics. *J Biol Chem,* 288(21), 14716–14726. doi:10.1074/jbc.M112.448258

Tripathy, B. C., & Oelmuller, R. 2012. Reactive oxygen species generation and signaling in plants. *Plant Signal Behav,* 7(12), 1621–1633. doi:10.4161/psb.22455

Van der Veen, A. G., Schorpp, K., Schlieker, C., et al. 2011. Role of the ubiquitin-like protein Urm1 as a noncanonical lysine-directed protein modifier. *Proc Natl Acad Sci U S A,* 108(5), 1763–1770. doi:10.1073/pnas.1014402108

Villeneuve, N. F., Tian, W., Wu, T., et al. 2013. USP15 negatively regulates Nrf2 through deubiquitination of Keap1. *Mol Cell,* 51(1), 68–79. doi:10.1016/j.molcel.2013.04.022

Wang, J., Yu, Y., Zhang, Z., et al. 2013. Arabidopsis CSN5B interacts with VTC1 and modulates ascorbic acid synthesis. *Plant Cell,* 25(2), 625–636. doi:10.1105/tpc.112.106880

Wee, S., Hetfeld, B., Dubiel, W., et al. 2002. Conservation of the COP9/signalosome in budding yeast. *BMC Genet,* 3, 15. Retrieved from http://www.ncbi.nlm.nih.gov/entrez/query.fcgi?cmd=Retrieve&db=PubMed&dopt=Citation&list_uids=12186635

Wei, N., Serino, G., & Deng, X. W. 2008. The COP9 signalosome: more than a protease. *Trends Biochem Sci,* 33(12), 592–600. Retrieved from http://www.ncbi.nlm.nih.gov/entrez/query.fcgi?cmd=Retrieve&db=PubMed&dopt=Citation&list_uids=18926707

Whitby, F. G., Xia, G., Pickart, C. M., et al. 1998. Crystal structure of the human ubiquitin-like protein NEDD8 and interactions with ubiquitin pathway enzymes. *J. Biol. Chem.,* 273(52), 34983–34991. doi:10.1074/jbc.273.52.34983

Yao, D., Gu, Z., Nakamura, T., et al. 2004. Nitrosative stress linked to sporadic Parkinson's disease: S-nitrosylation of parkin regulates its E3 ubiquitin ligase activity. *Proc Natl Acad Sci U S A,* 101(29), 10810–10814. doi:10.1073/pnas.0404161101

Yu, G., Liu, X., Tang, J., et al. 2019. NEDDylation facilitates the antiviral response in zebrafish. *Front Immunol,* 10, 1432. doi:10.3389/fimmu.2019.01432

Zhang, X., Zhang, Y. L., Qiu, G., et al. 2020. Hepatic NEDDylation targets and stabilizes electron transfer flavoproteins to facilitate fatty acid beta-oxidation. *Proc Natl Acad Sci U S A,* 117(5), 2473–2483. doi:10.1073/pnas.1910765117

Zhao, Y., Long, M. J. C., Wang, Y., et al. 2018. Ube2V2 is a Rosetta stone bridging redox and ubiquitin codes, coordinating DNA damage responses. *ACS Cent Sci,* 4(2), 246–259. doi:10.1021/acscentsci.7b00556

Zhou, W., Xu, J., Tan, M., et al. 2018. UBE2M is a stress-inducible dual E2 for NEDDylation and ubiquitylation that promotes targeted degradation of UBE2F. *Mol Cell,* 70(6), 1008–1024.e1006. doi:10.1016/j.molcel.2018.06.002

CHAPTER TEN

Structure, Function and Regulators of the 20S Proteasome

Tobias Jung and Annika Höhn

CONTENTS

10.1 INTRODUCTION

Metabolism is always accompanied by the formation of highly reactive particles able to virtually damage any cellular structure (Aprioku, 2013). Furthermore, pathological changes, redox shifts and oxidative stress are able to increase the formation of reactive particles, resulting in enhanced protein oxidation to which the cell needs to adapt and counteract. Consequently, cellular systems able to preserve functionality of the whole protein pool (proteome) are necessary. One of these systems is the "ubiquitin-proteasome system" (UPS), representing one of the two main cellular proteolytic machineries found in all three kingdoms of evolution (animals, plants and bacteria, while the latter are using prokaryotic ubiquitin-like (UBL) proteins [Barandun et al., 2017; Maupin-Furlow, 2014; Striebel et al., 2014]). The main task of the UPS is the recognition and proteolytic removal of damaged or no longer needed proteins. The process is so fundamental that the 20S proteolytic "core" is found even in the oldest known bacteria (*archaea*) (Becker and Darwin, 2017; Jastrab and Darwin, 2015). The UPS is composed of the

ubiquitin-conjugation system, several deubiquitinases, the 20S proteasome itself and the 19S regulator (Tanaka, 2009). Studies of the proteasome during the past decades provided far-reaching insights into proteasome structure and functions and the system of different regulators formed during evolution.

10.2 STRUCTURE, FUNCTION AND SUBSTRATES OF THE 20S COMPLEX

The 20S proteasome is a large proteolytic complex formed by 28 subunits overall (Kunjappu and Hochstrasser, 2014). The main function of 20S is the recognition and proteolytic degradation of oxidatively damaged proteins to maintain cellular functionality and to prevent accumulation of dysfunctional proteins (Pickering and Davies, 2012). The 20S proteasome degrades client proteins, generating oligopeptides ranging in length from 2 to 24 amino acid residues, while their abundance decreases with length according to a log-normal distribution (Kisselev et al., 1999). The resulting peptide products are then hydrolyzed to amino

DOI: 10.1201/9781003048138-10

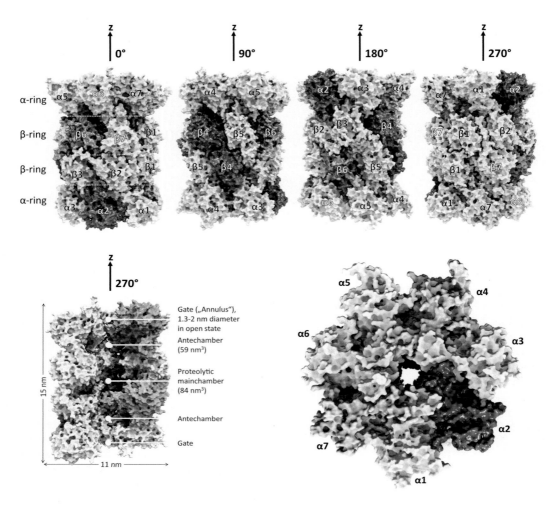

Figure 10.1 **Bovine 20S proteasome.** The upper row shows the bovine 20S proteasome, rotated in 90-degree steps. The α- and β-subunits are labeled accordingly. The lower row shows a sectional model of 20S with the foremost subunits removed (left side), exposing the two antechambers as well as the main chamber, harboring the active β-subunits ($β_1$, $β_2$, $β_5$). The right image of the bottom row shows a top view of 20S, the open central gate ("annulus") and the N-terminal ends of the α-subunits that regulate gate opening. (Adapted from Unno et al., 2002.)

acids by oligopeptidases and/or amino-carboxyl peptidases.

When viewed with an electron microscope, the 20S "core" proteasome is a cylindrical structure, formed of four stacked rings containing seven subunits, each as depicted in Figure 10.1. The four rings are arranged in the sequence α–β–β–α. α-Rings are composed of α-subunits only, and β-rings contain only β-subunits. The main function of the α-rings is substrate recognition and binding as well as strict regulation of substrate access to the inner proteolytic chamber formed by the β-rings.

The "ancient" archaea proteasome consists of a single type of both α- and β-subunit, thus both the α- and β-rings are homoheptamers ($α_7$ and $β_7$, respectively) (Kasahara and Flajnik, 2019). The proteolytic chamber of the archaeal 20S is also formed by two β-rings, containing 2 × 7 identical subunits, with their active centers facing the inner side of the proteasomal cylinder. In contrast, the eukaryotic mammalian proteasome is formed by seven different α ($α_1$–$α_7$) and β ($β_1$–$β_7$) subunits (Tanaka, 2009). Only three of those β-subunits, namely $β_1$, $β_2$ and $β_5$, carry specific proteolytic activities resulting to 2 × 3 different proteases in the proteolytic chamber. Furthermore, the proteolytic β-subunits have different proteolytic specificities: $β_1$ has a peptidyl-glutamyl-peptide-hydrolyzing-activity (caspase-like, cleaves after acidic amino acids, also termed as post-glutamyl-peptide hydrolytic-activity), $β_2$ provides a trypsin-like activity (cleaves after basic amino acids) and $β_5$ a chymotrypsin-like activity (cleaves after neutral amino acids). The individual proteolytic capacities are also different: $β_5 \gg β_2 \geq β_1$ (Jager et al., 1999).

All proteolytic subunits are threonine proteases, harboring an N-terminal Thr1 residue in their active center. The 28 subunits of an assembled 20S proteasome cumulate a molecular weight of about 700 kDa; the dimensions of the whole complex are about 15 × 11.5 nm (=150 × 115 Å) (Unno et al., 2002).

During evolution, the 20S core gained new functions through mutations of its subunits, development of new inducible subunits, as well as by formation of additional regulatory proteins and complexes that are able to modulate both activity and (substrate) specificity of the 20S proteasome. Different types of proteolytically active subunits can be induced, replacing the constitutive forms via *de novo* synthesis of new proteasomes. The "inducible" forms are iβ1 (also termed as PSMB9 or LMP2, for "low molecular weight protein 2"), iβ2 (PSMB10, LMP10 or MECL-1 for "multicatalytic endopeptidase complex-like 1") and iβ5 (PSMB8 or LMP7) that can be expressed in response to inflammation or infections, mediated by interferon-γ, TNF-α or lipopolysaccharides (interpret by the immune system as bacterial invasion). To discriminate between the two forms, the housekeeping constitutive proteasome is termed as "c20S" and the inducible form(s) as "i20S", or "immunoproteasome".

The inducible subunits are preferably incorporated into the assembling proteasome due to a higher affinity compared to the constitutive subunits. Thus, after several days of interferon-γ exposure, the whole c20S pool in murine liver can be replaced by i20S (Jung et al., 2009; Khan et al., 2001). In contrast to c20S that has a half-life of 10–12 days (Tanaka and Ichihara, 1989), the half-life of i20S is only 27 h, ensuring a fast removal of the inducible form. Short inflammatory stimuli shift the ratio of c20S/i20S in favor of i20S while during aging an increased share of i20S can also be observed (Fernando et al., 2019; Mishto et al., 2003). This shift in composition

can be explained by an increased amount of damaged proteins resulting from acute oxidative stress, or due to (age-related) pathologies associated with chronic inflammation, that are both able to induce enhanced formation of i20S. Compared to c20S, i20S exhibits higher trypsin-like and chymotrypsin-like and lower caspase-like activities (Hussong et al., 2010; Visekruna et al., 2006). Oligopeptides produced by immunoproteasomes are 8–10 residues long containing mainly hydrophobic or basic C-termini and appear to be more efficient in binding to MHC class I in order to be presented to the immune system on the cell's surface (Arellano-Garcia et al., 2014). Thus, immunoproteasomes are believed to enhance the generation of those antigenic peptides (Tanaka and Kasahara, 1998; Vigneron and Van den Eynde, 2012; F. Zhou, 2009).

According to Mishto et al. (2014), there are only minor differences between the pools of antigenic oligopeptides produced by i20S and c20S (Zanker et al., 2013), but significant differences in their proteolytic rates. Those results are still under discussion, since other authors state that i20S favors cleavage after hydrophobic and basic amino acids, resulting in oligopeptides better suited for MHC-I presentation (Hulpke and Tampe, 2013).

S-glutathionylation is an important redox-regulated posttranslational modification of 20S proteasomal cysteine residues. In yeast, the residues Cys76 and Cys221 of the α5-subunit were shown to be S-glutathionylated in a ROS-dependent manner, increasing the proteasomal activity via gate opening (Silva et al., 2012).

The composition of the proteasomal system is also redox-regulated. Slight oxidative stress induces an antioxidative Nrf2-mediated response augmenting proteasome activity via upregulation of the c20S and 11S regulator subunits expression (Pickering et al., 2012). In contrast, severe oxidative stress, mediated by NFκB, amplifies immunoproteasomal activity (Wright et al., 1995). The majority of mammalian cells express proteasomes containing both constitutive and inducible subunits, present in various amounts (Mishto and Liepe, 2017).

As mentioned above, substrate recognition, binding and "feeding" of a substrate into the proteolytic chamber are accomplished with the help of α-rings. In the center of each α-ring, the so-called proteasomal "gate" regulates the substrate access (Figure 10.1). The gate is formed by the

N-terminal ends of the α-subunits. $α_2$, $α_3$ and $α_4$ are mainly involved, containing the YDR-motif (Tyr_8-Asp_9-Arg_{10}), jointly inducing gate opening. Without any binding of substrate or regulators, the substrate access is blocked and the proteasome is inactive. Yeast deletion mutants of $α_3$ (missing the last 9 amino-residues) demonstrate a constitutively active proteasome, while deletion mutants of $α_7$ do not show any significant enhancement of the proteolytic activity.

Gate opening can be induced by substrates, binding of proteasome regulators or even exposure to low sodium dodecyl sulfate (SDS) concentrations, a denaturing detergent, leading to slight unfolding of the N-terminal ends and increased proteasome activity by a predominance of the "open gate" conformation (Osmulski et al., 2009).

The first type of 20S proteasome substrates is at least partially unfolded, misfolded and/or oxidatively damaged proteins that expose hydrophobic structures which are normally buried inside a soluble, natively folded protein. The second type of substrates are proteins whose unfolded regions are an intrinsic part of their native folding (Kumar Deshmukh et al., 2019), the intrinsically disordered proteins (IDPs). Proteolytic degradation of such partially unfolded substrates to oligopeptides by the 20S proteasome is ATP-independent.

There is adequate evidence for an allosteric behavior of the 20S complex, including interactions of its catalytic centers, as well as of the α- and the proteolytic β-subunits (Rechsteiner and Hill, 2005; Schmidtke et al., 2000). Additionally, allosteric coupling between the catalytic chamber and the 20S-19S regulator interface has been described.

10.3 UBIQUITINATION

For the recognition and proteolysis of a native substrate, special labeling is required, that is provided by the ubiquitin-conjugation system, a highly complex machinery involving thousands of different proteins, also responsible for substrate specificity (Chaugule and Walden, 2016; Iconomou and Saunders, 2016). Labeling of substrates occurs via covalent attachment of a short chain of ubiquitin (polyubiquitination), to the lysine residues of the target proteins (Audagnotto and Dal Peraro, 2017).

The proteins responsible for ubiquitination can be subdivided into four different classes of enzymes that are termed as **E1–E4**. The whole process is depicted in Figure 10.2.

To date, only eight different mammalian E1 enzymes are known, two of which are found in humans. **E1** proteins are responsible for ubiquitin activation where ubiquitin (a small protein of 76 amino acids, Ub) is attached via formation of a thioester-bond (C-terminal $glycine_{76}$ of Ub) to a cysteine in the center of E1. **E2** enzymes mediate the next step, namely ubiquitin-conjugation. About 50–75 mammalian forms of E2s are known (Ristic et al., 2014), 40 of them can be found in humans. **E3** enzymes mediate transfer of Ub to a substrate. An overall of about 650 E3 enzymes is known; 600 are classified as RING (Really Interesting New Gene)-type, and 50 as HECT (homologous to the E6-AP carboxyl terminus)-type (Ashizawa et al., 2012; Metzger et al., 2014). E3 enzymes deliver the substrate specificity of the UPS by recognition of a just few or even only a single specific substrate. The Ub-loaded E2 and the E3 that binds the substrate form a complex; in the case of a HECT-E3, the Ub is first transferred from E2 to E3 and then from E3 to the substrate, whereas in the case of a RING-E3, the Ub is directly transferred from E2 to the substrate.

In both cases, Ub is finally attached to the substrate via an amide bond, linking the C-terminus of Ub to a lysine of the target protein. After the first Ub is attached to a substrate, **E4** enzymes elongate the chain. A Ub_4 chain provides the strongest degradation signal for the 26S proteasome (Zhao and Ulrich, 2010). Nevertheless, few proteins can also be degraded by 26S after monobiquitination (like paired box 3 and syndecan 4) or after attachment of multiple single molecules of Ub (like the NF-κB precursor p105 or phospholipase D) (Kravtsova-Ivantsiv and Ciechanover, 2012). Moreover, some proteins can be degraded without any ubiquitination, such as the myeloid cell leukemia 1 (MCL1) (Stewart et al., 2010), the CCAAT/enhancer-binding protein δ (C/EBPδ) (S. Zhou and Dewille, 2007) or the ornithine decarboxylase (ODC) (Murakami et al., 1992).

How exactly a single substrate-attached Ub is extended to a short Ub chain is still discussed. While it is proposed that only a subset of E2 enzymes is able to attach Ub to another Ub molecule (Clague et al., 2015), other authors suggest that elongation is mediated by E4 enzymes, a subgroup of the E3 pool (Schwartz and Ciechanover, 2009).

PROTEOSTASIS AND PROTEOLYSIS

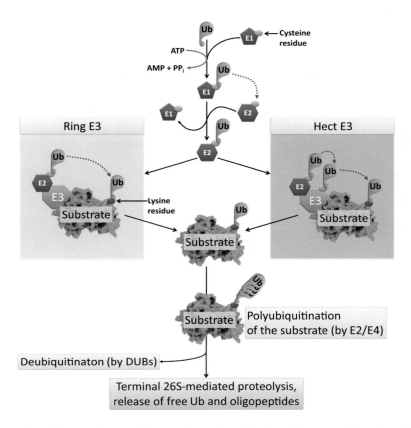

Figure 10.2 Function of the UPS. Ubiquitin (Ub, a small peptide used by the cell for posttranslational modification of proteins) is activated by E1 enzyme and transferred to E2 enzyme. E2 ligase binds the substrate-E3 ligasecomplex and the Ub bound to E2 enzyme is either transferred firstly to E3 and from E3 to the substrate (HECT E3 ubiquitination) or directly from E2 to the substrate (RING E3 ubiquitination). The Ub attached to the monoubiquitinated substrate is prolonged to a Ub chain (4 or more Ub-molecules) in a process mediated by E2 or E4 enzymes; this serves as a degradation signal for the 26S proteasome. After polyubiquitination, the substrate can either be deubiquitinated by appropriate enzymes (DUBs) or degraded by the 26S proteasome.

Importantly, polyubiquitination of a substrate does not result inevitably in 26S proteasome degradation, since there are also deubiquitinating enzymes (deubiquitinases, DUBs), able to remove an Ub chain from a potential 26S substrate. Consequently, DUBs represent another node of proteostasis control, determining the balance between synthesis and degradation (Ristic et al., 2014).

10.4 REGULATORS OF THE 20S PROTEASOME

Several different proteins and protein complexes are found in eukaryotic cells that can bind to one or both 20S α-rings. As mentioned, binding of the proteasome to its regulators can modulate both the 20S activity and specificity, as well as the characteristics of the oligopeptides released during protein degradation. All proteasome regulators are already present in the last eukaryotic common ancestor, namely the yeast.

10.4.1 The 19S Regulatory Complex (PA700)

The 19S regulator, one of the most prominent proteasome regulators, is a large protein complex containing at least 19 known subunits formed by 2 major sub-structures, a base and a lid. The whole complex is about 1 MDa. 20S can bind one or two 19S regulators. The 19S–20S complex (≈1.7 MDa) is termed "26S proteasome", the 19S-20S-19S complex (≈2.7 MDa) is technically a "30S proteasome", but in the literature, usually both configurations are termed as "26S". For a detailed depiction of 19S binding to the 20S proteasome, please refer

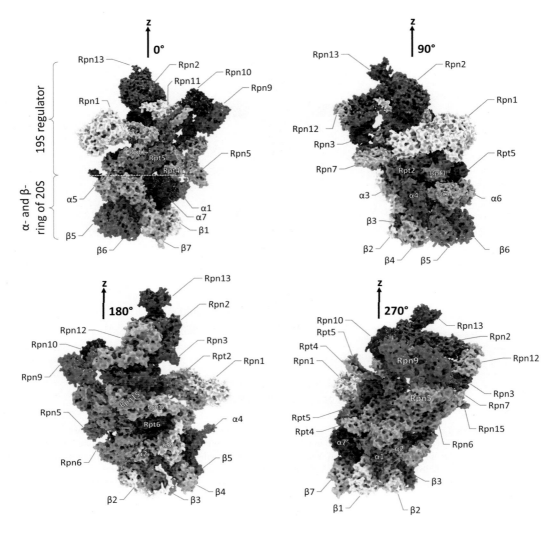

Figure 10.3 **The 19S regulator complex.** 19S regulator (above the line) bound to the human 20S proteasome (below the line, only the binding α- and attaching β-ring are shown). The single subunits of both 19S and 20S are labeled; the whole complex turns in 90-degree steps. The hexameric arranged ATPase subunits (Rpt1–6) of the 19S base bind to the 20S proteasomal α-ring and induce gate opening (by the subunits Rpt2, Rpt3 and Rpt5). The non-ATPase subunits Rpn1, Rpn2, Rpn10 and Rpn13 are also part of the base, while the lid is formed by Rpn3, Rpn5, Rpn7, Rpn8, Rpn9, Rpn11 and Rpn12. Substrate recognition and binding is done by the subunits Rpn10 and Rpn13. (Adapted from Unverdorben et al., 2014.)

to Figure 10.3. About 50% of the mammalian proteasomes in the cells are found uncapped, 30% are 19S-capped and the remaining 20% are associated with other regulators (Ben-Nissan and Sharon, 2014).

The "base", a hexameric ring formed by the ATPases Rpt1-6 (PSMC2, -1, -4, -6, -3, -5), and the non-ATPase-subunits PSMD1/Rpn2, PSMD2/Rpn1, PSMD4/Rpn10 and ADRM1/Rpn13, can bind to a proteasomal α-ring and induce gate opening, thus enhancing the 20S activity.

The "lid" contains the subunits PSMD3/Rpn3, PSMD6, -7, -8 (Rpn7, -8 and -12), PSMD11, -12, -13 and -14 (Rpn6, -5, -9 and -11). Substrates of the 26S proteasome are recognized and bound via the subunits Rpn10 and -13, while binding of 19S activates the 20S proteasome in a process involving the subunits Rpt2, -3 and 5 (Bedford et al., 2010).

Gate opening of the 20S "core" proteasome is induced via the HbYX-motif (Hb = a hydrophobic amino acid, Y = tyrosine, X = any amino acid)

(Kusmierczyk et al., 2011). The HbYX-motif is present at the C-termini of several proteasome regulators and enables binding of the regulator to cognate sites between the α-subunits and usually induces conformational changes of the α-ring like gate opening.

In contrast to uncapped 20S, the 26S proteasome is able to degrade natively folded and fully functional substrate proteins in an ATP-dependent manner. In the absence of ATP and Ca^{2+}, 19S complex detaches from the 20S complex (Aizawa et al., 1996; Cascio et al., 2002; Hohn et al., 2016; J. Y. Park et al., 2013). While NADPH binds and stabilizes the complex, the role of magnesium is still discussed (Tsvetkov et al., 2014). The ATP consumed by 26S for proteolysis of a natively folded substrate protein is not required for its fragmentation into oligopeptides, but rather for its unfolding.

As described in the previous section, the ubiquitin-conjugation system labels substrates by attaching short chains of polyubiquitin. If those single Ub molecules are linked via their Lys48 (K48), the substrate will be in most cases removed by the 26S, while Lys63 (K63) linkage usually results in lysosomal degradation. Both, (poly)-ubiquitination and deubiquitination of substrates, as well as the regulation of its balance are still poorly understood due to the high complexity of the UPS.

Recognition, binding and proteolysis of a polyubiquitinated substrate follow different steps (Grice and Nathan, 2016). First, the polyubiquitinated substrate is bound reversibly to 19S by the subunits Rpn10 and -13. Next, the substrate becomes deubiquitinated by 26S-associated enzymes such as Usp14, Uch37 or Rpn11 and is completely unfolded in an ATP-consuming manner, provided by the Rpt subunits of 19S. Finally, the unfolded substrate is translocated through the opened 20S gate to be degraded to oligopeptides.

Until now, six proteasome Ub receptors have been identified: Rpn1, Rpn10 and Rpn13 that are parts of the 19S complex, as well as three receptors that are reversibly associated with the 19S complex, namely Rad23, Dsk2 and Ddi1 (Jiang et al., 2018; Shi et al., 2016). Rpn1 is able to recognize both Ub chains as well as the UBL-binding sites of substrate-shuttling factors that deliver substrates to the 26S proteasome (Shi et al., 2016). Rpn1 recognizes and binds the UBL domain of the deubiquitinating enzyme Ubp6 and presents substrates to the proteasomal ATPases for unfolding.

Furthermore, the reversible phosphorylation of Rpn1 (at Ser_{361}) regulates 26S proteasomal assembly and function (Liu et al., 2020). Rpn10 binds both, K48- and K63-linked Ub chains, and plays a role in neurological diseases (Jiang et al., 2018). Deletion of Rpn10 in *Saccharomyces cerevisiae* influences the phenotype modestly without affecting viability (van Nocker et al., 1996). Like Rpn10, Rpn13 binds ubiquitin and the 26S proteasome (at the Rpn2 subunit of 19S) simultaneously. Rpn13 is considered as a drug target for proteasome inhibition, since Rpn13 inhibitors may lack the toxicity exhibited by common proteasome inhibitors targeting the proteolytically active 20S subunits (Jiang et al., 2018).

In *S. cerevisiae*, the Ub-receptor Rad23 is also involved in nucleotide excision repair, due to an interaction with the DNA damage recognition factor Rad4 (Wade and Auble, 2010). The N-terminal Ub-like (UBL) domain of this receptor can also physically interact with the Rpn1 subunit of the 19S complex (Kim et al., 2004). The two Ub-associated (UBA) domains of Rad23 can bind polyubiquitinated substrates, thus acting as a bridge to the 26S proteasome (Raasi et al., 2005). Proteins requiring Rad23 for degradation include many substrates determined for proteolysis by the ER-associated degradation machinery; other Rad23 targets are also involved in cell cycle regulation (Wilkinson et al., 2001). Consequently, loss of Rad23 shows phenotypic impact on protein quality control, DNA damage response, cellular metabolism and cell cycle (Wade and Auble, 2010).

Dsk2 (also termed as PLIC or ubiquilin) is also able to shuttle ubiquitinated substrates to the 26S proteasome (X. Chen et al., 2016). As in the other "shuttle" proteins (Rad23 and Ddi1), the UBA domain, a bundle of three α-helices, binds ubiquitinated substrates, while the UBL of Dsk2 binds to the Rpn10 26S subunit (Finley, 2009).

Ddi1 of *S. cerevisiae* and homologs in higher eukaryotes have been also proposed to be shuttling factors delivering polyubiquitinated substrates to the 26S proteasome. Although Ddi1 contains both, ubiquitin-binding UBA and proteasome-interacting UBL domains, the UBL domain is atypical, as it binds ubiquitin (Yip et al., 2020). In contrast to other "shuttling" proteins, Ddi1 contains a retroviral-like protease (RVP) domain, that is probably responsible for cleavage of the precursor of Nrf1 (a transcription factor) in higher eukaryotes, resulting in an upregulation of proteasome genes (Dirac-Svejstrup et al., 2020).

Rpn11 is a Zn^{2+}-dependent deubiquitinase (JAMM/MPN family) that removes the Ub chains attached to the substrate before entering the ring formed by the Rpt subunits where the targeted protein is unfolded and translocated into the 20S core (de Poot et al., 2017; Verma et al., 2002; T. Yao and Cohen, 2002). Rpn8 and Rpn11 form a heterodimer for substrate deubiquitination during 26S proteasome degradation (Worden et al., 2014). Rpn11 plays also a role in the mutual regulation of 26S proteasome and mitochondria; while mitochondrial stress induces disassembly of the 26S proteasome (Livnat-Levanon et al., 2014), mutations in Rpn11 induce fragmentation of the mitochondrial network (Rinaldi et al., 2004; Tar et al., 2014).

The energy necessary for substrate unfolding is provided by the Rpt1–6 heterohexameric "AAA+ ATPase motor" via ATP consumption.

Even if the 26S proteasome is able to unfold a native protein, the presence of an unstructured region, either at the terminus or as an internal flexible loop (20–30 residues), is a mandatory prerequisite for proteolysis (Bard et al., 2018). 26S-mediated degradation of a substrate is a process that may last from few minutes (e.g. the protein dihydrofolate dehydrogenase is degraded within 5 min) to several minutes (e.g. degradation of the I27 domain of titin takes about 40 min). The 12 ATPases in a 19S-20S-19S complex are estimated to consume about 300 molecules of ATP for the complete unfolding of the I27 domain (Henderson et al., 2011).

10.4.2 The 11S Regulatory Complex

The 11S (also termed as REG, PA28 or PA26 (in *Trypanosoma brucei*)) proteasome activator is found only in higher eukaryotes (Wang et al., 2020). Three different isoforms of the 11S subunits are known: α, β and γ. While α and β are found exclusively in the cytosol, γ is restricted to the nucleus (Stadtmueller and Hill, 2011). PA28αβ (heteroheptameric, formed of both α- and β-subunits) and PA28γ (homoheptamer, only γ-subunits) are the main regulatory complexes. The structure of a homoheptameric PA28α$_7$ complex bound to 20S is depicted in Figure 10.4.

According to Raule et al. (2014), binding of the PA28αβ-complex to the 20S core proteasome does not increase the overall proteolytic rate compared to the uncapped 20S, but rather it reduces the size of its degradation products and increases their hydrophilicity. Such fragments are suited for MHC-I antigen presentation on the cell's surface to cytotoxic CD8[+] T lymphocytes. In contrast, other publications suggest another central role of the 11S that involves the enhanced removal of damaged proteins upon oxidative stress, due to the higher i20S proteasome activity, further enhanced following 11S binding that induces gate opening (Di Cola, 1992; Dubiel et al., 1992; Kuehn and Dahlmann, 1996).

Both i20S and 11S (PA28αβ) are induced by interferon-γ (Pickering et al., 2010), so that in case of a viral infection, enhanced proteolytic capacity provides more antigenic fragments of viral proteins that can be presented on the cells' surface to enhance the immune response.

Interferon-γ is induced via a receptor-mediated stimulation or as response to early produced cytokines (like interleukins and IFN-α/β) or stimulated by T-cell receptors (TCRs) or natural killer (NK) cell receptors; INF's signal via transmembrane receptors is mainly transduced via Jak-Stat pathways, among others (Malmgaard, 2004).

The nuclear isoform of PA28 (PA28γ, PSME3) forms the homoheptameric complex PA28γ$_7$. PA28αβ-complexes, PA28γ$_7$, can also bind to either 20S or 26S complexes. However, most of PA28γ$_7$ is found unbound possibly performing proteasome-unrelated functions. Its role in regulation of nuclear bodies, trafficking splicing factors and chromatin dynamics have already been shown (Wang et al., 2020).

Previously it was assumed that PA28γ may degrade only short peptides (Ma et al., 1992), but recent evidence suggests that intact proteins can also be PA28γ targets (X. Chen et al., 2007; Li et al., 2006, 2007). How natively folded proteins are unfolded and translocated into the 20S core proteasome in an ATP-independent manner is still poorly understood (Mao et al., 2008).

10.4.3 The PA200 Regulator

The mammalian PA200 (proteasome activator 200 kDa, Figure 10.5, Blm10 – bleomycin resistance 10 or Blm10p in yeast, Figure 10.6) is a dome-shaped nuclear monomeric protein that binds the α-rings of 20S core inducing a conformational change (gate opening) via its C-terminal HbYX-motif and a loop

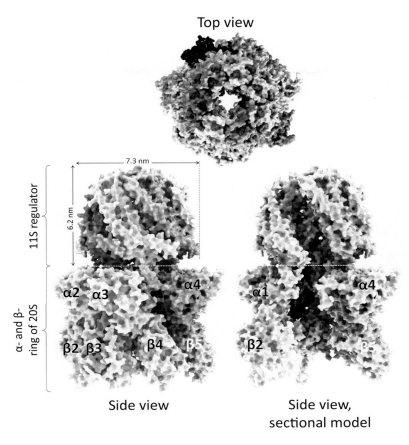

Top view

7.3 nm

6.2 nm

11S regulator

α- and β-
ring of 20S

α2 α3   α4   α1   α4

β2 β3   β4   β5   β2   β5

Side view

Side view,
sectional model

Figure 10.4 The 11 S regulator complex. The upper row displays a top view of the 11S regulator (here: PA28α$_7$) bound to the 20S proteasome. The right image of the second row depicts the complex with some 11S and 20S subunits removed, exposing the channel of 11S leading into the inner chamber of 20S, while the left image shows the whole complex.

(T652 to K574) that contacts the N-termini of the α5, α6, α1 and α2 20S subunits.

The human PA200 has a molecular weight of 200 kDa; three different PA200 isoforms have been identified (PA200i/ii/iii), but only PA200i seems to be associated with the 20S core (Blickwedehl et al., 2007).

Previously, a 23-Å-resolution electron microscopic study suggested that only six of the seven 20S α subunits interact with PA200, while the α$_7$ subunit does not (Ortega et al., 2005). According to Guan et al. (2020), α$_1$, α$_6$ and α$_7$ subunits have a smaller contact area than the other four subunits, while α$_7$ interacts with PA200 via its N-terminal loop only after a conformational change induced following its binding.

Two main anchor regions were identified by Toste Rego and da Fonseca (2019); one involves the C-terminus of PA200 (residues 1838-1844) including an extended peptide sequence upstream of the HbYX motif also found in other 20S binding proteins. The second one comprises an analogous binding pocket at the α1-α2 interface, docking an extended loop of PA200, formed by the residues 561–576.

Whether PA200 can facilitate substrate entrance into the 20S is still discussed (Fort et al., 2015), although according to Guan et al. (2020), PA200 reveals two openings formed by numerous positively charged residues that are probably the gates enabling the passage of substrates into the 20S core.

According to cryo-EM data by Guan et al. (2020), binding of PA200 to 20S is quite similar to Blm10, 11S and 19S complexes, suggesting a uniform type of interaction between those regulators and the 20S core that ultimately induces gate opening. However, PA200 has two different openings distinct from Blm10, one in the center and one located at the edge of this regulator.

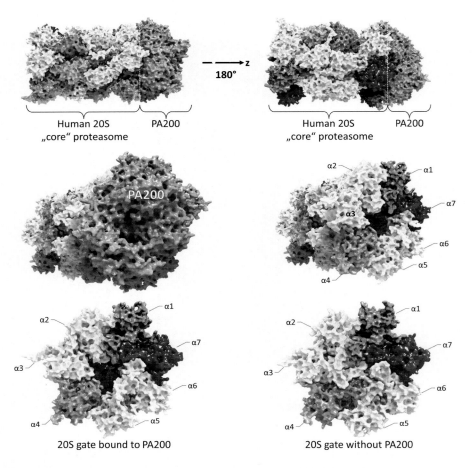

Figure 10.5 The proteasome activator PA200. The top row shows the human 20S proteasome bound to the proteasomal regulator PA200 (dome-shaped protein on the right side of 20S) and the same structure rotated by 180-degree. The second row depicts 20S bound to PA200 (left) and the opened structure of the proteasomal gate while bound to PA200 (right). The bottom row shows the "activated" and opened proteasomal gate by binding to PA200 (left image) and the closed gate of the "uncapped" 20S proteasome (right image). Binding to PA200 induces the main conformational changes of gate opening in the α2–α5 subunits. (Right image: Adapted from Guan et al., 2020.)

Ninety percent of the proteasomes isolated from testicular tissue are bound to either one or two PA200 regulators while in muscle it is only 7.9% (Qian et al., 2013). PA200 is found in three different complexes with the 20S complex: as PA200–20S–PA200, as PA200–20S and as PA200–20S–19S. The relative proportions are different in distinct organs and tissues (Guan et al., 2020; Qian et al., 2013). *In vitro* incubation of PA200 and 20S proteasome at different ratios results also in different proportions of the possible complexes; at a PA200:20S ratio of 4.4:1, 33.3% PA200–20S–PA200 complexes and 51% PA200–20S complexes are found, while 15.7% of 20S proteasome remains unbound. At a ratio of 8.8:1, there is a shift to 47.6% PA200–20S–PA200 complexes, 40.2% PA200–20S complexes and 12.2% unbound 20S proteasome, respectively.

Deletion of PA200 in mice abolishes acetylation-dependent proteolytic degradation of histones during DNA double-strand breaks and delays the removal of core histones in elongated spermatids. PA200 enhances the ATP-independent proteolysis of acetylated core histones but does not affect the degradation of polyubiquitinated substrates (Mandemaker et al., 2018). Furthermore, mitochondrial inhibitors such as rotenone and antimycin A are able to induce a redistribution of PA200 in the nucleus, resulting in withdrawal from certain gene promoters and binding to others. This

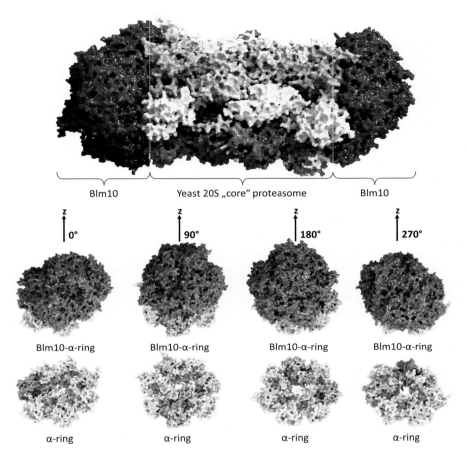

Figure 10.6 **The yeast Blm10 regulator.** The upper row displays the yeast 20S proteasome double capped with the Blm10 proteasomal regulator (dome-shaped monomer) attached to both of the 20S proteasomal α rings. The single proteasomal subunits are labeled (Sadre-Bazzaz et al., 2010). The middle row of images shows Blm10 rotated in 90-degree steps (from left to right), the lower line shows the bound α ring of 20S only, as well as the opened central gate.

suggests a correlation of the mitochondrial status with the PA200–chromatin interaction (Douida et al., 2020).

Welk et al. (2019) investigated the role of PA200 in myofibroblast differentiation and fibrotic tissue remodeling, showing both PA200 and the formation of PA200–proteasome complexes upregulated in hyperplastic basal cells and myofibroblasts of fibrotic lungs from patients with idiopathic pulmonary fibrosis.

The yeast Blm10 is about 250 kDa, revealing a difference of 301 residues compared to the human PA200 (Guan et al., 2020). Blm10 may facilitate gate opening via its penultimate C-terminal tyrosine/phenylalanine, displacing the Pro17 reverse turns of the proteasomal α subunits (Guan et al., 2020). Yeast lacking Blm10 is more susceptible

to DNA-damaging agents; in mammals PA200 is associated with DNA double-strand breaks repair, chromatin remodeling, acetylation-regulated histone degradation, maintenance of mitochondrial inheritance, spermatogenesis as well as proteasome maturation and assembly (Jung et al., 2009).

Recently, a promoting role of PA200/Blm10 was shown in the acetylation-dependent degradation of core histones during DNA repair and spermiogenesis (L. B. Chen et al., 2020). Elevated transcription of Blm10 reduced the amount of acetylated core histones during aging and prolonged the replicative cellular lifespan, suggesting that cells can counteract aging by an upregulation of PA200/Blm10. Moreover, reduced expression of PA200/Blm10 was identified as the leading cause of declining proteasomal activity during aging.

10.5 TISSUE SPECIFIC PROTEASOMES

Another tissue specific form of the proteasome is the so-called thymoproteasome (t20S), first identified by Murata et al. (2007). The thymoproteasome contains iβ1, iβ5 and tβ5 (PSMB11), a subunit only found in cortical thymic epithelial cells (cTECs). In cTECs, t20S plays an important role in the positive selection of T-cells (CD8$^+$T, which recognize and kill infected and aberrant cells), while expression of β5 or iβ5 is not sufficient to compensate for a lack of tβ5 (Murata et al., 2007; Nitta et al., 2010; Xing et al., 2013).

Another isoform of a 20S proteasomal subunit is the testis-specific α_4s (PSMA8, the paralog of PSMA7 (α_4)), essential for male fertility and expressed in spermatocytes from the pachytene stage (Zhang et al., 2019). The spermatoproteasomes contain the regular active subunits (β_1, β_2 and β_5) as well as the PA200 regulator; the subunit α_4s stimulates the acetylation-dependent degradation of core histones during spermiogenesis and somatic DNA repair (Gomez et al., 2019). Deletion of α_4s blocks both degradation of the core histones at DNA damage loci and meiotic DNA repair, resulting in meiotic arrest at metaphase I in spermatocytes (Zhang et al., 2019). α_4s is essential for meiotic DNA repair, meiosis and fertility in male at least partially by promoting histone degradation (Gomez et al., 2019).

10.6 CONCLUSIONS

The 20S proteasome, initially a necessary complex to maintain proteostasis, has been expanded in mammals to include numerous regulators, inducible subunits and a system for highly specific polyubiquitination of substrates. The UPS is virtually involved in every cellular function, the fine regulation of the protein pool while it co-evolved along with the immune system, one of the most complex systems of higher life. Following understanding of its function, it has become a highly interesting therapeutic target. For example, proteasomal inhibitors like bortezomib are applied in multiple myeloma therapy. Nutlin-3a or idasanutlin that inhibits MDM2 (it polyubiquitinates the tumor suppressor p53 and induces its proteasomal degradation) is applied for treatment of glioblastoma (preclinical trials), diffuse large B cell lymphoma, multiple myeloma, prostate cancer (phase I/II clinical trial) and acute myeloid leukemia (in a phase III clinical trial combined with chemotherapy) (J. Park et al., 2020). MLN4924 (Pevonedistat) inhibits cullin-ring ligases, the largest E3 family that promotes polyubiquitination of about 20% of cellular proteins doomed to proteasomal degradation (Soucy et al., 2009), thereby suppressing cancer cell growth (phase I clinical trials for certain solid tumors as well as hematologic malignancies) and acute myelogenous leukemia (Swords et al., 2010), while it is also known to inhibit angiogenesis during tumor development (W. T. Yao et al., 2014).

REFERENCES

Aizawa, H., Kawahara, H., Tanaka, K., et al. 1996. Activation of the proteasome during Xenopus egg activation implies a link between proteasome activation and intracellular calcium release. *Biochem Biophys Res Commun*, 218(1), 224–228. doi:10.1006/bbrc.1996.0039

Aprioku, J. S. 2013. Pharmacology of free radicals and the impact of reactive oxygen species on the testis. *J Reprod Infertil*, 14(4), 158–172. Retrieved from https://www.ncbi.nlm.nih.gov/pubmed/24551570

Arellano-Garcia, M. E., Misuno, K., Tran, S. D., et al. 2014. Interferon-gamma induces immunoproteasomes and the presentation of MHC I-associated peptides on human salivary gland cells. *PLoS One*, 9(8), e102878. doi:10.1371/journal.pone.0102878

Ashizawa, A., Higashi, C., Masuda, K., et al. 2012. The ubiquitin system and Kaposi's sarcoma-associated herpesvirus. *Front Microbiol*, 3, 66. doi:10.3389/fmicb.2012.00066

Audagnotto, M., & Dal Peraro, M. 2017. Protein post-translational modifications: in silico prediction tools and molecular modeling. *Comput Struct Biotechnol J*, 15, 307–319. doi:10.1016/j.csbj.2017.03.004

Barandun, J., Damberger, F. F., Delley, C. L., et al. 2017. Prokaryotic ubiquitin-like protein remains intrinsically disordered when covalently attached to proteasomal target proteins. *BMC Struct Biol*, 17(1), 1. doi:10.1186/s12900-017-0072-1

Bard, J. A. M., Goodall, E. A., Greene, E. R., et al. 2018. Structure and function of the 26S proteasome. *Annu Rev Biochem*, 87, 697–724. doi:10.1146/annurev-biochem-062917-011931

Becker, S. H., & Darwin, K. H. 2017. Bacterial proteasomes: mechanistic and functional insights. *Microbiol Mol Biol Rev*, 81(1). doi:10.1128/MMBR.00036-16

Bedford, L., Paine, S., Sheppard, P. W., et al. 2010. Assembly, structure, and function of the 26S proteasome. Trends Cell Biol, 20(7), 391–401. doi:10.1016/j.tcb.2010.03.007

Ben-Nissan, G., & Sharon, M. 2014. Regulating the 20S proteasome ubiquitin-independent degradation pathway. Biomolecules, 4(3), 862–884. doi:10.3390/biom4030862

Blickwedehl, J., McEvoy, S., Wong, I., et al. 2007. Proteasomes and proteasome activator 200 kDa (PA200) accumulate on chromatin in response to ionizing radiation. Radiat Res, 167(6), 663–674. doi:10.1667/RR0690.1

Cascio, P., Call, M., Petre, B. M., et al. 2002. Properties of the hybrid form of the 26S proteasome containing both 19S and PA28 complexes. EMBO J, 21(11), 2636–2645. doi:10.1093/emboj/21.11.2636

Chaugule, V. K., & Walden, H. 2016. Specificity and disease in the ubiquitin system. Biochem Soc Trans, 44(1), 212–227. doi:10.1042/BST20150209

Chen, X., Barton, L. F., Chi, Y., et al. 2007. Ubiquitin-independent degradation of cell-cycle inhibitors by the REGgamma proteasome. Mol Cell, 26(6), 843–852. doi:10.1016/j.molcel.2007.05.022

Chen, X., Randles, L., Shi, K., et al. 2016. Structures of Rpn1 T1:Rad23 and hRpn13:hPLIC2 reveal distinct binding mechanisms between substrate receptors and shuttle factors of the proteasome. Structure, 24(8), 1257–1270. doi:10.1016/j.str.2016.05.018

Chen, L. B., Ma, S., Jiang, T. X., et al. 2020. Transcriptional upregulation of proteasome activator Blm10 antagonizes cellular aging. Biochem Biophys Res Commun, 532(2), 211–218. doi:10.1016/j.bbrc.2020.07.003

Clague, M. J., Heride, C., & Urbe, S. 2015. The demographics of the ubiquitin system. Trends Cell Biol, 25(7), 417–426. doi:10.1016/j.tcb.2015.03.002

de Poot, S. A. H., Tian, G., & Finley, D. 2017. Meddling with fate: the proteasomal deubiquitinating enzymes. J Mol Biol, 429(22), 3525–3545. doi:10.1016/j.jmb.2017.09.015

Di Cola, D. 1992. Human erythrocyte contains a factor that stimulates the peptidase activities of multicatalytic proteinase complex. Ital J Biochem, 41(4), 213–224. Retrieved from https://www.ncbi.nlm.nih.gov/pubmed/1428780

Dirac-Svejstrup, A. B., Walker, J., Faull, P., et al. 2020. DDI2 is a ubiquitin-directed endoprotease responsible for cleavage of transcription factor NRF1. Mol Cell, 79(2), 332–341.e337. doi:10.1016/j.molcel.2020.05.035

Douida, A., Batista, F., Robaszkiewicz, A., et al. 2020. The proteasome activator PA200 regulates expression of genes involved in cell survival upon selective mitochondrial inhibition in neuroblastoma cells. J Cell Mol Med, 24(12), 6716–6730. doi:10.1111/jcmm.15323

Dubiel, W., Pratt, G., Ferrell, K., et al. 1992. Purification of an 11S regulator of the multicatalytic protease. J Biol Chem, 267(31), 22369–22377. Retrieved from https://www.ncbi.nlm.nih.gov/pubmed/1429590

Fernando, R., Drescher, C., Nowotny, K., et al. 2019. Impaired proteostasis during skeletal muscle aging. Free Radic Biol Med, 132, 58–66. doi:10.1016/j.freeradbiomed.2018.08.037

Finley, D. 2009. Recognition and processing of ubiquitin-protein conjugates by the proteasome. Annu Rev Biochem, 78, 477–513. doi:10.1146/annurev.biochem.78.081507.101607

Fort, P., Kajava, A. V., Delsuc, F., et al. 2015. Evolution of proteasome regulators in eukaryotes. Genome Biol Evol, 7(5), 1363–1379. doi:10.1093/gbe/evv068

Gomez, H. L., Felipe-Medina, N., Condezo, Y. B., et al. 2019. The PSMA8 subunit of the spermatoproteasome is essential for proper meiotic exit and mouse fertility. PLoS Genet, 15(8), e1008316. doi:10.1371/journal.pgen.1008316

Grice, G. L., & Nathan, J. A. 2016. The recognition of ubiquitinated proteins by the proteasome. Cell Mol Life Sci, 73(18), 3497–3506. doi:10.1007/s00018-016-2255-5

Guan, H., Wang, Y., Yu, T., et al. 2020. Cryo-EM structures of the human PA200 and PA200-20S complex reveal regulation of proteasome gate opening and two PA200 apertures. PLoS Biol, 18(3), e3000654. doi:10.1371/journal.pbio.3000654

Henderson, A., Erales, J., Hoyt, M. A., et al. 2011. Dependence of proteasome processing rate on substrate unfolding. J Biol Chem, 286(20), 17495–17502. doi:10.1074/jbc.M110.212027

Hohn, A., Konig, J., & Jung, T. 2016. Metabolic syndrome, redox state, and the proteasomal system. Antioxid Redox Signal, 25(16), 902–917. doi:10.1089/ars.2016.6815

Hulpke, S., & Tampe, R. 2013. The MHC I loading complex: a multitasking machinery in adaptive immunity. Trends Biochem Sci, 38(8), 412–420. doi:10.1016/j.tibs.2013.06.003

Hussong, S. A., Kapphahn, R. J., Phillips, S. L., et al. 2010. Immunoproteasome deficiency alters retinal proteasome's response to stress. J Neurochem, 113(6), 1481–1490. doi:10.1111/j.1471-4159.2010.06688.x

Iconomou, M., & Saunders, D. N. 2016. Systematic approaches to identify E3 ligase substrates. Biochem J, 473(22), 4083–4101. doi:10.1042/BCJ20160719

Jager, S., Groll, M., Huber, R., et al. 1999. Proteasome beta-type subunits: unequal roles of propeptides in core particle maturation and a hierarchy of active site function. J Mol Biol, 291(4), 997–1013. doi:10.1006/jmbi.1999.2995

Jastrab, J. B., & Darwin, K. H. 2015. Bacterial proteasomes. Annu Rev Microbiol, 69, 109–127. doi:10.1146/annurev-micro-091014-104201

Jiang, T. X., Zhao, M., & Qiu, X. B. 2018. Substrate receptors of proteasomes. Biol Rev Camb Philos Soc, 93(4), 1765–1777. doi:10.1111/brv.12419

Jung, T., Catalgol, B., & Grune, T. 2009. The proteasomal system. Mol Aspects Med, 30(4), 191–296. doi:10.1016/j.mam.2009.04.001

Kasahara, M., & Flajnik, M. F. 2019. Origin and evolution of the specialized forms of proteasomes involved in antigen presentation. Immunogenetics, 71(3), 251–261. doi:10.1007/s00251-019-01105-0

Khan, S., van den Broek, M., Schwarz, K., et al. 2001. Immunoproteasomes largely replace constitutive proteasomes during an antiviral and antibacterial immune response in the liver. J Immunol, 167(12), 6859–6868. doi:10.4049/jimmunol.167.12.6859

Kim, I., Mi, K., & Rao, H. 2004. Multiple interactions of rad23 suggest a mechanism for ubiquitylated substrate delivery important in proteolysis. Mol Biol Cell, 15(7), 3357–3365. doi:10.1091/mbc.e03-11-0835

Kisselev, A. F., Akopian, T. N., Woo, K. M., et al. 1999. The sizes of peptides generated from protein by mammalian 26 and 20 S proteasomes: implications for understanding the degradative mechanism and antigen presentation. J Biol Chem, 274(6), 3363–3371. doi:10.1074/jbc.274.6.3363

Kravtsova-Ivantsiv, Y., & Ciechanover, A. 2012. Noncanonical ubiquitin-based signals for proteasomal degradation. J Cell Sci, 125(Pt 3), 539–548. doi:10.1242/jcs.093567

Kuehn, L., & Dahlmann, B. 1996. Proteasome activator PA28 and its interaction with 20S proteasomes. Arch Biochem Biophys, 329(1), 87–96. doi:10.1006/abbi.1996.0195

Kumar Deshmukh, F., Yaffe, D., Olshina, M. A., et al. 2019. The contribution of the 20S proteasome to proteostasis. Biomolecules, 9(5). doi:10.3390/biom9050190

Kunjappu, M. J., & Hochstrasser, M. 2014. Assembly of the 20S proteasome. Biochim Biophys Acta, 1843(1), 2–12. doi:10.1016/j.bbamcr.2013.03.008

Kusmierczyk, A. R., Kunjappu, M. J., Kim, R. Y., et al. 2011. A conserved 20S proteasome assembly factor requires a C-terminal HbYX motif for proteasomal precursor binding. Nat Struct Mol Biol, 18(5), 622–629. doi:10.1038/nsmb.2027

Li, X., Lonard, D. M., Jung, S. Y., et al. 2006. The SRC-3/AIB1 coactivator is degraded in a ubiquitin- and ATP-independent manner by the REGgamma proteasome. Cell, 124(2), 381–392. doi:10.1016/j.cell.2005.11.037

Li, X., Amazit, L., Long, W., et al. 2007. Ubiquitin- and ATP-independent proteolytic turnover of p21 by the REGgamma-proteasome pathway. Mol Cell, 26(6), 831–842. doi:10.1016/j.molcel.2007.05.028

Liu, X., Xiao, W., Zhang, Y., et al. 2020. Reversible phosphorylation of Rpn1 regulates 26S proteasome assembly and function. Proc Natl Acad Sci U S A, 117(1), 328–336. doi:10.1073/pnas.1912531117

Livnat-Levanon, N., Kevei, E., Kleifeld, O., et al. 2014. Reversible 26S proteasome disassembly upon mitochondrial stress. Cell Rep, 7(5), 1371–1380. doi:10.1016/j.celrep.2014.04.030

Ma, C. P., Slaughter, C. A., & DeMartino, G. N. 1992. Identification, purification, and characterization of a protein activator (PA28) of the 20 S proteasome (macropain). J Biol Chem, 267(15), 10515–10523. Retrieved from https://www.ncbi.nlm.nih.gov/pubmed/1587832

Malmgaard, L. 2004. Induction and regulation of IFNs during viral infections. J Interferon Cytokine Res, 24(8), 439–454. doi:10.1089/1079990041689665

Mandemaker, I. K., Geijer, M. E., Kik, I., et al. 2018. DNA damage-induced replication stress results in PA200-proteasome-mediated degradation of acetylated histones. EMBO Rep, 19(10). doi:10.15252/embr.201745566

Mao, I., Liu, J., Li, X., et al. 2008. REGgamma, a proteasome activator and beyond? Cell Mol Life Sci, 65(24), 3971–3980. doi:10.1007/s00018-008-8291-z

Maupin-Furlow, J. A. 2014. Prokaryotic ubiquitin-like protein modification. Annu Rev Microbiol, 68, 155–175. doi:10.1146/annurev-micro-091313-103447

Metzger, M. B., Pruneda, J. N., Klevit, R. E., et al. 2014. RING-type E3 ligases: master manipulators of E2 ubiquitin-conjugating enzymes and ubiquitination. Biochim Biophys Acta, 1843(1), 47–60. doi:10.1016/j.bbamcr.2013.05.026

Mishto, M., & Liepe, J. 2017. Post-translational peptide splicing and T-cell responses. Trends Immunol, 38(12), 904–915. doi:10.1016/j.it.2017.07.011

Mishto, M., Santoro, A., Bellavista, E., et al. 2003. Immunoproteasomes and immunosenescence. Ageing Res Rev, 2(4), 419–432. doi:10.1016/s1568-1637(03)00030-8

Mishto, M., Liepe, J., Textoris-Taube, K., et al. 2014. Proteasome isoforms exhibit only quantitative differences in cleavage and epitope generation. Eur J Immunol, 44(12), 3508–3521. doi:10.1002/eji.201444902

Murakami, Y., Matsufuji, S., Kameji, T., et al. 1992. Ornithine decarboxylase is degraded by the 26S proteasome without ubiquitination. Nature, 360(6404), 597–599. doi:10.1038/360597a0

Murata, S., Sasaki, K., Kishimoto, T., et al. 2007. Regulation of CD8+ T cell development by thymus-specific proteasomes. Science, 316(5829), 1349–1353. doi:10.1126/science.1141915

Nitta, T., Murata, S., Sasaki, K., et al. 2010. Thymoproteasome shapes immunocompetent repertoire of CD8+ T cells. Immunity, 32(1), 29–40. doi:10.1016/j.immuni.2009.10.009

Ortega, J., Heymann, J. B., Kajava, A. V., et al. 2005. The axial channel of the 20S proteasome opens upon binding of the PA200 activator. J Mol Biol, 346(5), 1221–1227. doi:10.1016/j.jmb.2004.12.049

Osmulski, P. A., Hochstrasser, M., & Gaczynska, M. 2009. A tetrahedral transition state at the active sites of the 20S proteasome is coupled to opening of the alpha-ring channel. Structure, 17(8), 1137–1147. doi:10.1016/j.str.2009.06.011

Park, J. Y., Jang, S. Y., Shin, Y. K., et al. 2013. Calcium-dependent proteasome activation is required for axonal neurofilament degradation. Neural Regen Res, 8(36), 3401–3409. doi:10.3969/j.issn.1673-5374.2013.36.005

Park, J., Cho, J., & Song, E. J. 2020. Ubiquitin-proteasome system (UPS) as a target for anticancer treatment. Arch Pharm Res. doi:10.1007/s12272-020-01281-8

Pickering, A. M., & Davies, K. J. 2012. Degradation of damaged proteins: the main function of the 20S proteasome. Prog Mol Biol Transl Sci, 109, 227–248. doi:10.1016/B978-0-12-397863-9.00006-7

Pickering, A. M., Koop, A. L., Teoh, C. Y., et al. 2010. The immunoproteasome, the 20S proteasome and the PA28alphabeta proteasome regulator are oxidative-stress-adaptive proteolytic complexes. Biochem J, 432(3), 585–594. doi:10.1042/BJ20100878

Pickering, A. M., Linder, R. A., Zhang, H., et al. 2012. Nrf2-dependent induction of proteasome and Pa28alphabeta regulator are required for adaptation to oxidative stress. J Biol Chem, 287(13), 10021–10031. doi:10.1074/jbc.M111.277145

Qian, M. X., Pang, Y., Liu, C. H., et al. 2013. Acetylation-mediated proteasomal degradation of core histones during DNA repair and spermatogenesis. Cell, 153(5), 1012–1024. doi:10.1016/j.cell.2013.04.032

Raasi, S., Varadan, R., Fushman, D., et al. 2005. Diverse polyubiquitin interaction properties of ubiquitin-associated domains. Nat Struct Mol Biol, 12(8), 708–714. doi:10.1038/nsmb962

Raule, M., Cerruti, F., Benaroudj, N., et al. 2014. PA28alphabeta reduces size and increases hydrophilicity of 20S immunoproteasome peptide products. Chem Biol, 21(4), 470–480. doi:10.1016/j.chembiol.2014.02.006

Rechsteiner, M., & Hill, C. P. 2005. Mobilizing the proteolytic machine: cell biological roles of proteasome activators and inhibitors. Trends Cell Biol, 15(1), 27–33. doi:10.1016/j.tcb.2004.11.003

Rinaldi, T., Pick, E., Gambadoro, A., et al. 2004. Participation of the proteasomal lid subunit Rpn11 in mitochondrial morphology and function is mapped to a distinct C-terminal domain. Biochem J, 381(Pt 1), 275–285. doi:10.1042/BJ20040008

Ristic, G., Tsou, W. L., & Todi, S. V. 2014. An optimal ubiquitin-proteasome pathway in the nervous system: the role of deubiquitinating enzymes. Front Mol Neurosci, 7, 72. doi:10.3389/fnmol.2014.00072

Sadre-Bazzaz, K., Whitby, F. G., Robinson, H., et al. 2010. Structure of a Blm10 complex reveals common mechanisms for proteasome binding and gate opening. Mol Cell, 37(5), 728–735. doi:10.1016/j.molcel.2010.02.002

Schmidtke, G., Emch, S., Groettrup, M., et al. 2000. Evidence for the existence of a non-catalytic modifier site of peptide hydrolysis by the 20 S proteasome. J Biol Chem, 275(29), 22056–22063. doi:10.1074/jbc.M002513200

Schwartz, A. L., & Ciechanover, A. 2009. Targeting proteins for destruction by the ubiquitin system: implications for human pathobiology. Annu Rev Pharmacol Toxicol, 49, 73–96. doi:10.1146/annurev.pharmtox.051208.165340

Shi, Y., Chen, X., Elsasser, S., et al. 2016. Rpn1 provides adjacent receptor sites for substrate binding and deubiquitination by the proteasome. Science, 351(6275). doi:10.1126/science.aad9421

Silva, G. M., Netto, L. E., Simoes, V., et al. 2012. Redox control of 20S proteasome gating. Antioxid Redox Signal, 16(11), 1183–1194. doi:10.1089/ars.2011.4210

Soucy, T. A., Smith, P. G., Milhollen, M. A., et al. 2009. An inhibitor of NEDD8-activating enzyme as a new approach to treat cancer. Nature, 458(7239), 732–736. doi:10.1038/nature07884

Stadtmueller, B. M., & Hill, C. P. 2011. Proteasome activators. Mol Cell, 41(1), 8–19. doi:10.1016/j.molcel.2010.12.020

Stewart, D. P., Koss, B., Bathina, M., et al. 2010. Ubiquitin-independent degradation of anti-apoptotic MCL-1. *Mol Cell Biol*, 30(12), 3099–3110. doi:10.1128/MCB.01266-09

Striebel, F., Imkamp, F., Ozcelik, D., et al. 2014. Pupylation as a signal for proteasomal degradation in bacteria. *Biochim Biophys Acta*, 1843(1), 103–113. doi:10.1016/j.bbamcr.2013.03.022

Swords, R. T., Kelly, K. R., Smith, P. G., et al. 2010. Inhibition of NEDD8-activating enzyme: a novel approach for the treatment of acute myeloid leukemia. *Blood*, 115(18), 3796–3800. doi:10.1182/blood-2009-11-254862

Tanaka, K. 2009. The proteasome: overview of structure and functions. *Proc Jpn Acad Ser B Phys Biol Sci*, 85(1), 12–36. doi:10.2183/pjab.85.12

Tanaka, K., & Ichihara, A. 1989. Half-life of proteasomes (multiprotease complexes) in rat liver. *Biochem Biophys Res Commun*, 159(3), 1309–1315. doi:10.1016/0006-291x(89)92253-5

Tanaka, K., & Kasahara, M. 1998. The MHC class I ligand-generating system: roles of immunoproteasomes and the interferon-gamma-inducible proteasome activator PA28. *Immunol Rev*, 163, 161–176. doi:10.1111/j.1600-065x.1998.tb01195.x

Tar, K., Dange, T., Yang, C., et al. 2014. Proteasomes associated with the Blm10 activator protein antagonize mitochondrial fission through degradation of the fission protein Dnm1. *J Biol Chem*, 289(17), 12145–12156. doi:10.1074/jbc.M114.554105

Toste Rego, A., & da Fonseca, P. C. A. 2019. Characterization of fully recombinant human 20S and 20S-PA200 proteasome complexes. *Mol Cell*, 76(1), 138–147.e135. doi:10.1016/j.molcel.2019.07.014

Tsvetkov, P., Myers, N., Eliav, R., et al. 2014. NADH binds and stabilizes the 26S proteasomes independent of ATP. *J Biol Chem*, 289(16), 11272–11281. doi:10.1074/jbc.M113.537175

Unno, M., Mizushima, T., Morimoto, Y., et al. 2002. The structure of the mammalian 20S proteasome at 2.75 A resolution. *Structure*, 10(5), 609–618. doi:10.1016/s0969-2126(02)00748-7

Unverdorben, P., Beck, F., Sledz, P., et al. 2014. Deep classification of a large cryo-EM dataset defines the conformational landscape of the 26S proteasome. *Proc Natl Acad Sci U S A*, 111(15), 5544–5549. doi:10.1073/pnas.1403409111

van Nocker, S., Sadis, S., Rubin, D. M., et al. 1996. The multiubiquitin-chain-binding protein Mcb1 is a component of the 26S proteasome in Saccharomyces cerevisiae and plays a nonessential, substrate-specific role in protein turnover. *Mol Cell Biol*, 16(11), 6020–6028. doi:10.1128/mcb.16.11.6020

Verma, R., Aravind, L., Oania, R., et al. 2002. Role of Rpn11 metalloprotease in deubiquitination and degradation by the 26S proteasome. *Science*, 298(5593), 611–615. doi:10.1126/science.1075898

Vigneron, N., & Van den Eynde, B. J. 2012. Proteasome subtypes and the processing of tumor antigens: increasing antigenic diversity. *Curr Opin Immunol*, 24(1), 84–91. doi:10.1016/j.coi.2011.12.002

Visekruna, A., Joeris, T., Seidel, D., et al. 2006. Proteasome-mediated degradation of IkappaBalpha and processing of p105 in Crohn disease and ulcerative colitis. *J Clin Invest*, 116(12), 3195–3203. doi:10.1172/JCI28804

Wade, S. L., & Auble, D. T. 2010. The Rad23 ubiquitin receptor, the proteasome and functional specificity in transcriptional control. *Transcription*, 1(1), 22–26. doi:10.4161/trns.1.1.12201

Wang, X., Meul, T., & Meiners, S. 2020. Exploring the proteasome system: a novel concept of proteasome inhibition and regulation. *Pharmacol Ther*, 107526. doi:10.1016/j.pharmthera.2020.107526

Welk, V., Meul, T., Lukas, C., et al. 2019. Proteasome activator PA200 regulates myofibroblast differentiation. *Sci Rep*, 9(1), 15224. doi:10.1038/s41598-019-51665-0

Wilkinson, C. R., Seeger, M., Hartmann-Petersen, R., et al. 2001. Proteins containing the UBA domain are able to bind to multi-ubiquitin chains. *Nat Cell Biol*, 3(10), 939–943. doi:10.1038/ncb1001-939

Worden, E. J., Padovani, C., & Martin, A. 2014. Structure of the Rpn11-Rpn8 dimer reveals mechanisms of substrate deubiquitination during proteasomal degradation. *Nat Struct Mol Biol*, 21(3), 220–227. doi:10.1038/nsmb.2771

Wright, K. L., White, L. C., Kelly, A., et al. 1995. Coordinate regulation of the human TAP1 and LMP2 genes from a shared bidirectional promoter. *J Exp Med*, 181(4), 1459–1471. doi:10.1084/jem.181.4.1459

Xing, Y., Jameson, S. C., & Hogquist, K. A. 2013. Thymoproteasome subunit-beta5T generates peptide-MHC complexes specialized for positive selection. *Proc Natl Acad Sci U S A*, 110(17), 6979–6984. doi:10.1073/pnas.1222244110

Yao, T., & Cohen, R. E. 2002. A cryptic protease couples deubiquitination and degradation by the proteasome. *Nature*, 419(6905), 403–407. doi:10.1038/nature01071

Yao, W. T., Wu, J. F., Yu, G. Y., et al. 2014. Suppression of tumor angiogenesis by targeting the protein NEDDylation pathway. *Cell Death Dis*, 5, e1059. doi:10.1038/cddis.2014.21

Yip, M. C. J., Bodnar, N. O., & Rapoport, T. A. 2020. Ddi1 is a ubiquitin-dependent protease. *Proc Natl Acad Sci U S A*, 117(14), 7776–7781. doi:10.1073/pnas.1902298117

Zanker, D., Waithman, J., Yewdell, J. W., et al. 2013. Mixed proteasomes function to increase viral peptide diversity and broaden antiviral CD8+ T cell responses. *J Immunol*, 191(1), 52–59. doi:10.4049/jimmunol.1300802

Zhang, Q., Ji, S. Y., Busayavalasa, K., et al. 2019. Meiosis I progression in spermatogenesis requires a type of testis-specific 20S core proteasome. *Nat Commun*, 10(1), 3387. doi:10.1038/s41467-019-11346-y

Zhao, S., & Ulrich, H. D. 2010. Distinct consequences of posttranslational modification by linear versus K63-linked polyubiquitin chains. *Proc Natl Acad Sci U S A*, 107(17), 7704–7709. doi:10.1073/pnas.0908764107

Zhou, F. 2009. Molecular mechanisms of IFN-gamma to up-regulate MHC class I antigen processing and presentation. *Int Rev Immunol*, 28(3-4), 239–260. doi:10.1080/08830180902978120

Zhou, S., & Dewille, J. W. 2007. Proteasome-mediated CCAAT/enhancer-binding protein delta (C/EBPdelta) degradation is ubiquitin-independent. *Biochem J*, 405(2), 341–349. doi:10.1042/BJ20070082

CHAPTER ELEVEN

Cellular Responses to Proteasome Impairment

Maja Studencka-Turski and Elke Krüger

CONTENTS

11.1 INTRODUCTION

The ubiquitin-proteasome system (UPS) controls the proteolysis of the vast majority of intracellular proteins in mammalian cells. Therefore, UPS dysfunction may lead to several pathological disorders including systemic autoinflammation, neurodevelopmental or neurodegenerative disorders as well as aging-associated pathologies (Llamas et al., 2020).

The most commonly observed decline in proteasome activity has been shown to be associated with physiological changes occurring in aged organisms. Remarkably, this age-related loss of function may affect various levels of proteasome-mediated degradation pathways, such as reduced proteasome gene expression, reduced proteolytic activity, disassembly of proteasomes or inactivation by interacting with protein aggregates (Chondrogianni et al., 2003; Hwang et al., 2007; Vernace et al., 2007; Thibaudeau et al., 2018).

Moreover, proteasome dysfunction is not only restricted to aging, since aberrations in the UPS system have been implicated, either as a primary cause or secondary consequence, in the pathogenesis of neurodegenerative diseases (Figueiredo-Pereira et al., 2014; Llamas et al., 2020; Schmidt and Finley, 2014). For instance, insufficient proteasome function leading to chronic expression of unfolded, misfolded or damaged proteins contributes to abnormal deposition of protein aggregates and development of various proteinopathies, such as Alzheimer's disease, Parkinson's disease, polyglutamine (PolyQ)-related diseases and many others (Fan et al., 2014; Powers et al., 2009). In this regard, a recent study has led to the identification of a mechanistic link between proteasomal activity decline and induction of inflammation in microglial cells and supported a role for the proteasome impairment in triggering neuroinflammation, which is a key component of neurodegenerative diseases (Ebstein et al., 2019; Studencka-Turski et al., 2019).

Other noteworthy, however rare, causes of proteasome dysfunction are genetic alterations in genes encoding proteasomal subunits. Interestingly, the spectrum of cellular events triggered upon functional impairment of proteasome and its implication for human pathologies depends on the localization of these mutations. Specifically, 20S mutations cause a systemic autoinflammation syndrome referred to as chronic atypical neutrophilic dermatosis with lipodystrophy and elevated temperature (CANDLE) (Torrelo et al., 2010). In addition, further symptoms caused by 20S mutations are nowadays also known under other terms,

DOI: 10.1201/9781003048138-11

such as joint contractures – muscle atrophy – microcytic anemia – panniculitis-induced lipodystrophy syndrome (JMP), Nakajo–Nishimura syndrome (NKJO), proteasome-associated auto-inflammatory syndrome (PRAAS) and POMP-related autoinflammation and immune dysregulation disease (PRAID) (Agarwal et al., 2010; Brehm et al., 2015; Kitamura et al., 2011; Poli et al., 2018). For the sake of simplicity, the name CANDLE/PRAAS syndrome will be used as a generic term in this chapter. On molecular level, CANDLE/PRAAS patients display accumulation of ubiquitinated proteins which is very often accompanied by elevated levels of proinflammatory cytokines and chemokines (IL-6, IL-8, IP-10) as well as type I interferons (IFN) (Agarwal et al., 2010; Brehm et al., 2015; de Jesus et al., 2019; Kitamura et al., 2011; Poli et al., 2018). However, when mutations affect the 19S regulatory complex, subjects do not exhibit visible signs of autoinflammation but instead display mainly syndromic neurodevelopmental features such as early-onset severe deafness, congenital cataracts, intellectual disability, feeding difficulties and subtle dysmorphic facial abnormalities (Küry et al., 2017; Kröll-Hermi et al., 2020).

Finally, proteasome impairment and accompanied cellular responses may as well be intentionally induced as in the case of cancer therapies. Inhibition of the proteasome became an area of interest in cancer research following observations that malignant cells of many types frequently exhibit elevated levels of 26S proteasomes as well as unusually high proteasome activity (Arlt et al., 2009; Chen and Madura, 2005). Since then, numerous types of inhibitors have been designed interfering either with the activity of standard proteasomes (SP) or immunoproteasomes (IP) or both (Crawford et al., 2011; Ettari et al., 2018).

In fact, IP were considered for a very long time as proteolytic complexes essentially specialized in generation of short oligopeptides for MHC-I presentation. Nevertheless, further studies verified this concept but pointed out the active role of IP in the regulation of the unfolded protein response (UPR), integrated stress response (ISR) or mTOR pathway and the control of immune signaling (Angeles et al., 2012; Ebstein et al., 2012; Groettrup et al., 2010; Zhao et al., 2016). IP are abundantly and constitutively expressed in immune cells, but when an increased protein turnover is required, their cytokine-stimulated expression can be triggered in almost all cell types. Therefore, specific impairment of IP aims to overcome toxicities associated with pan-proteasome inhibition which is targeting SP as well as IP (Ettari et al., 2018). In this context, it is important to note that the first founder mutation of CANDLE/PRAAS has been identified in PSMB8 encoding the IP subunit β5i/LMP7 (Torrelo et al., 2010).

11.2 CONSEQUENCES OF DIMINISHED PROTEASOME ACTIVITY ON THE ER HOMEOSTASIS

Alterations in the proteostasis network caused either by proteasome dysfunction or inhibition trigger activation of several signaling pathways. The main cellular sensor of disturbed balance between protein synthesis and degradation is the endoplasmic reticulum (ER) (Figure 11.1). Accumulation of misfolded or damaged proteins in the ER leads to activation of the ER stress condition and initiation of quality control mechanisms: the UPR as well as the ER-associated degradation (ERAD) (Bhattacharya and Qi, 2019; Hwang and Qi, 2018). At a steady state, nascent proteins are translocated into the ER lumen, which is a major folding compartment for secreted and/or plasma membrane proteins. During folding, ERAD pathway responds to the presence of misfolded proteins in the ER lumen to induce transport of the proteins to the cytosol where they can be targeted for proteolysis by the UPS (Figure 11.1). In the course of ER stress induced by proteasome impairment, ERAD regulators are retained in the ER, which leads to either ERAD overload or failure (Harding et al., 2003; Schröder and Kaufman, 2005). When ERAD is blocked, aberrant proteins accumulate within the ER and trigger activation of the UPR (Hwang and Qi, 2018; Schröder and Kaufman, 2005). The UPR signaling consists of three branches that together assure a coordinated response to proteasome impairment, restore protein-folding homeostasis or activate apoptosis when damage is irreversible (Bravo et al., 2013; Hetz et al., 2015). The branches are activated by three different stress sensors: inositol-requiring protein 1α (IRE-1α), protein kinase RNA-like endoplasmic reticulum kinase (PERK) and activating transcription factor 6 (ATF6) (Figure 11.1). Under normal conditions, luminal domains of these ER stress sensors are associated with the

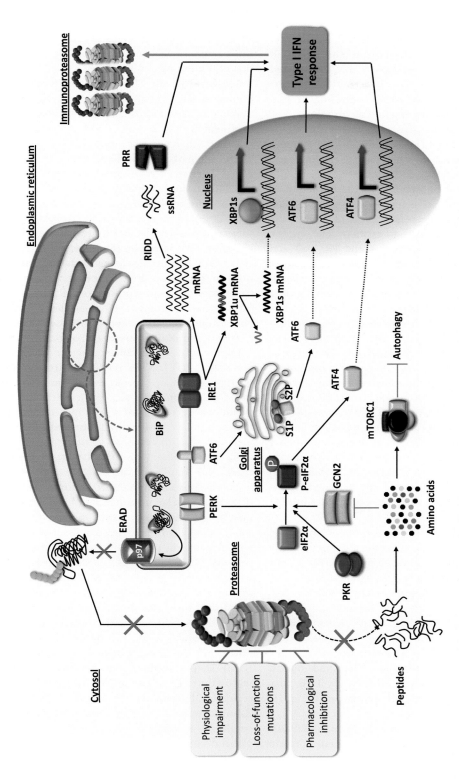

Figure 11.1 **Schematic representation of signaling pathways induced upon proteasome impairment.** Proteasome impairment blocks the ER-associated degradation of misfolded proteins, causing their accumulation in the ER lumen and induction of the UPR. As a result, the UPR-induced sXBP1, ATF6 and ATF4 transcription factors along with IRE1-activated RIDD pathway trigger type I interferon response. The decreased proteolytic activity of proteasomes also leads to an amino acid deficiency which is sensed by the integrative stress response (ISR) and affects mTOR signaling.

chaperone protein immunoglobulin binding protein (BiP), which keeps them in an inactive monomeric state. When proteasome dysfunction results in an accumulation of proteins in the ER lumen, BiP dissociates from PERK, IRE1α and ATF6, thereby triggering their activation and reestablishment of proteostasis by controlling protein translation and increasing folding capacity.

As a result of BiP dissociation from the luminal domain of PERK, this stress sensor undergoes oligomerization, trans-autophosphorylation and subsequent activation (Bertolotti et al., 2000). Activated PERK phosphorylates the α-subunit of eukaryotic translation initiation factor-2 (eIF2α) at Ser51, which decreases levels of active eIF2 and attenuates protein translation. The decline of protein synthesis reduces the ER load and allows to counterbalance disrupted proteostasis. Furthermore, PERK-mediated phosphorylation of eIF2α triggers transcriptional activation via a translational upregulation of transcripts consisting of short inhibitory upstream open reading frames (uORFs) (Vattem and Wek, 2004). The main preferentially translated gene transcript is ATF4-encoding basic leucine zipper transcription factor 4 (ATF4) that increases expression of genes involved in metabolism, alleviation of oxidative stress as well as activation of ERAD. Nevertheless, when the ER stress cannot be resolved, ATF4 may eventually induce the CCAAT-enhancer-binding protein homologous protein (CHOP), a transcription factor that triggers apoptosis by binding to the death receptor pathway and upregulating the expression of death receptors 4 and 5 (DR4, DR5) (Hu et al., 2018).

Similar to PERK, IRE1α also oligomerizes and undergoes trans-autophosphorylation in response to accumulation of unfolded proteins. These conformational changes trigger its additional effector function, the endoribonuclease activity. This activity serves for the endonucleolytic cleavage of a transcript encoding a basic leucine zipper (bZIP)-type transcription factor – X-box-binding protein (XBP1). This generates an active XBP1 which translocates into the nucleus and induces transcription of genes involved in UPR (e.g. CHOP) and ERAD (Madhusudhan et al., 2015; Yang et al., 2017).

The last ER stress sensor activated upon BiP dissociation is the type II transmembrane protein ATF6. This facilitates translocation of ATF6 from the ER to the Golgi apparatus and subsequent cleavage by the Golgi-resident site-1 and -2 proteases (S1P and S2P). The processed fragment of ATF6 moves into the nucleus and binds to ER stress response elements to activate gene expression of, for example, BiP and XBP1.

Activation of the UPR following proteasome inhibition was initially reported in studies of multiple myeloma cells subjected to bortezomib treatment (Lee et al., 2003) and later confirmed in other cell types (Bao et al., 2016; Cybulsky, 2013). Remarkably, further studies revealed a link between activation of the UPR following proteasome dysfunction and neuroinflammation (Pintado et al., 2017; Studencka-Turski et al., 2019). Consistently with these observations, subjects displaying decreased proteasome activity also show UPR activation (Brehm et al., 2015; Poli et al., 2018). For instance, subjects carrying mutations in POMP, encoding a proteasome assembly factor coordinating incorporation of proteasome subunits into 20S complexes, exhibit elevated phosphorylation of eIF2α as well as increased levels of ATF6, BiP and sXBP1 transcripts (Poli et al., 2018).

Importantly, PERK kinase and the transcription factors ATF4, ATF6 and sXBP1 are also part of the coordinated gene expression program called ISR (Harding et al., 2003). ISR is another evolutionarily conserved signaling network that becomes induced following proteasomal impairment but also in response to nutrient deprivation, viral infection or oxidative stress (García et al., 2007; Harding et al., 2003; Ye et al., 2010). It is worth mentioning that nutrient deprivation, in particular insufficiency of amino acids, may in fact originate from the disrupted proteasome activity since proteasome-mediated protein degradation provides a majority of cellular peptides (Suraweera et al., 2012). Moreover, ER stress caused by proteasome impairment can also activate the ISR (Harding et al., 1999). All various disturbances of cellular homeostasis can be sensed by four kinases: PERK, the general control nonderepressible 2 (GCN2), the protein kinase R (PKR) and the heme-regulated inhibitor kinase (HRI). They all mediate phosphorylation of eIF2α within the eIF2–eIF2B complex, which is a central regulatory hub for the ISR (Figure 11.1) (Pakos-Zebrucka et al., 2016; Rzymski and Harris, 2007). As already mentioned, phosphorylation causes a global decrease of protein synthesis and promotes the translation of specific mRNAs, such as ATF4 (Rzymski et al., 2010). Thus, alike UPR, ISR helps to reestablish protein homeostasis by attenuating general mRNA translation and adjusting the synthesis of several transcription factors that trigger activation of stress-specific genes.

11.3 ACTIVATION OF THE TCF11/NRF1 IN RESPONSE TO PROTEASOME IMPAIRMENT

Another crucial compensatory mechanism induced upon proteasome dysfunction involves the processing of the long isoform of the nuclear factor, erythroid 2-Like 1 (Nrf1) TCF11/Nrf1. TCF11 transcription factor was originally described as a key regulator of 26S proteasome formation in human cells that is critical for restoring proteostasis following transient proteasome impairment (Steffen et al., 2010; Radhakrishnan et al., 2014; Vangala et al., 2016). TCF11/Nrf1 belongs to NF-E2-related factors, which represent the Cap'n'collar (Cnc) bZIP transcription factor family that plays a central role in the regulation of genes involved in antioxidant stress responses by binding to the antioxidant response elements (AREs) (Chan et al., 1995; Liu et al., 2019). At steady-state conditions, TCF11/Nrf1 resides in the ER membrane, where it is targeted to an ERAD-mediated protein degradation involving the E3 ubiquitin ligase HRD1 and the AAA ATPase p97 (Steffen et al., 2010) (Figure 11.2). A decreased proteasome activity typically leads to accumulation of misfolded and otherwise damaged proteins, which in turn triggers de-N-glycosylation via N-glycanase 1 (NGLY1) (Tomlin et al., 2017) and proteolytic cleavage of TCF11/Nrf1 by the aspartyl protease DDI2 (Steffen et al., 2010; Koizumi et al., 2016; Dirac-Svejstrup et al., 2020) (Figure 11.2). These posttranslational modifications enable nuclear translocation of the TCF11/Nrf1 C-terminal fragment from the ER, association with Maf proteins and activation of proteasome gene expression following binding to AREs (Johnsen et al., 1996; Steffen et al., 2010; Radhakrishnan et al., 2014) (Figure 11.2).

Similar to UPR, it has been demonstrated that CANDLE/PRAAS patients display elevated expression of processed TCF11/Nrf1, which suggests activation of the compensatory mechanisms in cells with either decreased or absent proteasomal activity (Sotzny et al., 2016; Poli et al., 2018). Consistently, fibroblasts with mutations in PSMC3, encoding an ATPase protein from the 19S regulatory complex, also exhibit activation of the TCF11/Nrf1 transcriptional pathway, emphasizing once more the resemblance of both syndromes (Kröll-Hermi et al., 2020). Surprisingly, a recent study showed that murine Nrf1 can be induced in response to IP inhibition, suggesting a link between this transcription factor and the regulation of the inducible isoforms of the proteasome (Studencka-Turski et al., 2019).

11.4 RESPONSE OF MTOR SIGNALING TO DECREASED PROTEASOME ACTIVITY

Originally, the link between disrupted proteasome activity and mTOR signaling was made following observations that bortezomib treatment leads to dephosphorylation and inactivation of the evolutionarily conserved target of rapamycin complex 1 (TORC1) in mantle cell lymphoma and multiple myeloma (Decaux et al., 2010; Hutter et al., 2012) (Figure 11.1). Later studies revealed that this crosstalk is based on homeostatic regulation of energy and metabolic status, in particular availability of free amino acids for protein synthesis (Adegoke et al., 2019; Zhang and Manning, 2015; Zhao et al., 2015). As already stated, under normal conditions, SP as well as IP are responsible for breakdown of the vast majority of proteins, thereby being the major source of cellular peptides, which are further degraded to amino acids. Once proteasome activity is disrupted, the amino acids' pool decreases, which not only activates the ISR pathway but also interferes with mTORC1 activity (Meul et al., 2020; Zhang et al., 2014; Zhang and Manning, 2015; Zhao et al., 2016; Zhao and Goldberg, 2016). mTORC1 is induced by growth factor signaling pathways under conditions of intracellular nutrients abundance and in the presence of energy. Once activated, mTORC1 stimulates synthesis of proteins, lipids and nucleotides and at the same time inhibits autophagy (Howell et al., 2013; Singh and Cuervo, 2011). Initially, it was believed that under conditions of nutrient depletion, mTORC1 becomes inactivated resulting in stimulation of autophagy, which then provides cells with amino acids for the protein synthesis or for energy production (Russell et al., 2014). However, studies of Zhao et al. have shown that following nutrient starvation, mTORC1 inhibition triggers the coordinated protein degradation by autophagy as well as proteasomes (Zhao et al., 2015; Zhao and Goldberg, 2016). Therefore,

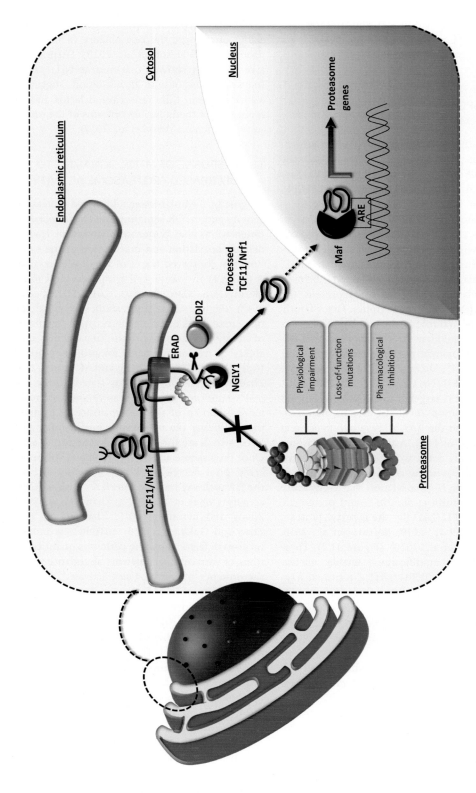

Figure 11.2 **Mechanism of the TCF11/Nrf1-mediated transcriptional activation of proteasome genes.** Proteasome impairment prevents degradation of the TCF11/Nrf1 protein and triggers de-glycosylation by NGLY1 as well as proteolytic cleavage by DDI2. Processed TCF11/Nrf1 translocates into the nucleus where it heterodimerizes with small Maf proteins and induces the transcription of proteasome and mitophagy genes.

proteasomal-mediated proteolysis not only assures quality control during protein synthesis and folding but also, alike autophagy, delivers amino acids available for the synthesis of new proteins. Consistently with that notion, a study of Zhang et al. demonstrated that mTORC1 signaling assures constant supply of amino acids via activation of TCF11/Nrf1, thereby increasing cellular proteasome levels (Zhang et al., 2014).

11.5 INDUCTION OF THE INNATE IMMUNE RESPONSE UPON PROTEASOME IMPAIRMENT

The initial observation that CANDLE/PRAAS patients exhibit constitutive activation of type I IFN clearly links the disrupted activity of proteasome with activation of the innate immune response. Nevertheless, the specific drivers that trigger this response are still not completely characterized. Given that the observed response shares the characteristics of a sterile inflammation, the most common approach in understanding the underlying mechanism is to look for molecules that are generated upon proteasome impairment and might be considered as danger-associated molecular patterns (DAMPS). Such DAMPS are subsequently recognized by pattern recognition receptors (PRR) and activate coordinated response of the immune system. Interestingly, a recent study revealed that induction of UPR following proteasome impairment plays a direct role in generation of DAMPS (Studencka-Turski et al., 2019). Specifically, it has been shown that initiation of the type I IFN response in myeloid cells upon proteasome inhibition is mediated by the endoribonuclease activity of the ER stress-induced IRE1α. IRE1α triggers a pathway known as IRE1-dependent decay (RIDD), which degrades mRNAs encoding ER-targeted proteins (Coelho and Domingos, 2014; Hollien et al., 2009). The generated short single-stranded RNA fragments may be recognized by a cytosolic PRR-RIG-I (retinoic acid inducible gene I), which in turn triggers a TBK1/IRF3-mediated induction of type I IFN (Hollien et al., 2009; Eckard et al., 2014) (Figure 11.1). Actually, the hypothesis that ER stress may trigger sterile inflammation was developed much earlier following the observation that treatment of human embryonic kidney cells with agents inducing ER stress led to accumulation of proteins in the ER lumen and subsequent nuclear translocation of

the transcription factor NF-κB (Pahl and Baeuerle, 1995). This notion was later supported by the analyses of knockout mice models for UPR components or the use of inhibitors disrupting activity of specific UPR branches, which all together confirmed that induction of ER stress induces sterile inflammation by activating NF-κB signaling (Jiang et al., 2003; Yamazaki et al., 2009; Hu et al., 2018). Interestingly, induction of the NF-κB signaling has also been reported in several cases of multiple myeloma or lymphoma patients subjected to bortezomib treatment (Sung et al., 2008; Hideshima et al., 2009; Juvekar et al., 2011; Vrábel et al., 2019). Correspondingly, CANDLE/PRAAS patients commonly express elevated levels of IL-6, an induction which is mainly regulated by the NF-κB pathway (Brehm et al., 2015; de Jesus et al., 2019; Kanazawa, 2012; McDermott et al., 2013; Poli et al., 2018). Furthermore, constitutively induced NF-κB signaling has been observed during aging and neurodegeneration, both processes often associated with proteasomal defects (Kriete and Mayo, 2009; Shih et al., 2015). Proteasome impairment-induced NF-κB signaling is an unprecedented phenomenon, given that proteasomal activity is required for activation of both the canonical and non-canonical NF-κB pathways. Nevertheless, this unusual activity is known as an atypical NF-κB signaling which is induced specifically following disrupted proteasomal proteolysis and therefore does not require proteasome for nuclear translocation of p65/RelA subunit of the NF-κB transcription factor complex (Cullen et al., 2010).

11.6 CONCLUDING REMARKS

Dysregulation of the UPS function caused either by pharmacological inhibition, genetic alteration or in the course of organismal aging triggers a broad spectrum of complex molecular events, with detrimental effects on cellular homeostasis. Multiple lines of evidence indicate that UPR as well as ISR are the major sentinels of insufficient proteasomal activity and play a direct role in induction of the innate immune response. Moreover, both of these pathways eventually interfere with the mTOR signaling, suggesting that proteasome impairment also leads to a broader metabolic dysregulation. Given that UPS plays a part in numerous intracellular processes, further efforts in this field are needed to better understand complex mechanisms underlying pathology of proteasome dysfunction and to dissect drivers of inflammation induced upon proteasome impairment.

REFERENCES

Adegoke, O.A.J., Beatty, B.E., Kimball, S.R., et al., 2019. Interactions of the super complexes: when mTORC1 meets the proteasome. Int. J. Biochem. Cell. Biol. 117, 105638. doi:10.1016/j.biocel.2019.105638

Agarwal, A.K., Xing, C., DeMartino, G.N., et al., 2010. PSMB8 encoding the β5i proteasome subunit is mutated in joint contractures, muscle atrophy, microcytic anemia, and panniculitis-induced lipodystrophy syndrome. Am. J. Hum. Genet. 87, 866–872. doi:10.1016/j.ajhg.2010.10.031

Angeles, A., Fung, G., Luo, H., 2012. Immune and non-immune functions of the immunoproteasome. Front. Biosci. (Landmark Ed.) 17, 1904–1916. doi:10.2741/4027

Arlt, A., Bauer, I., Schafmayer, C., et al., 2009. Increased proteasome subunit protein expression and proteasome activity in colon cancer relate to an enhanced activation of nuclear factor E2-related factor 2 (Nrf2). Oncogene 28, 3983–3996. doi:10.1038/onc.2009.264

Bao, W., Gu, Y., Ta, L., et al., 2016. Induction of autophagy by the MG-132 proteasome inhibitor is associated with endoplasmic reticulum stress in MCF-7 cells. Mol. Med. Rep. 13, 796–804. doi:10.3892/mmr.2015.4599

Bertolotti, A., Zhang, Y., Hendershot, L.M., et al., 2000. Dynamic interaction of BiP and ER stress transducers in the unfolded-protein response. Nat. Cell Biol. 2, 326–332. doi:10.1038/35014014

Bhattacharya, A., Qi, L., 2019. ER-associated degradation in health and disease – from substrate to organism. J. Cell. Sci. 132. doi:10.1242/jcs.232850

Bravo, R., Parra, V., Gatica, D., et al., 2013. Endoplasmic reticulum and the unfolded protein response: dynamics and metabolic integration. Int. Rev. Cell. Mol. Biol. 301, 215–290. doi:10.1016/B978-0-12-407704-1.00005-1

Brehm, A., Liu, Y., Sheikh, A., et al., 2015. Additive loss-of-function proteasome subunit mutations in CANDLE/PRAAS patients promote type I IFN production. J. Clin. Investig. 125, 16. doi: 10.1172/JCI81260

Chan, J.Y., Cheung, M.C., Moi, P., et al., 1995. Chromosomal localization of the human NF-E2 family of bZIP transcription factors by fluorescence in situ hybridization. Hum. Genet. 95, 265–269. doi:10.1007/bf00225191

Chen, L., Madura, K., 2005. Increased proteasome activity, ubiquitin-conjugating enzymes, and eEF1A translation factor detected in breast cancer tissue. Cancer Res. 65, 5599–5606. doi:10.1158/0008-5472.CAN-05-0201

Chondrogianni, N., Stratford, F.L.L., Trougakos, I.P., et al., 2003. Central role of the proteasome in senescence and survival of human fibroblasts: induction of a senescence-like phenotype upon its inhibition and resistance to stress upon its activation. J. Biol. Chem. 278, 28026–28037. doi:10.1074/jbc.M301048200

Coelho, D.S., Domingos, P.M., 2014. Physiological roles of regulated Ire1 dependent decay. Front. Genet. 5. doi:10.3389/fgene.2014.00076

Crawford, L.J., Walker, B., Irvine, A.E., 2011. Proteasome inhibitors in cancer therapy. J. Cell. Commun. Signal. 5, 101–110. doi:10.1007/s12079-011-0121-7

Cullen, S.J., Ponnappan, S., Ponnappan, U., 2010. Proteasome inhibition up-regulates inflammatory gene transcription induced by an atypical pathway of NF-κB activation. Biochem. Pharmacol. 79, 706–714. doi:10.1016/j.bcp.2009.10.006

Cybulsky, A.V., 2013. The intersecting roles of endoplasmic reticulum stress, ubiquitin–proteasome system, and autophagy in the pathogenesis of proteinuric kidney disease. Kidney Int. 84, 25–33. doi:10.1038/ki.2012.390

de Jesus, A.A., Brehm, A., VanTries, R., et al., 2019. Novel proteasome assembly chaperone mutations in PSMG2/PAC2 cause the autoinflammatory interferonopathy CANDLE/PRAAS4. J. Allergy Clin. Immunol. doi:10.1016/j.jaci.2018.12.1012

Decaux, O., Clement, M., Magrangeas, F., et al., 2007. Bortezomib inhibits mTOR pathway in multiple myeloma cell lines via induced expression of REDD1. Blood 110, 242–242. doi:10.1182/blood.V110.11.242.242

Dirac-Svejstrup, A.B., Walker, J., Faull, P., et al., 2020. DDI2 is a ubiquitin-directed endoprotease responsible for cleavage of transcription factor NRF1. Mol. Cell 79, 332–341.e7. doi:10.1016/j.molcel.2020.05.035

Ebstein, F., Kloetzel, P.-M., Krüger, E., et al., 2012. Emerging roles of immunoproteasomes beyond MHC class I antigen processing. Cell. Mol. Life Sci. 69, 2543–2558. doi:10.1007/s00018-012-0938-0

Ebstein, F., Poli Harlowe, M.C., Studencka-Turski, M., et al., 2019. Contribution of the unfolded protein response (UPR) to the pathogenesis of proteasome-associated autoinflammatory syndromes (PRAAS). Front Immunol 10. doi:10.3389/fimmu.2019.02756

Eckard, S.C., Rice, G.I., Fabre, A., et al., 2014. The SKIV2L RNA exosome limits activation of the RIG-I-like receptors. Nat. Immunol. 15, 839–845. doi:10.1038/ni.2948

Ettari, R., Zappalà, M., Grasso, S., et al., 2018. Immunoproteasome-selective and non-selective inhibitors: a promising approach for the treatment of multiple myeloma. Pharmacol. Ther. 182, 176–192. doi:10.1016/j.pharmthera.2017.09.001

Fan, H.-C., Ho, L.-I., Chi, C.-S., et al., 2014. Polyglutamine (PolyQ) diseases: genetics to treatments. Cell Transplant. 23, 441–458. doi:10.3727/096368914X678454

Figueiredo-Pereira, M.E., Rockwell, P., Schmidt-Glenewinkel, T., et al., 2014. Neuroinflammation and J2 prostaglandins: linking impairment of the ubiquitin-proteasome pathway and mitochondria to neurodegeneration. Front. Mol. Neurosci. 7, 104. doi:10.3389/fnmol.2014.00104

García, M.A., Meurs, E.F., Esteban, M., 2007. The dsRNA protein kinase PKR: virus and cell control. Biochimie 89, 799–811. doi:10.1016/j.biochi.2007.03.001

Groettrup, M., Kirk, C.J., Basler, M., 2010. Proteasomes in immune cells: more than peptide producers? Nat. Rev. Immunol. 10, 73–78. doi:10.1038/nri2687

Harding, H.P., Zhang, Y., Ron, D., 1999. Protein translation and folding are coupled by an endoplasmic-reticulum-resident kinase. Nature 397, 271–274. doi:10.1038/16729

Harding, H.P., Calfon, M., Urano, F., et al., 2002. Transcriptional and translational control in the mammalian unfolded protein response. Annu. Rev. Cell Dev. Biol. 18, 575–599. doi:10.1146/annurev.cellbio.18.011402.160624

Harding, H.P., Zhang, Y., Zeng, H., et al., 2003. An integrated stress response regulates amino acid metabolism and resistance to oxidative stress. Mol. Cell 11, 619–633. doi:10.1016/s1097-2765(03)00105-9

Hetz, C., Chevet, E., Oakes, S.A., 2015. Proteostasis control by the unfolded protein response. Nat. Cell Biol. 17, 829–838. doi:10.1038/ncb3184

Hideshima, T., Ikeda, H., Chauhan, D., et al., 2009. Bortezomib induces canonical nuclear factor-kappaB activation in multiple myeloma cells. Blood 114, 1046–1052. doi:10.1182/blood-2009-01-199604

Hollien, J., Lin, J.H., Li, H., et al., 2009. Regulated Ire1-dependent decay of messenger RNAs in mammalian cells. J. Cell Biol. 186, 323–331. doi:10.1083/jcb.200903014

Howell, J.J., Ricoult, S.J.H., Ben-Sahra, I., et al., 2013. A growing role for mTOR in promoting anabolic metabolism. Biochem. Soc. Trans. 41, 906–912. doi:10.1042/BST20130041

Hu, F., Yu, X., Wang, H., et al., 2011. ER stress and its regulator X-box-binding protein-1 enhance polyIC-induced innate immune response in dendritic cells. Eur. J. Immunol. 41, 1086–1097. doi:10.1002/eji.201040831

Hu, H., Tian, M., Ding, C., et al., 2019. The C/EBP homologous protein (CHOP) transcription factor functions in endoplasmic reticulum stress-induced apoptosis and microbial infection. Front. Immunol. 9. doi:10.3389/fimmu.2018.03083

Hutter, G., Zimmermann, Y., Rieken, M., et al., 2012. Proteasome inhibition leads to dephosphorylation and downregulation of protein expression of members of the Akt/mTOR pathway in MCL. Leukemia 26, 2442–2444. doi:10.1038/leu.2012.118

Hwang, J., Qi, L., 2018. Quality control in the endoplasmic reticulum: crosstalk between ERAD and UPR pathways. Trends Biochem Sci 43, 593–605. doi:10.1016/j.tibs.2018.06.005

Hwang, J.S., Hwang, J.S., Chang, I., et al., 2007. Age-associated decrease in proteasome content and activities in human dermal fibroblasts: restoration of normal level of proteasome subunits reduces aging markers in fibroblasts from elderly persons. J. Gerontol. A Biol. Sci. Med. Sci. 62, 490–499. doi:10.1093/gerona/62.5.490

Jiang, H.-Y., Wek, S.A., McGrath, B.C., et al., 2003. Phosphorylation of the alpha subunit of eukaryotic initiation factor 2 is required for activation of NF-kappaB in response to diverse cellular stresses. Mol. Cell. Biol. 23, 5651–5663. doi:10.1128/mcb.23.16.5651-5663.2003

Johnsen, O., Skammelsrud, N., Luna, L., et al., 1996. Small Maf proteins interact with the human transcription factor TCF11/Nrf1/LCR-F1. Nucleic Acids Res. 24, 4289–4297. doi:10.1093/nar/24.21.4289

Juvekar, A., Manna, S., Ramaswami, S., et al., 2011. Bortezomib induces nuclear translocation of I B resulting in gene-specific suppression of NF-B-dependent transcription and induction of apoptosis in CTCL. Mol. Cancer Res. 9, 183–194. doi:10.1158/1541-7786.MCR-10-0368

Kanazawa, N., 2012. Nakajo-Nishimura syndrome: an autoinflammatory disorder showing Pernio-like rashes and progressive partial lipodystrophy. Allerg. Int. 61, 197–206. doi:10.2332/allergolint.11-RAI-0416

Kitamura, A., Maekawa, Y., Uehara, H., et al., 2011. A mutation in the immunoproteasome subunit PSMB8 causes autoinflammation and lipodystrophy in humans. J. Clin. Invest. 121, 4150–4160. doi:10.1172/JCI58414

Koizumi, S., Irie, T., Hirayama, S., et al., 2016. The aspartyl protease DDI2 activates Nrf1 to compensate for proteasome dysfunction. eLife 5, e18357. doi:10.7554/eLife.18357

Kriete, A., Mayo, K.L., 2009. Atypical pathways of NF-kappaB activation and aging. Exp. Gerontol. 44, 250–255. doi:10.1016/j.exger.2008.12.005

Kröll-Hermi, A., Ebstein, F., Stoetzel, C., et al., 2020. Proteasome subunit PSMC3 variants cause neurosensory syndrome combining deafness and cataract due to proteotoxic stress. EMBO Mol Med 12, e11861. doi:10.15252/emmm.201911861

Küry, S., Besnard, T., Ebstein, F., et al., 2017. De novo disruption of the proteasome regulatory subunit PSMD12 causes a syndromic neurodevelopmental disorder. Am. J. Hum. Genet. 100, 689. doi:10.1016/j.ajhg.2017.03.003

Lee, A.-H., Iwakoshi, N.N., Anderson, K.C., et al., 2003. Proteasome inhibitors disrupt the unfolded protein response in myeloma cells. PNAS 100, 9946–9951. doi:10.1073/pnas.1334037100

Liu, P., Kerins, M.J., Tian, W., et al., 2019. Differential and overlapping targets of the transcriptional regulators NRF1, NRF2, and NRF3 in human cells. J. Biol. Chem. 294, 18131–18149. doi:10.1074/jbc.RA119.009591

Llamas, E., Alirzayeva, H., Loureiro, R., et al., 2020. The intrinsic proteostasis network of stem cells. Curr. Opin. Cell Biol. 67, 46–55. doi:10.1016/j.ceb.2020.08.005

Madhusudhan, T., Wang, H., Dong, W., et al., 2015. Defective podocyte insulin signalling through p85-XBP1 promotes ATF6-dependent maladaptive ER-stress response in diabetic nephropathy. Nat. Commun. 6, 6496. doi:10.1038/ncomms7496

McDermott, A., de Jesus, A.A., Liu, Y., et al., 2013. A case of proteasome-associated auto-inflammatory syndrome with compound heterozygous mutations in PSMB8. J. Am. Acad. Dermatol. 69, e29–e32. doi:10.1016/j.jaad.2013.01.015

Meul, T., Berschneider, K., Schmitt, S., et al., 2020. Mitochondrial regulation of the 26S proteasome. Cell Rep. 32, 108059. doi:10.1016/j.celrep.2020.108059

Pahl, H.L., Baeuerle, P.A., 1995. A novel signal transduction pathway from the endoplasmic reticulum to the nucleus is mediated by transcription factor NF-kappa B. EMBO J. 14, 2580–2588. doi: 10.1002/j.1460-2075.1995.tb07256.x.

Pakos-Zebrucka, K., Koryga, I., Mnich, K., et al., 2016. The integrated stress response. EMBO Rep. 17, 1374–1395. doi:10.15252/embr.201642195

Pintado, C., Macías, S., Domínguez-Martín, H., et al., 2017. Neuroinflammation alters cellular proteostasis by producing endoplasmic reticulum stress, autophagy activation and disrupting ERAD activation. Sci. Rep. 7, 8100. doi:10.1038/s41598-017-08722-3

Poli, M.C., Ebstein, F., Nicholas, S.K., et al., 2018. Heterozygous truncating variants in POMP escape nonsense-mediated decay and cause a unique immune dysregulatory syndrome. Am. J. Hum. Genet. 102, 1126–1142. doi:10.1016/j.ajhg.2018.04.010

Powers, E.T., Morimoto, R.I., Dillin, A., et al., 2009. Biological and chemical approaches to diseases of proteostasis deficiency. Annu. Rev. Biochem. 78, 959–991. doi:10.1146/annurev.biochem.052308.114844

Radhakrishnan, S.K., Lee, C.S., Young, P., et al., 2010. Transcription factor Nrf1 mediates the proteasome recovery pathway after proteasome inhibition in mammalian cells. Mol. Cell 38, 17–28. doi:10.1016/j.molcel.2010.02.029

Russell, R.C., Yuan, H.-X., Guan, K.-L., 2014. Autophagy regulation by nutrient signaling. Cell. Res. 24, 42–57. doi:10.1038/cr.2013.166

Rzymski, T., Harris, A.L., 2007. The unfolded protein response and integrated stress response to anoxia. Clin. Cancer Res. 13, 2537–2540. doi:10.1158/1078-0432.CCR-06-2126

Rzymski, T., Milani, M., Pike, L., et al., 2010. Regulation of autophagy by ATF4 in response to severe hypoxia. Oncogene 29, 4424–4435. doi:10.1038/onc.2010.191

Schmidt, M., Finley, D., 2014. Regulation of proteasome activity in health and disease. Biochim. Biophys. Acta 1843. doi:10.1016/j.bbamcr.2013.08.012

Schröder, M., Kaufman, R.J., 2005. The mammalian unfolded protein response. Annu. Rev. Biochem. 74, 739–789. doi:10.1146/annurev.biochem.73.011303.074134

Shih, R.-H., Wang, C.-Y., Yang, C.-M., 2015. NF-kappaB signaling pathways in neurological inflammation: a mini review. Front. Mol. Neurosci. 8. doi:10.3389/fnmol.2015.00077

Singh, R., Cuervo, A.M., 2011. Autophagy in the cellular energetic balance. Cell Metab. 13, 495–504. doi:10.1016/j.cmet.2011.04.004

Sotzny, F., Schormann, E., Kühlewindt, I., et al., 2016. TCF11/Nrf1-mediated induction of proteasome expression prevents cytotoxicity by rotenone. Antioxid. Redox Signal. 25, 870–885. doi:10.1089/ars.2015.6539

Steffen, J., Seeger, M., Koch, A., et al., 2010. Proteasomal degradation is transcriptionally controlled by TCF11 via an ERAD-dependent feedback loop. Mol. Cell 40, 147–158. doi:10.1016/j.molcel.2010.09.012

Studencka-Turski, M., Çetin, G., Junker, H., et al., 2019. Molecular insight into the IRE1α-mediated type I interferon response induced by proteasome impairment in myeloid cells of the brain. Front. Immunol. 10, 2900. doi:10.3389/fimmu.2019.02900

Sung, M.-H., Bagain, L., Chen, Z., et al., 2008. Dynamic effect of bortezomib on NF-κB activity and gene expression in tumor cells. Mol. Pharmacol. 74, 1215–1222. doi:10.1124/mol.108.049114

Suraweera, A., Münch, C., Hanssum, A., et al., 2012. Failure of amino acid homeostasis causes cell death following proteasome inhibition. Mol. Cell 48, 242–253. doi:10.1016/j.molcel.2012.08.003

Thibaudeau, T.A., Anderson, R.T., Smith, D.M., 2018. A common mechanism of proteasome impairment by neurodegenerative disease-associated oligomers. Nat. Commun. 9, 1–14. doi:10.1038/s41467-018-03509-0

Tomlin, F.M., Gerling-Driessen, U.I.M., Liu, Y.-C., et al., 2017. Inhibition of NGLY1 inactivates the transcription factor Nrf1 and potentiates proteasome inhibitor cytotoxicity. ACS Cent. Sci. 3, 1143–1155. doi:10.1021/acscentsci.7b00224

Torrelo, A., Patel, S., Colmenero, I., et al., 2010. Chronic atypical neutrophilic dermatosis with lipodystrophy and elevated temperature (CANDLE) syndrome. J. Am. Acad. Dermatol. 62, 489–495. doi:10.1016/j.jaad.2009.04.046

Vangala, J.R., Sotzny, F., Krüger, E., et al., 2016. Nrf1 can be processed and activated in a proteasome-independent manner. Curr. Biol. 26, R834–R835. doi:10.1016/j.cub.2016.08.008

Vattem, K.M., Wek, R.C., 2004. Reinitiation involving upstream ORFs regulates ATF4 mRNA translation in mammalian cells. Proc. Natl. Acad. Sci. 101, 11269–11274. doi:10.1073/pnas.0400541101

Vernace, V.A., Arnaud, L., Schmidt-Glenewinkel, T., et al., 2007. Aging perturbs 26S proteasome assembly in Drosophila melanogaster. FASEB J. 21, 2672–2682. doi:10.1096/fj.06-6751com

Vrábel, D., Pour, L., Ševčíková, S., 2019. The impact of NF-κB signaling on pathogenesis and current treatment strategies in multiple myeloma. Blood Rev. 34, 56–66. doi:10.1016/j.blre.2018.11.003

Yamazaki, H., Hiramatsu, N., Hayakawa, K., et al., 2009. Activation of the Akt-NF-kappaB pathway by subtilase cytotoxin through the ATF6 branch of the unfolded protein response. J. Immunol. 183, 1480–1487. doi:10.4049/jimmunol.0900017

Yang, Y., Liu, L., Naik, I., et al., 2017. Transcription factor C/EBP homologous protein in health and diseases. Front. Immunol. 8. doi:10.3389/fimmu.2017.01612

Ye, J., Kumanova, M., Hart, L.S., et al., 2010. The GCN2-ATF4 pathway is critical for tumour cell survival and proliferation in response to nutrient deprivation. EMBO J. 29, 2082–2096. doi:10.1038/emboj.2010.81

Zhang, Y., Manning, B.D., 2015. mTORC1 signaling activates NRF1 to increase cellular proteasome levels. Cell Cycle 14, 2011–2017. doi:10.1080/15384101.2015.1044188

Zhang, Y., Nicholatos, J., Dreier, J.R., et al., 2014. Coordinated regulation of protein synthesis and degradation by mTORC1. Nature 513, 440–443. doi:10.1038/nature13492

Zhao, J., Goldberg, A.L., 2016. Coordinate regulation of autophagy and the ubiquitin proteasome system by MTOR. Autophagy 12, 1967–1970. doi:10.1080/15548627.2016.1205770

Zhao, J., Zhai, B., Gygi, S.P., et al., 2015. mTOR inhibition activates overall protein degradation by the ubiquitin proteasome system as well as by autophagy. Proc. Natl. Acad. Sci. U. S. A. 112, 15790–15797. doi:10.1073/pnas.1521919112

Zhao, J., Garcia, G.A., Goldberg, A.L., 2016. Control of proteasomal proteolysis by mTOR. Nature 529, E1–2. doi:10.1038/nature16472

CHAPTER TWELVE

Proteasome Fate in Aging and Proteinopathies

Mary A. Vasilopoulou, Nikoletta Papaevgeniou and Niki Chondrogianni

CONTENTS

12.1 INTRODUCTION

The term "protein" derived from the Greek word "πρώτειος" (proteios), which means "primary", reflects the importance of the proteins (Olzscha, 2019). The term proteostasis reveals the network of mechanisms responsible for proteome maintenance (Chondrogianni et al., 2015b) Most of the newly synthesized cellular proteins must fold in a specific stable three-dimensional structure that should be maintained over their lifetime to be biologically functional and not to disturb the function of other molecules (Anfinsen, 1973). Protein folding is error-prone (during mRNA transcription/maturation/translation, folding etc.). Thus, misfolded proteins must be recognized and degraded to maintain a healthy proteome without aberrant interactions that can lead to toxic protein aggregates. Moreover, proteostatic mechanisms ensure that proteins are not only folded in the right way but also at the right time, quantity and location (Balchin et al., 2016).

The balance between the different proteostatic mechanisms is pivotal to maintain the proteome integrity within individual cells, tissues and organs. However, various conditions promote the disruption of this balance leading to the accumulation of misfolded and damaged proteins and the accumulation of toxic protein aggregates. Two models may explain this disequilibrium: (a) continuous expression of misfolded proteins (due to abnormally enhanced expression of mutated proteins, incorrect protein folding and/or problematic degradation)

DOI: 10.1201/9781003048138-12

that ultimately defeat the guardians of the proteome or progressively disrupt the normal proteostasis function (i.e. in pathological conditions) or
(b) gradual deterioration of the proteostatic mechanisms (synthesis, folding or degradation) during
aging progression or due to environmental factors
(such as various kinds of stress) that lead to accelerated protein aggregation (Hartl, 2017).

In this chapter, we briefly present the mechanisms of protein quality control, and we mainly
focus on the ubiquitin–proteasome system (UPS).
We initially describe the various UPS components
and we present studies demonstrating that proteasome activation is a valuable anti-aging strategy.
In the second part of the chapter, we discuss the
four major proteinopathies, namely Alzheimer's
(AD), Parkinson's (PD), Huntington's (HD) disease
and amyotrophic lateral sclerosis (ALS) and the
UPS implication in their pathology. Interventions
based on proteasome activation are also presented
as a promising therapeutic approach.

12.2 PROTECTIVE MECHANISMS OF PROTEIN QUALITY CONTROL

To maintain proteostasis, a proteostasis network
(PN) with components distributed in three pro-
cess-related branches (synthesis, folding and degradation) exists (Chondrogianni et al., 2015b).
Molecular chaperones are the key factors that act
as liaisons between these processes. A summary
of PN implications in health and disease is illustrated in Figure 12.1.

12.2.1 Synthesis

Protein synthesis is a high energy-consuming
process related to environmental conditions,
such as nutrients uptake and stress (Mayer and
Grummt, 2006). Therefore, regulation of the core
constituents of the mRNA translational machinery is vital. Despite its importance, protein synthesis remains an error-prone procedure; 1 in

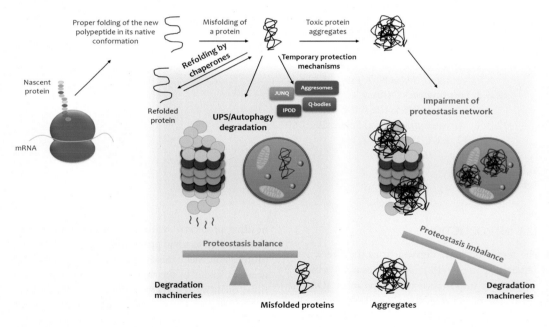

Figure 12.1 **The proteostasis network in health and disease.** Newly synthesized proteins are folded to acquire their final
native conformation, which is thermodynamically stable. This process is under the strict surveillance of various molecular
chaperones that assure the proper protein folding and prevent aberrant intramolecular interactions. If not bound by chaperones, fully or partially unfolded proteins expose their hydrophobic amino acid residues or get increased; both conditions favor
protein aggregation. Under normal conditions, the cellular degradation machineries maintain protein homeostasis via elimination of the damaged proteins either through sequestration of PN components that aim to refold these proteins (chaperones) or
through the UPS- or autophagy-mediated degradation. Aggresomes, JUNQ, Q-bodies and IPOD are also serving as protective quality-control compartments to which misfolded/aggregated proteins are sequestered as a temporary protective cellular
response to a failure of the other proteostatic mechanisms (left blue panel). However, the age-related dysfunction of the PN
components disrupts the proteostasis balance. The decreased function of the UPS and autophagy facilitates the protein aggregation process leading to the accumulation of toxic protein aggregates and the subsequent progression of PMDs (right red panel).

20 protein molecules produced will be defective, potentially causing misfolding or reduced stability (Zaher and Green, 2009). Apart from the core translation components, protein synthesis also incorporates various chaperones and components of the UPS. Chaperones act directly on the ribosomes to assist the folding procedure and prevent premature misfolding of the nascent chain. Upon error sensing, the ribosome-associated quality-control pathway is activated and UPS factors cooperate with chaperones to remove defective/unrepaired chains (Joazeiro, 2017).

12.2.2 Protein Folding

Proper translation should be followed by proper folding of the nascent polypeptide. Chaperones are key PN effectors ensuring the proper folding of the new polypeptide at the right time and its conformational maintenance (Dunker et al., 2008), during which the polypeptide chain acquires the minimum energy conformation. Although the process is very efficient, errors may still occur, resulting in a non-native folding state (Hinault et al., 2006). The structural stability of these misfolded proteins depends on the residues exposed to the cytosol; hydrophobic residues trigger a cataract of proteostatic mechanisms. If this system fails to repair the defective proteins, the exposed hydrophobic motifs interact with cellular macromolecules resulting in oligomerization and consequently aggregation leading to cytotoxicity (Olzscha et al., 2011).

Upon accumulation of misfolded proteins and toxic aggregates, the PN triggers a cytosolic response to stress (heat-shock response [HSR]) regulated by the heat-shock factors (HSFs). HSFs bind to heat-shock elements in promoter regions to initiate the transcription of various genes encoding chaperones and other PN factors.

If all proteostatic mechanisms fail to eliminate the misfolded/aggregated proteins, PN sequesters the defective proteins in distinct compartments as the last alternative defensive strategy. These compartments function as temporary storage until the misfolded proteins can be refolded or degraded. Four distinct compartments have been described so far; the Q-body/Cyto-Q compartment, the juxtanuclear quality-control compartment (JUNQ), the insoluble protein deposit (IPOD) and the microtubule-dependent inclusion bodies (IBs), namely aggresomes (Hartl, 2017).

12.2.3 Protein Degradation

Proteins that cannot fold properly must be cleared to prevent their accumulation and further aggregation. Their disposal is achieved through degradation by the two major cellular proteolytic systems, namely the UPS and the autophagy-lysosomal pathway (ALP). UPS mainly takes part in the degradation of individual proteins, whereas ALP can also mediate the disposal of aggregates or membrane-associated proteins (Ciechanover and Kwon, 2017; Cuervo and Wong, 2014).

During aging and various pathological conditions, UPS and ALP functions decrease leading to PN impairment (Cuervo and Wong, 2014). Consequently, widespread protein aggregation occurs, especially in post-mitotic cells, such as muscle and neurons, leading to various diseases (Walther et al., 2015).

12.2.3.1 Proteasome

Eukaryotic proteasomes are multi-subunit megacomplexes consisting of regulatory (RP) and catalytic (CP) particles. The CP degrades ~80% of the proteome (Ciechanover and Kwon, 2017), including short-lived, oxidized and misfolded proteins in an ATP-dependent or -independent manner (Collins and Goldberg, 2017). Proteasome degradation is involved in several cellular functions such as cell cycle, apoptosis, differentiation, response to stress, immune responses, genetic regulation and metabolism, among others.

The CP is a barrel-like structure, also known as the 20S core. It is produced through the assembly of four rings; the two outer rings are formed by seven different α subunits (α_{1-7}), the two inner rings consist of seven different β subunits (β_{1-7}) in an $\alpha_{1-7}\beta_{1-7}\beta_{1-7}\alpha_{1-7}$ arrangement. The N-terminal domains of the α-type subunits regulate substrate access to the inner cavity. Three of the β-type subunits (β1, caspase-like activity/C-L; β2, trypsin-like activity/T-L; β5, chymotrypsin-like activity/CT-L) consist the proteolytic centers (Rousseau and Bertolotti, 2018). The 20S CP can stand alone to degrade small polypeptides and oxidized proteins in an ATP-independent manner (Pickering and Davies, 2012). In mammals, there is also another form of proteasome, known as the immunoproteasome (Ebstein et al., 2012). This form consists of β1i, β2i and β5i immunosubunits (substituting β1, β2 and β5 constitutive subunits, respectively) following *de novo* synthesis and assembly (Kniepert and Groettrup, 2014). The immunosubunits exhibit differences in the cleavage sites, thus generating peptides more appropriate for antigen presentation by the major histocompatibility complex (MHC) class I (Kloetzel, 2004). Apart from the role of immunoproteasomes in the immune response, they have a distinct role in cellular adaptation to oxidative stress through the selective degradation of oxidized proteins (Pickering et al., 2010).

The 26S proteasome consists of the 20S CP and one 19S RP forming a large multi-subunit protease. If two 19S complexes bind at the two ends of the 20S, the 30S proteasome is formed. Degradation of protein substrates through the 26S/30S proteasomes occurs in an ATP-dependent manner. 19S RP is divided into two major structures; the base and the lid. The base is formed by six ATPases [RP triple-A protein 1 (Rpt1)–Rpt6] that form an AAA ATPase ring and four non-ATPases [RP non-ATPase 1 (Rpn1), Rpn2, Rpn10 and Rpn13]. Nine non-ATPase subunits form the lid (Rpn3, Rpn5-9, Rpn11, Rpn12 and Rpn15/Sem1/DSS1). The 19S RP is responsible for the recognition, binding, deubiquitination and unfolding of the protein substrate as well as for its entrance to the inner CP cavity through opening of the α-gated channel (Rousseau and Bertolotti, 2018). Table 12.1 summarizes the proteasome subunits and the human genes encoding them.

Apart from the 19S RP, other proteasome regulators have been identified to bind to the 20S CP giving rise to alternative proteasome forms (Stadtmueller and Hill, 2011). 11S regulator complex (also known as PA28 or REG) is a proteasome activator that exists in three isoforms; the heteroheptamer PA28α and PA28β and the homoheptamer PA28γ. The 11S regulator preferentially binds to the immunoproteasome to increase its activity and to facilitate MHC class I antigen presentation. The 11S regulator is also implicated in cellular adaptation to oxidative stress by enhancing immunoproteasome selectivity for oxidatively modified proteins (Pickering and Davies, 2012).

UPS-mediated degradation requires substrate tagging with ubiquitin (Ub), a small evolutionary conserved protein. Ub is incorporated into the client protein and facilitates ATP-dependent degradation. Ubiquitination involves three enzyme-catalyzed steps; (a) the ATP-consuming activation of Ub by the E1 enzyme (Ub-activating enzyme), (b) the conjugation of Ub to the E2 enzyme (Ub-conjugating enzyme) and (c) the Ub transfer from the E2 to the E3 enzyme (ligase) to the protein substrate or directly to the substrate that is anchored on the E3 enzyme. Finally, a poly-Ub chain is formed onto the targeted protein through the repeated action of E1, E2 and E3 enzymes. Upon poly-Ub chain formation, the substrate is transferred to the proteasome for degradation. Before substrate entrance into the CP, Ub molecules are released by deubiquitinating enzymes [DUBs; Rpn11, Ub-specific protease 14 (USP14),

TABLE 12.1

Human genes encoding the 20S and 19S proteasome subunits

20S proteasome				19S regulatory particle			
α-Subunits	Gene	β-Subunits	Gene	Base subunits	Gene	Lid subunits	Gene
α1	PSMA1	β1	PSMB6	Rpt1	PSMC2	Rpn3	PSMD3
α2	PSMA2	β2	PSMB7	Rpt2	PSMC1	Rpn5	PSMD12
α3	PSMA3	β3	PSMB3	Rpt3	PSMC4	Rpn6	PSMD11
α4	PSMA4	β4	PSMB2	Rpt4	PSMC6	Rpn7	PSMD6
α5	PSMA5	β5	PSMB5	Rpt5	PSMC3	Rpn8	PSMD7
α6	PSMA6	β6	PSMB1	Rpt6	PSMC5	Rpn9	PSMD13
α7	PSMA7	β7	PSMB4	Rpn1	PSMD2	Rpn11	PSMD14
				Rpn2	PSMD1	Rpn12	PSMD8
				Rpn10	PSMD4	Sem1	SEM1
				Rpn13	ADRM1		

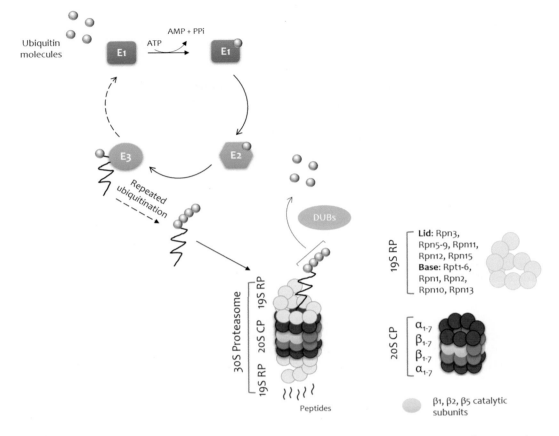

Ubiquitin
molecules

AMP + PPi

ATP

E1

E1

E3

Repeated
ubiquitination

E2

DUBs

Lid: Rpn3,
Rpn5-9, Rpn11,
Rpn12, Rpn15
Base: Rpt1-6,
Rpn1, Rpn2,
Rpn10, Rpn13

19S RP

30S Proteasome

19S RP

20S CP

19S RP

α_{1-7}
β_{1-7}
β_{1-7}
α_{1-7}

20S CP

Peptides

$\beta1$, $\beta2$, $\beta5$ catalytic
subunits

Figure 12.2 **The ubiquitin–proteasome system (UPS).** Ub is incorporated into the protein substrate via the concerted action of three enzymes; E1 Ub-activating, E2 Ub-conjugating and E3 Ub ligase enzymes. A three-step repeated action of these enzymes produces a poly-Ub chain (the standard form of which is tetra-Ub) onto the targeted protein. The ubiquitinated protein is thus recognized by the 19S RP. After its recognition, the DUBs remove the Ub molecules from the protein substrate that then enters the 20S CP. DUBs also ensure Ub recycling, which will participate in another ubiquitination cycle. The structure of the 30S proteasome is also illustrated. It consists of the 20S CP and two 19S RP. $\beta1$, $\beta2$ and $\beta5$ 20S proteasome subunits are the catalytic centers of the CP.

UCHL37] for recycling (Figure 12.2) (Rousseau and Bertolotti, 2018).

Substrate unfolding is a prerequisite for its entrance in the inner-CP cavity for degradation. Thus, the proteasome is unable to degrade aggregated proteins. Therefore, chaperones that act as disaggregases are required for the degradation of proteasome-targeted aggregated proteins. In general, chaperones retain the unsuccessfully (re) folded proteins in a state proper for Ub labeling (Esser et al., 2004).

The functional decline of the UPS during aging results in the accumulation and consequently the aggregation of non-degraded proteins. Among others, a decline in the proteasome amount and/ or activity is observed. This decline may be due

to changes in subunits expression and/or modifications of proteasomal subunits (Papaevgeniou and Chondrogianni, 2016). The age-related proteasomal impairment is followed by increased levels of oxidatively damaged proteins and protein aggregation (Pickering and Davies, 2012). Aggregated proteins anchor directly to the proteasome clogging its catalytic centers leading to further impairment and more protein aggregates. As a result, toxic inclusions accumulate leading to degeneration (Thibaudeau et al., 2018).

Deceleration of the progression of aging and age-related diseases has been achieved through proteasome activation via genetic manipulation or compound treatment (Papaevgeniou and Chondrogianni, 2016) (Figure 12.3). Several

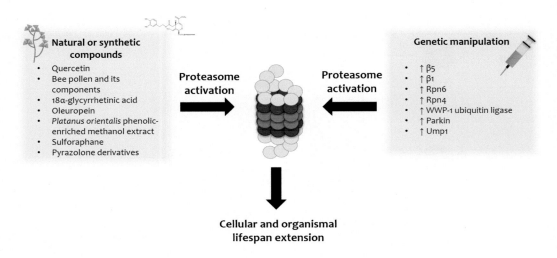

Natural or synthetic compounds

- Quercetin
- Bee pollen and its components
- 18α-glycyrrhetinic acid
- Oleuropein
- *Platanus orientalis* phenolic-enriched methanol extract
- Sulforaphane
- Pyrazolone derivatives

Proteasome activation

Proteasome activation

Genetic manipulation

- ↑ β5
- ↑ β1
- ↑ Rpn6
- ↑ Rpn4
- ↑ WWP-1 ubiquitin ligase
- ↑ Parkin
- ↑ Ump1

Cellular and organismal lifespan extension

Figure 12.3 **Genetic- and compound-mediated UPS modulation in aging.** UPS activation can be achieved through genetic means or through natural or synthetic compounds leading to cellular and/or organismal lifespan extension.

compounds have been identified as proteasome activators with simultaneous anti-aging properties. Studies in cellular and organismal aging models have revealed such compounds, like quercetin, bee pollen and its components, 18α-glycyrrhetinic acid (18α-GA), oleuropein, *Platanus orientalis* phenolic-enriched methanol extract, pyrazolone derivatives and sulforaphane, as reviewed by Chondrogianni and colleagues (2020). Dietary restriction through food deprivation or genetic manipulation, a known anti-aging strategy, is also characterized by increased proteasome activities in multiple animal models (*Saccharomyces cerevisiae, Caenorhabditis elegans* and rats; reviewed in Papaevgeniou and Chondrogianni, 2016). Overexpression of β5 subunit leads to proteasome activation and decelerates the progression of the age-related phenotype *in vitro* and *in vivo* (Chondrogianni et al., 2005; Chondrogianni et al., 2015a; Munkácsy et al., 2019; Nguyen et al., 2019). This is also the case for β1 overexpression in mammalian cells (Chondrogianni et al., 2005). Rpn11 overexpression rescues the proteasome activity decline observed in old flies while it promotes lifespan extension (Tonoki et al., 2009). RPN-6 overexpression exhibits similar results in *C. elegans* (Vilchez et al., 2012). Likewise, reduced degradation of the transcription factor of many proteasome genes, namely Rpn4/Ufd5, due to loss of UBR2 and MUB1 ligases, enhances proteasome activity and confers lifespan extension in *S. cerevisiae* (Kruegel et al., 2011). Overexpression of the WWP-1 Ub ligase enhances the degradation rates and promotes longevity in human fibroblasts

and *C. elegans* (Cao et al., 2011; Carrano et al., 2009). Additionally, overexpression of parkin, a critical RING domain Ub ligase, leads to increased proteasome function and lifespan extension in *Drosophila melanogaster* (Um et al., 2010a). Overexpression of Ump1, a chaperone necessary for 20S assembly in yeast, also leads to enhanced degradation rates and chronological lifespan elongation (Chen et al., 2006).

12.2.3.2 Autophagy

Under basal conditions, soluble proteins can also be degraded through chaperone-mediated autophagy (CMA) (Kaushik and Cuervo, 2018). Misfolded proteins may not always avoid aggregation, thus forming larger insoluble inclusions and inducing proteotoxicity. Although aggregated proteins are poor proteasome substrates, they are better ALP targets (Lamark and Johansen, 2012). In contrast to CMA, inclusions or aggregates are engulfed by double-membrane vesicles, the autophagosomes. Autophagosomes sequester their content to deliver it into the lysosome through fusion (He and Klionsky, 2009). Impaired lysosomal clearance has been suggested to promote degenerative disorders (Nixon, 2013).

12.3 PROTEINOPATHIES

During aging, gradual proteostasis deterioration provokes various toxic protein aggregates-related pathological conditions (Labbadia and Morimoto,

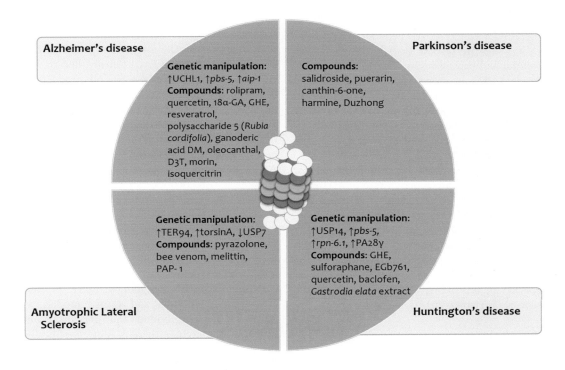

Alzheimer's disease

Genetic manipulation: ↑UCHL1, ↑pbs-5, ↑aip-1
Compounds: rolipram, quercetin, 18α-GA, GHE, resveratrol, polysaccharide 5 (*Rubia cordifolia*), ganoderic acid DM, oleocanthal, D3T, morin, isoquercitrin

Parkinson's disease

Compounds: salidroside, puerarin, canthin-6-one, harmine, Duzhong

Amyotrophic Lateral Sclerosis

Genetic manipulation: ↑TER94, ↑torsinA, ↓USP7
Compounds: pyrazolone, bee venom, melittin, PAP-1

Huntington's disease

Genetic manipulation: ↑USP14, ↑pbs-5, ↑rpn-6.1, ↑PA28γ
Compounds: GHE, sulforaphane, EGb761, quercetin, baclofen, *Gastrodia elata* extract

Figure 12.4 **Genetic- and compound-mediated UPS modulation in PMDs.** UPS modulation through genetic means or through the administration of natural or synthetic compounds can reduce the levels of misfolded proteins and alleviate the symptoms of the major neurodegenerative diseases, such as AD, PD, HD and ALS.

2015). Post-mitotic cell types, such as neurons, are mostly affected by these aggregates leading to the disease phenotype (Sala et al., 2017). Consequently, toxic protein aggregates' formation is a hallmark of many pathologies, termed as proteinopathies or protein misfolding diseases (PMDs). One or two species of aggregated proteins often drive the progression of each disease. Proteinopathies are categorized depending on the content and structure of the involved aggregates; amyloidosis, tauopathy, synucleinopathy (Bayer, 2015) and poly-glutaminopathy (Walker, 2007). Neurodegenerative diseases are the most prominent type of proteinopathies that affect a large part of the population, especially the elderly. AD, PD, HD and ALS are disease examples reflecting the heterogeneity of proteinopathies. Most patients manifest a PMD in a stochastic and age-dependent manner (Hartl, 2017), highlighting the pivotal role of maintaining proteostasis with age. Since UPS inhibition is a common characteristic for most proteinopathies, we will mainly focus on the proteolytic part of proteostasis and predominately on UPS that is impaired upon PMD

progression. We will also present potential therapeutic approaches based on proteasome modulation, as summarized in Figure 12.4.

12.3.1 Alzheimer's Disease

Two main types of protein aggregates characterize AD; the extracellular amyloid-β (Aβ) peptide that accumulates in plaques and the intracellular microtubule-associated protein tau that accumulates in neurofibrillary tangles (NFTs) in the cerebral cortex (Masters et al., 2015).

APP is the first gene identified to play an essential role in AD. It encodes the β-amyloid precursor protein (APP), a transmembrane, receptor-like glycoprotein expressed ubiquitously in neural cell types (Masters et al., 2015). APP undergoes post-translational proteolytic processing through two distinct pathways; the amyloidogenic and the non-amyloidogenic one. In the amyloidogenic pathway, APP is cleaved by β-secretase, leading to a large soluble ectodomain of APP and a C-terminal peptide (C99). The subsequent cleavage of C99 by γ-secretase gives rise to Aβ peptide.

The γ-secretase cleaves APP at different positions generating a variety of Aβ peptides. The precise cleavage sites are pivotal for Aβ self-aggregation potential and pathogenicity. The 42 amino acids peptide ($A\beta_{42}$) is more aggregation-prone and neurotoxic than the predominant $A\beta_{40}$ peptide. In the non-amyloidogenic pathway, APP is cleaved by α- and γ-secretase generating $A\beta_{17-40}$ and $A\beta_{17-42}$ peptides (p3 peptides), which do not form aggregates (Jeong, 2017).

Although the amyloid hypothesis argues for the Aβ peptide as the primary cause of AD, many pathological features in patients, such as cognitive impairment, correlate with tau aggregation. Tau is a microtubule-associated protein normally located on the axon contributing to axonal transport through binding and stabilization of the microtubules. In AD, tau undergoes hyperphosphorylation, misfolding and aggregation, giving rise to NFTs (Jeong, 2017).

12.3.1.1 UPS in AD

Detection of Ub in senile plaques was the first indication for the link between UPS and AD. Ub was then found accumulated in both plaques and tangles in AD patients, while UBB[+1] (a Ub mutant protein that lacks the glycine 76 conjugating residue) was elevated in these structures. UBB[+1] is expressed in neurons and binds to proteasome targets in competition with normal Ub. It inhibits the Ub-dependent proteolysis in neuronal cells acting as a detrimental factor leading to misfolded protein aggregation and further neurotoxicity (reviewed in Huang and Figueiredo-Pereira, 2010).

$A\beta_{40}$ and $A\beta_{42}$ peptides interact directly and inhibit the 20S proteasome (Oddo, 2008). As reviewed in Y. Zhang and colleagues (2017), decreased proteasome activities are detected in brain samples of AD patients and transgenic AD mice, while tau and Ub co-localization occurs in NFTs. Moreover, upon proteasome inhibition, tau degradation is impaired. There is a positive correlation between increased aggregated tau protein levels and progressive 26S activity decline in the brain of transgenic mice expressing tau. Conversely, enhanced cAMP-protein kinase A (PKA) phosphorylation via rolipram induces proteasome activity, subsequently protecting against tau toxicity. Rolipram improves tauopathy-induced cognitive impairment in the same AD mouse model, albeit only in the early-stage disease.

Dysfunction of Ub carboxyl-terminal hydrolase L1 (UCHL1) DUB has been revealed in AD. UCHL1 overexpression reduces Aβ production, decreases plaque formation and improves memory loss in transgenic AD mice (M. Zhang et al., 2014). Increased proteasome content and activity through pbs-5 overexpression delays the paralysis in a C. elegans transgenic strain expressing human $A\beta_{1-42}$ peptide while it reduces both total and oligomeric Aβ peptide levels (Chondrogianni et al., 2015a). Likewise, overexpression of aip-1, an inducible 19S subunit, decreases Aβ accumulation in a nematode AD model and significantly prevents Aβ toxicity (Hassan et al., 2009), advocating for the protective effect of proteasome activation against Aβ proteotoxicity.

Various natural compounds exert positive effects on AD cellular and animal models, as reviewed in Panagiotidou and Chondrogianni (2020) and references therein. More specifically, the flavonoid quercetin and 18α-GA, a triterpenoid isolated from Glycyrrhiza radix, both identified as potent proteasome activators, delay $A\beta_{1-42}$-induced paralysis and lower Aβ aggregation in an AD nematode model. Reduced Aβ toxicity is also observed in human and murine cells treated with 18α-GA. Similar results are obtained with guarana hydroalcoholic extract (GHE) from Paullinia cupana var. sorbilis and resveratrol. Treatment with resveratrol promotes Aβ peptide clearance through enhanced proteasomal degradation in APP695-transfected HEK293 cells, while in an AD transgenic mouse model, resveratrol improves cognitive impairment and increases proteasome content and function. Cells treated with polysaccharide 5 from Rubia cordifolia and ganoderic acid DM isolated from Ganoderma lucidum exhibit lower Aβ aggregate levels, an effect probably linked to enhanced proteasome function. Oleocanthal, one of the active components of extra virgin olive oil, attenuates Aβ and tau aggregation via enhancement of their clearance and upregulates Psmd1 (encodes Rpn2) and Psmβ4 (encodes β7) subunits expression in SH-SY5Y cells under normal and stress conditions. 3H-1,2-dithiole-3-thione (D3T), known to induce proteasome subunits' expression and to enhance proteasome activity in an Nrf2-dependent manner, decreases Aβ peptide levels in various AD models, implying a potential link between its beneficial effects and its

PROTEOSTASIS AND PROTEOLYSIS

proteasome activating properties. Finally, two fla-vonols, namely morin and isoquercitrin, impede β- and γ-secretase activities to prevent $A\beta_{25-35}$ aggregation and to disaggregate the already existing amyloid fibrils *in vitro*. They also reduce ROS production and prevent the CT-L activity decline induced by oxidative stress.

12.3.2 Parkinson's Disease

PD is the second most common neurodegenerative disease after AD. Cognitive impairment and disability of patients to perform their daily routine autonomously are among its pathological features. The most common PD characteristics are dyskinesia, tremor and rigidity (Poewe et al., 2017).

Loss of the dopaminergic neurons in the substantia nigra pars compacta located in the midbrain and presence of Lewy bodies and Lewy neurites (insoluble cytoplasmic inclusions of α-synuclein fibrils) in the remaining survival nigral neurons are the two main PD hallmarks. Lewy neurites are mainly located in the amygdala and striatum and are more abundant than Lewy bodies. However, the latter seems to be more critical for PD progression, as they are detected in the later disease stages (Rocha et al., 2018).

The major constituent of Lewy bodies and Lewy neurites is α-synuclein (encoded by the *SNCA* gene), a soluble 140 amino acid protein with a relatively unknown normal function. It is expressed in various tissues, but its expression in the nervous system, especially in the presynaptic nerve terminals, is enriched. It consists of three regions, one of which comprises a repeated hydrophobic and aggregation-prone sequence. Soluble α-synuclein monomers form oligomers and then protofibrils, leading to the formation of insoluble α-synuclein fibrils. Mutations in the *SNCA* gene, post-translational modifications or proteostatic mechanisms impairment are few examples that could trigger the pathological form of this protein and its subsequent accumulation and aggregation (Poewe et al., 2017; Rocha et al., 2018).

12.3.2.1 UPS in PD

Accumulation of α-synuclein is strongly associated with UPS malfunction. All 20S activities are impaired in the brain samples of PD patients. Moreover, the α-type proteasome subunits are reduced or lost in dopaminergic neurons in the

same samples. α-Synuclein overexpression in PC12 cells promotes proteasome impairment; soluble α-synuclein oligomers are associated with the 26S complex and the resulting proteasome inhibition further increases the levels of these forms (reviewed in Opattova et al., 2015). α-Synuclein soluble oligomers possess a three-dimensional structure that inhibits the 26S proteasome by binding to α-subunits, thus preventing gate opening (Thibaudeau et al., 2018). Microinjection of lactacystin, a proteasome inhibitor, above the substantia nigra of C57Bl/6 mice leads to a PD-like phenotype with locomotor impairment and α-synuclein accumulation (Savolainen et al., 2017). Mutations in the *PARKIN* gene (that encodes E3 parkin Ub ligase) are strongly associated with PD pathology. Parkin stimulates the 26S activity through interaction with the 19S subunits, while its knockdown reduces 26S activity in mice and flies (Um et al., 2010b).

Salidroside, an extract from the *Rhodiola rosea* L., induces α-synuclein clearance by decreasing the high molecular weight ubiquitinated proteins and increasing the levels of free Ub, UCHL1 and parkin in the 6-hydroxydopamine (6-OHDA)-induced PD model *in vitro* (Li et al., 2018). Puerarin, an isoflavone isolated from *Pueraria lobata*, reverses the 1-methyl-4-phenylpyridinium (MPP+)-induced reduction of CT-L activity in SH-SY5Y cells, accompanied by reduced levels of α-synuclein and Ub-conjugated proteins (Cheng et al., 2009). Canthin-6-one, an indole alkaloid, decreases wild-type and mutant α-synuclein levels *in vitro* through UPS activation attributed to *Psmd1* (encodes Rpn2) increased mRNA levels due to PKA activation (Yuan et al., 2019). The naturally derived alkaloid harmine exhibits similar results (Cai et al., 2019). Oral administration of Duzhong (derived from the bark of *Eucommia ulmoides* Oliv.) prevents the 1-methyl-4-phenyl-1,2,3,6-tetrahydropyridine (MPTP)-induced bradykinesia in mice and protects against MPTP-induced behavioral dysfunctions. Duzhong also reverses the MPP+-induced reduction of proteasome activity in SH-SY5Y cells (Guo et al., 2015).

12.3.3 Huntington's Disease

HD is an autosomal dominant progressive neurodegenerative disease caused by the (CAG)n trinucleotide repeat expansion, encoding glutamine, in the exon 1 of huntingtin (HTT) gene. Huntingtin is mainly expressed in the brain and testes, but its normal function is not fully understood. Impairment of movement, cognitive decline, emotional and behavioral disorders are the main characteristics of HD (Bates et al., 2015).

The normal HTT alleles have 6–36 trinucleotide repeats. Repeats between 36 and 39 show reduced penetrance, while individuals with above 40 repeats expansions will manifest HD. There is a correlation between the number of expansions and HD onset, as the higher the amount of the repeats, the earlier the age of the disease onset (Walker, 2007). Due to abnormal CAG expansions, mutant huntingtin (mHTT) contains a long polyglutamine (polyQ) tract at its N-terminus that is prone to form fibrils and aggregates and consequently neurotoxicity. IBs of mHTT aggregates are nuclear or cytoplasmic in the striatum, cerebral cortex, cerebellum and spinal cord. Several studies attribute a protective role to IB (Ramdzan et al., 2017), while others support their lethal toxicity (Arrasate and Finkbeiner, 2012; Bates et al., 2015).

12.3.3.1 UPS in HD

Aggregates of mHTT and IBs exhibit positive Ub or proteasome subunits labeling, revealing the UPS implication in HD (Juenemann et al., 2018). HTT filaments interact directly with the 19S subunits, while soluble oligomeric mHTT impedes 26S function through gate-opening inhibition (Thibaudeau et al., 2018). These studies suggest that UPS is recruited to eliminate the HTT aggregates, but this is a double-edged sword; as the UPS struggles to reduce mHTT aggregates, the expanded polyQ tracts possibly clog the proteasome. In support, many studies highlight the impairment of proteasome function in the presence of mHTT, while UPS inhibition has been linked to the accumulation of mHTT aggregates (reviewed in Fernandez-Cruz and Reynaud, 2020).

The cytoplasmic USP19 stimulates polyQ-expanded protein aggregation in cells through the enhancement of their protein levels. This involves the interaction of Hsp90 with the N-terminus of HTT and the subsequent recruitment of USP19. Inhibition of HSP90 disrupts its binding to HTT, leading to HTT cleavage by the UPS (He et al., 2017). Moreover, overexpression of USP14 in cells overexpressing mHTT reduces the mHTT aggregates *in vitro* through the UPS by preventing

IRE1α phosphorylation that is induced by mHTT (Hyrskyluoto et al., 2014).

Overexpression of the *pbs-5* or *rpn-6.1* proteasome subunits confers resistance to polyQ proteotoxicity, delays paralysis progression and improves motility in HD nematode models (Chondrogianni et al., 2015a; Vilchez et al., 2012). Overexpression of PA28γ proteasome activator increases proteasome activity and improves the survival of striatal neurons expressing mHTT (Seo et al., 2007). Also, PA28γ gene transfer in HD transgenic mice enhances C-L activity and improves the recorded behavioral abnormalities (Jeon et al., 2016). In contrast, a mutant *S. cerevisiae* strain, which prevents S119-Rpt6 phosphorylation, exhibits reduced proteasome activity and significantly larger polyQ HTT aggregates (Marquez-Lona et al., 2017).

A *Ginkgo biloba* leaf extract (EGb 761) and GHE from *P. cupana* reduce polyQ aggregation through proteasome activity enhancement in HD cellular and nematode models, respectively (Boasquivis et al., 2018; Stark and Behl, 2014). Likewise, sulforaphane reduces mHTT cytotoxicity and increases mHTT degradation *in vitro* in a proteasome-dependent manner (Liu et al., 2014). Quercetin-treated mHTT-transfected Neuro2a cells exhibit increased CT-L activity, resulting in reduced mHTT aggregates (Chakraborty et al., 2015). Baclofen, a GABAB receptor agonist, increases the CT-L activity in mHTT-expressing striatal cells and in an HD mouse model. Baclofen-injected HD mice possess a decreased number of Ub-positive neuronal intranuclear inclusions and present improvement of behavioral abnormalities (Kim and Seo, 2014). Furthermore, *Gastrodia elata* extract prevents mHTT aggregates' formation through PKA-mediated proteasome activation in mHTT-transfected PC12 cells (Huang et al., 2011).

Regarding the potential protective role of IBs, recent studies revealed that proteasomes maintain their activity and are dynamically and reversibly recruited into them. In agreement, only the HTT filaments that have not been recruited into IBs can interact with and inhibit the 26S proteasome (reviewed in Harding and Tong, 2018).

12.3.4 Amyotrophic Lateral Sclerosis

ALS is a highly heterogeneous neurodegenerative disorder characterized by the death of both upper (neurons projected from the cortex to the brainstem and spinal cord) and lower (neurons projected from the brainstem or spinal cord to the muscle) motor neurons. The degeneration usually begins with asymmetric involvement of muscle tissues and results in fatal paralysis, mainly through failure of neuromuscular respiratory function (Sibilla and Bertolotti, 2017).

ALS is considered a PMD due to the accumulation of ubiquitinated proteinaceous inclusions in motor neurons and neural accessory cells. Mutations in *TARDBP* and *SOD1* genes have been implicated in ALS. *TARDBP* gene encodes the TAR DNA-binding protein 43 (TDP-43), which is the main constituent of the skein-like inclusions found in most ALS patients. Normal TDP-43 protein is a nuclear RNA/DNA-binding protein participating in RNA metabolism. Contrary to the nuclear localization of normal TDP-43, misfolded TDP-43 aggregates are cytoplasmic. Mutations in *SOD1* are the most common mutations in familial ALS patients. Inclusions in *SOD1*-associated ALS patients contain aggregates of the mutant superoxide dismutase-1 protein (mSOD1), a major cytoplasmic antioxidant enzyme (Hardiman et al., 2017; Sibilla and Bertolotti, 2017).

12.3.4.1 UPS in ALS

Studies have revealed that the soluble TDP-43 oligomers are preferentially degraded by the UPS, while UPS inhibition results in elevated levels of ubiquitinated TDP-43 aggregates (Hergesheimer et al., 2019). Moreover, the Ub-conjugating enzyme UBE2E3 promotes ubiquitination of a specific TDP-43 mutant, while silencing of the Ub isopeptidase Y (UBPY) results in accumulation of insoluble high molecular weight TDP-43 and Ub species and enhancement of eye neurodegeneration in *D. melanogaster* (Hans et al., 2014). Knockout of Rpt3 19S subunit in mice results in locomotor dysfunction and progressive motor neuron loss (Tashiro et al., 2012). Ubiquilin is a molecular chaperon with a Ub-binding domain, actively involved in UPS and autophagy and an ALS-related protein, while Ub chaperone ubiquilin 2 (UBQLN2) mutations have been linked to the X-linked forms of ALS with frontotemporal dementia. Knockdown of *dUbqn* (the homolog of ubiquilin) in *D. melanogaster* increases the levels of Ub chains and poly-ubiquitinated proteins, implying a proteostasis defect. *dUbqn* knockdown in neurons and motor neurons of the flies afflicts locomotion and memory and increases the soluble

cytoplasmic Ub+-TBPH/TDP-43. Restoration of Ub-dependent proteolysis through overexpression of the chaperone VCP, namely TER94 (a member of the AAA+ ATPase family), reduces the soluble Ub+-TBPH/TDP-43 but does not restore its nuclear localization (Jantrapirom et al., 2018).

Concerning *SOD1*-associated ALS, proteasome activity or expression of UPS components is reduced in the presence of mSOD1 protein in both *in vitro* and *in vivo* ALS models. In agreement, thoracic spinal cord samples from ALS patients display decreased levels of β5 proteasome subunit, while CT-L activity inhibition increases mSOD1 levels, indicating a proteasome-dependent mSOD1 turnover as reviewed by Fernandez-Cruz and Reynaud (2020) and references therein. In a recent study, knockdown of USP7 decreases SOD1[G85R] protein levels *in vitro* through activation of the transforming growth factor β (TGFβ)-SMAD pathway. More specifically, loss of *USP7* decreases the levels of the E3 Ub ligase NEDD4L that orchestrates the degradation of SMAD2, the transcription modulator of the TGFβ pathway. This, in turn, leads to enhanced clearance of misfolded proteins through autophagy (T. Zhang et al., 2020). Pyrazolone induces proteasome activation in PC12-SOD1 (G93A) cells; proteomic analysis revealed the 26S proteasome subunits PSMC1 (Rpt2) and PSMC4 (Rpt3) as its targets (Trippier et al., 2014). Bee venom exerts neuroprotective effects by improving motor activity in an ALS mouse model. Melittin, a 26 amino acid polypeptide component of bee venom, improves motor activity of hSOD1[G93A] mice through proteasome activity restoration (Yang et al., 2011). Overexpression of torsinA, an AAA+ family member with a chaperone-like activity, attenuates mSOD1-induced endoplasmic reticulum (ER) stress by driving mSOD1 to proteasomal elimination in an ALS nematode model (Thompson et al., 2014). Finally, the peptide proteasome-activating peptide 1 (PAP-1), a conformational proteasome activator, increases proteasome activity in SH-SY5Y cells expressing mSOD1 and induces the clearance of protein aggregates *in vitro* (Dal Vechio et al., 2014).

12.4 CONCLUSIONS

Various studies point to proteostasis impairment as a crucial contributor to aging and PMD pathogenesis (Labbadia and Morimoto, 2015). The formation of large aggregates along with their dysfunctional removal by the PN components results in their accumulation and ultimately in cytotoxicity. Proteasome-mediated degradation is one of the key processes to ensure cell survival and longevity, while proteasome dysfunction has been linked to aging and PMD onset and progression. Understanding the mechanisms implicated in proteasomal degradation may facilitate the design of intervention strategies that target UPS functional enhancement as a means to reduce or even reverse aging and PMD pathology.

ACKNOWLEDGMENTS

We apologize to all the authors whose work we could not cite due to space limitations. NC lab is currently funded by the European Union and Greek national funds through the Operational Program Competitiveness, Entrepreneurship and Innovation under the call RESEARCH − CREATE − INNOVATE (MIS 5031214 and 5030851). We also acknowledge funding by Greece and the European Union (European Social Fund [ESF]) through the Operational Program "Human Resources Development, Education and Lifelong Learning 2014–2020" in the context of the project "Proteostasis" (MIS 5050341).

REFERENCES

Anfinsen, C.B., 1973. Principles that govern the folding of protein chains. Science (80−) 181, 223–230. doi:10.1126/science.181.4096.223

Arrasate, M., Finkbeiner, S., 2012. Protein aggregates in Huntington's disease. Exp Neurol 238, 1–11. doi:10.1016/j.expneurol.2011.12.013

Balchin, D., Hayer-Hartl, M., Hartl, F.U., 2016. In vivo aspects of protein folding and quality control. Science (80−) 353, aac4354. doi:10.1126/science.aac4354

Bates, G.P., Dorsey, R., Gusella, J.F., et al., 2015. Huntington disease. Nat Rev Dis Prim 1, 15005. doi:10.1038/nrdp.2015.5

Bayer, T.A., 2015. Proteinopathies, a core concept for understanding and ultimately treating degenerative disorders? Eur Neuropsychopharmacol 25, 713–724. doi:10.1016/j.euroneuro.2013.03.007

Boasquivis, P.F., Silva, G.M.M., Paiva, F.A., et al., 2018. Guarana (Paullinia cupana) extract protects Caenorhabditis elegans models for Alzheimer disease and Huntington disease through activation of antioxidant and protein degradation pathways. Oxid Med Cell Longev 2018, 9241308. doi:10.1155/2018/9241308

Cai, C.Z., Zhou, H.F., Yuan, N.N., et al., 2019. Natural alkaloid harmine promotes degradation of alpha-synuclein via PKA-mediated ubiquitin-proteasome system activation. Phytomedicine 61, 152842. doi:10.1016/j.phymed.2019.152842

Cao, X., Xue, L., Han, L., et al., 2011. WW domain-containing E3 ubiquitin protein ligase 1 (WWP1) delays cellular senescence by promoting p27(Kip1) degradation in human diploid fibroblasts. J Biol Chem 286, 33447–33456. doi:10.1074/jbc.M111.225565

Carrano, A.C., Liu, Z., Dillin, A., et al., 2009. A conserved ubiquitination pathway determines longevity in response to diet restriction. Nature 460, 396–399. doi:10.1038/nature08130

Chakraborty, J., Rajamma, U., Jana, N., et al., 2015. Quercetin improves the activity of the ubiquitin-proteasomal system in 150Q mutated huntingtin-expressing cells but exerts detrimental effects on neuronal survivability. J Neurosci Res 93, 1581–1591. doi:10.1002/jnr.23618

Chen, Q., Thorpe, J., Dohmen, J.R., et al., 2006. Ump1 extends yeast lifespan and enhances viability during oxidative stress: central role for the proteasome? Free Radic Biol Med 40, 120–126. doi:10.1016/j.freeradbiomed.2005.08.048

Cheng, Y.F., Zhu, G.Q., Wang, M., et al., 2009. Involvement of ubiquitin proteasome system in protective mechanisms of Puerarin to MPP(+)-elicited apoptosis. Neurosci Res 63, 52–58. doi:10.1016/j.neures.2008.10.009

Chondrogianni, N., Tzavelas, C., Pemberton, A.J., et al., 2005. Overexpression of proteasome β5 subunit increases the amount of assembled proteasome and confers ameliorated response to oxidative stress and higher survival rates. J Biol Chem 280, 11840–11850. doi:10.1074/jbc.M413007200

Chondrogianni, N., Georgila, K., Kourtis, N., et al., 2015a. 20S proteasome activation promotes life span extension and resistance to proteotoxicity in Caenorhabditis elegans. FASEB J 29, 611–622. doi:10.1096/fj.14-252189

Chondrogianni, N., Voutetakis, K., Kapetanou, M., et al., 2015b. Proteasome activation: an innovative promising approach for delaying aging and retarding age-related diseases. Ageing Res Rev doi:10.1016/j.arr.2014.12.003

Chondrogianni, N., Vasilopoulou, M.A., Kapetanou, M., et al., 2020. Proteasome modulation: a way to delay aging?, in: Rattan, S.I.S. (Ed.), Encyclopedia of Biomedical Gerontology. Academic Press, Oxford, pp. 92–104. doi:https://doi.org/10.1016/B978-0-12-801238-3.11461-8

Ciechanover, A., Kwon, Y.T., 2017. Protein quality control by molecular chaperones in neurodegeneration. Front Neurosci 11, 185. doi:10.3389/fnins.2017.00185

Collins, G.A., Goldberg, A.L., 2017. The logic of the 26S proteasome. Cell 169, 792–806. doi:10.1016/j.cell.2017.04.023

Cuervo, A.M., Wong, E., 2014. Chaperone-mediated autophagy: roles in disease and aging. Cell Res doi:10.1038/cr.2013.153

Dal Vechio, F.H., Cerqueira, F., Augusto, O., et al., 2014. Peptides that activate the 20S proteasome by gate opening increased oxidized protein removal and reduced protein aggregation. Free Radic Biol Med 67, 304–313. doi:10.1016/j.freeradbiomed.2013.11.017

Dunker, A.K., Silman, I., Uversky, V.N., et al., 2008. Function and structure of inherently disordered proteins. Curr Opin Struct Biol 18, 756–764. doi:10.1016/j.sbi.2008.10.002

Ebstein, F., Kloetzel, P.M., Krüger, E., et al., 2012. Emerging roles of immunoproteasomes beyond MHC class I antigen processing. Cell Mol Life Sci doi:10.1007/s00018-012-0938-0

Esser, C., Alberti, S., Höhfeld, J., 2004. Cooperation of molecular chaperones with the ubiquitin/proteasome system. Biochim Biophys Acta – Mol Cell Res doi:10.1016/j.bbamcr.2004.09.020

Fernandez-Cruz, I., Reynaud, E., 2020. Proteasome subunits involved in neurodegenerative diseases. Arch Med Res doi:10.1016/j.arcmed.2020.09.007

Guo, H., Shi, F., Li, M., et al., 2015. Neuroprotective effects of Eucommia ulmoides Oliv. and its bioactive constituent work via ameliorating the ubiquitin-proteasome system. BMC Complement Altern Med 15, 151. doi:10.1186/s12906-015-0675-7

Hans, F., Fiesel, F.C., Strong, J.C., et al., 2014. UBE2E ubiquitin-conjugating enzymes and ubiquitin isopeptidase y regulate TDP-43 protein ubiquitination. J Biol Chem 289, 19164–19179. doi:10.1074/jbc.M114.561704

Hardiman, O., Al-Chalabi, A., Chio, A., et al., 2017. Amyotrophic lateral sclerosis. Nat Rev Dis Prim 3, 17071. doi:10.1038/nrdp.2017.71

Harding, R.J., Tong, Y.F., 2018. Proteostasis in Huntington's disease: disease mechanisms and therapeutic opportunities. Acta Pharmacol Sin 39, 754–769. doi:10.1038/aps.2018.11

Hartl, F.U., 2017. Protein misfolding diseases. Annu Rev Biochem 86, 21–26. doi:10.1146/annurev-biochem-061516-044518

Hassan, W.M., Merin, D.A., Fonte, V., et al., 2009. AIP-1 ameliorates beta-amyloid peptide toxicity in a Caenorhabditis elegans Alzheimer's disease model. Hum Mol Genet 18, 2739–2747. doi:10.1093/hmg/ddp209

He, C., Klionsky, D.J., 2009. Regulation mechanisms and signaling pathways of autophagy. Annu Rev Genet doi:10.1146/annurev-genet-102808-114910

He, W.T., Xue, W., Gao, Y.G., et al., 2017. HSP90 recognizes the N-terminus of huntingtin involved in regulation of huntingtin aggregation by USP19. Sci Rep 7. doi:10.1038/s41598-017-13711-7

Hergesheimer, R.C., Chami, A.A., de Assis, D.R., et al., 2019. The debated toxic role of aggregated TDP-43 in amyotrophic lateral sclerosis: a resolution in sight? Brain 142, 1176–1194. doi:10.1093/brain/awz078

Hinault, M.P., Ben-Zvi, A., Goloubinoff, P., 2006. Chaperones and proteases: cellular fold-controlling factors of proteins in neurodegenerative diseases and aging. J Mol Neurosci 30, 249–265. doi:10.1385/JMN:30:3:249

Huang, Q., Figueiredo-Pereira, M.E., 2010. Ubiquitin/proteasome pathway impairment in neurodegeneration: therapeutic implications. Apoptosis 15, 1292–1311. doi:10.1007/s10495-010-0466-z

Huang, C.L., Yang, J.M., Wang, K.C., et al., 2011. Gastrodia elata prevents huntingtin aggregations through activation of the adenosine A(2)A receptor and ubiquitin proteasome system. J Ethnopharmacol 138, 162–168. doi:10.1016/j.jep.2011.08.075

Hyrskyluoto, A., Bruelle, C., Lundh, S.H., et al., 2014. Ubiquitin-specific protease-14 reduces cellular aggregates and protects against mutant huntingtin-induced cell degeneration: involvement of the proteasome and ER stress-activated kinase IRE1α. Hum Mol Genet 23, 5928–5939. doi:10.1093/hmg/ddu317

Jantrapirom, S., Lo Piccolo, L., Yoshida, H., et al., 2018. Depletion of ubiquilin induces an augmentation in soluble ubiquitinated Drosophila TDP-43 to drive neurotoxicity in the fly. Biochim Biophys Acta Mol Basis Dis 1864, 3038–3049. doi:10.1016/j.bbadis.2018.06.017

Jeon, J., Kim, W., Jang, J., et al., 2016. Gene therapy by proteasome activator, PA28gamma, improves motor coordination and proteasome function in Huntington's disease YAC128 mice. Neuroscience 324, 20–28. doi:10.1016/j.neuroscience.2016.02.054

Jeong, S., 2017. Molecular and cellular basis of neurodegeneration in Alzheimer's disease. Mol Cells 40, 613–620. doi:10.14348/molcells.2017.0096

Joazeiro, C.A.P., 2017. Ribosomal stalling during translation: providing substrates for ribosome-associated protein quality control. Annu Rev Cell Dev Biol 33, 343–368. doi:10.1146/annurev-cellbio-111315-125249

Juenemann, K., Jansen, A.H.P., van Riel, L., et al., 2018. Dynamic recruitment of ubiquitin to mutant huntingtin inclusion bodies. Sci Rep 8, 1405. doi:10.1038/s41598-018-19538-0

Kaushik, S., Cuervo, A.M., 2018. The coming of age of chaperone-mediated autophagy. Nat Rev Mol Cell Biol doi:10.1038/s41580-018-0001-6

Kim, W., Seo, H., 2014. Baclofen, a GABAB receptor agonist, enhances ubiquitin-proteasome system functioning and neuronal survival in Huntington's disease model mice. Biochem Biophys Res Commun 443, 706–711. doi:10.1016/j.bbrc.2013.12.034

Kloetzel, P.M., 2004. Generation of major histocompatibility complex class I antigens: functional interplay between proteasomes and TPPII. Nat Immunol doi:10.1038/ni1090

Kniepert, A., Groettrup, M., 2014. The unique functions of tissue-specific proteasomes. Trends Biochem Sci doi:10.1016/j.tibs.2013.10.004

Kruegel, U., Robison, B., Dange, T., et al., 2011. Elevated proteasome capacity extends replicative lifespan in Saccharomyces cerevisiae. PLoS Genet 7. doi:10.1371/journal.pgen.1002253

Labbadia, J., Morimoto, R.I., 2015. The biology of proteostasis in aging and disease. Annu Rev Biochem 84, 435–464. doi:10.1146/annurev-biochem-060614-033955

Lamark, T., Johansen, T., 2012. Aggrephagy: selective disposal of protein aggregates by macroautophagy. Int J Cell Biol doi:10.1155/2012/736905

Li, T., Feng, Y., Yang, R., et al., 2018. Salidroside promotes the pathological alpha-synuclein clearance through ubiquitin-proteasome system in SH-SY5Y cells. Front Pharmacol 9, 377. doi:10.3389/fphar.2018.00377

Liu, Y., Hettinger, C.L., Zhang, D., et al., 2014. Sulforaphane enhances proteasomal and autophagic activities in mice and is a potential therapeutic reagent for Huntington's disease. J Neurochem 129, 539–547. doi:10.1111/jnc.12647

Marquez-Lona, E.M., Torres-Machorro, A.L., Gonzales, F.R., et al., 2017. Phosphorylation of the 19S regulatory particle ATPase subunit, Rpt6, modifies susceptibility to proteotoxic stress and protein aggregation. PLoS One 12, e0179893. doi:10.1371/journal.pone.0179893

Masters, C.L., Bateman, R., Blennow, K., et al., 2015. Alzheimer's disease. Nat Rev Dis Prim 1, 15056. doi:10.1038/nrdp.2015.56

Mayer, C., Grummt, I., 2006. Ribosome biogenesis and cell growth: mTOR coordinates transcription by all three classes of nuclear RNA polymerases. Oncogene 25, 6384–6391. doi:10.1038/sj.onc.1209883

Munkácsy, E., Chocron, E.S., Quintanilla, L., et al., 2019. Neuronal-specific proteasome augmentation via Prosβ5 overexpression extends lifespan and reduces age-related cognitive decline. Aging Cell 18. doi:10.1111/acel.13005

Nguyen, N.N., Rana, A., Goldman, C., et al., 2019. Proteasome β5 subunit overexpression improves proteostasis during aging and extends lifespan in Drosophila melanogaster. Sci Rep 9, 3170. doi:10.1038/s41598-019-39508-4

Nixon, R.A., 2013. The role of autophagy in neurodegenerative disease. Nat Med 19, 983–997. doi:10.1038/nm.3232

Oddo, S., 2008. The ubiquitin-proteasome system in Alzheimer's disease. J Cell Mol Med 12, 363–373. doi:10.1111/j.1582-4934.2008.00276.x

Olzscha, H., 2019. Posttranslational modifications and proteinopathies: how guardians of the proteome are defeated. Biol Chem 400, 895–915. doi:10.1515/hsz-2018-0458

Olzscha, H., Schermann, S.M., Woerner, A.C., et al., 2011. Amyloid-like aggregates sequester numerous metastable proteins with essential cellular functions. Cell 144, 67–78. doi:10.1016/j.cell.2010.11.050

Opattova, A., Cente, M., Novak, M., et al., 2015. The ubiquitin proteasome system as a potential therapeutic target for treatment of neurodegenerative diseases. Gen Physiol Biophys 34, 337–352. doi:10.4149/gpb_2015024

Panagiotidou, E., Chondrogianni, N., 2020. We are what we eat: ubiquitin-proteasome system (UPS) modulation through dietary products. Adv Exp Med Biol 1233, 329–348. doi:10.1007/978-3-030-38266-7_15

Papaevgeniou, N., Chondrogianni, N., 2016. UPS activation in the battle against aging and aggregation-related diseases: an extended review, in: Methods in Molecular Biology. Humana Press Inc., New York, doi:10.1007/978-1-4939-3756-1_1

Pickering, A.M., Davies, K.J.A., 2012. Degradation of damaged proteins: the main function of the 20S proteasome, in: Progress in Molecular Biology and Translational Science. Elsevier B.V., Amsterdam, pp. 227–248. doi:10.1016/B978-0-12-397863-9.00006-7

Pickering, A.M., Koop, A.L., Teoh, C.Y., et al., 2010. The immunoproteasome, the 20S proteasome and the PA28αβ proteasome regulator are oxidative-stress-adaptive proteolytic complexes. Biochem J 432, 585–594. doi:10.1042/BJ20100878

Poewe, W., Seppi, K., Tanner, C.M., et al., 2017. Parkinson disease. Nat Rev Dis Prim 3, 17013. doi:10.1038/nrdp.2017.13

Ramdzan, Y.M., Trubetskov, M.M., Ormsby, A.R., et al., 2017. Huntingtin inclusions trigger cellular quiescence, deactivate apoptosis, and lead to delayed necrosis. Cell Rep 19, 919–927. doi:10.1016/j.celrep.2017.04.029

Rocha, E.M., De Miranda, B., Sanders, L.H., 2018. Alpha-synuclein: pathology, mitochondrial dysfunction and neuroinflammation in Parkinson's disease. Neurobiol Dis 109, 249–257. doi:10.1016/j.nbd.2017.04.004

Rousseau, A., Bertolotti, A., 2018. Regulation of proteasome assembly and activity in health and disease. Nat Rev Mol Cell Biol 19, 697–712. doi:10.1038/s41580-018-0040-z

Sala, A.J., Bott, L.C., Morimoto, R.I., 2017. Shaping proteostasis at the cellular, tissue, and organismal level. J Cell Biol 216, 1231–1241. doi:10.1083/jcb.201612111

Savolainen, M.H., Albert, K., Airavaara, M., et al., 2017. Nigral injection of a proteasomal inhibitor, lactacystin, induces widespread glial cell activation and shows various phenotypes of Parkinson's disease in young and adult mouse. Exp Brain Res 235, 2189–2202. doi:10.1007/s00221-017-4962-z

Seo, H., Sonntag, K.C., Kim, W., et al., 2007. Proteasome activator enhances survival of Huntington's disease neuronal model cells. PLoS One 2, e238. doi:10.1371/journal.pone.0000238

Sibilla, C., Bertolotti, A., 2017. Prion properties of SOD1 in amyotrophic lateral sclerosis and potential therapy. Cold Spring Harb Perspect Biol 9. doi:10.1101/cshperspect.a024141

Stadtmueller, B.M., Hill, C.P., 2011. Proteasome activators. Mol Cell doi:10.1016/j.molcel.2010.12.020

Stark, M., Behl, C., 2014. The Ginkgo biloba extract EGb 761 modulates proteasome activity and polyglutamine protein aggregation. Evid-Based Complement Altern Med 2014. doi:10.1155/2014/940186

Tashiro, Y., Urushitani, M., Inoue, H., et al., 2012. Motor neuron-specific disruption of proteasomes, but not autophagy, replicates amyotrophic lateral sclerosis. J Biol Chem 287, 42984–42994. doi:10.1074/jbc.M112.417600

Thibaudeau, T.A., Anderson, R.T., Smith, D.M., 2018. A common mechanism of proteasome impairment by neurodegenerative disease-associated oligomers. Nat Commun 9. doi:10.1038/s41467-018-03509-0

Thompson, M.L., Chen, P., Yan, X., et al., 2014. TorsinA rescues ER-associated stress and locomotive defects in C. elegans models of ALS. Dis Model Mech 7, 233–243. doi:10.1242/dmm.013615

Tonoki, A., Kuranaga, E., Tomioka, T., et al., 2009. Genetic evidence linking age-dependent attenuation of the 26S proteasome with the aging process. Mol Cell Biol 29, 1095–1106. doi:10.1128/mcb.01227-08

Trippier, P.C., Zhao, K.T., Fox, S.G., et al., 2014. Proteasome activation is a mechanism for pyrazolone small molecules displaying therapeutic potential in amyotrophic lateral sclerosis. ACS Chem Neurosci 5, 823–829. doi:10.1021/cn500147v

Um, J.W., Im, E., Lee, H.J., et al., 2010. Parkin directly modulates 26S proteasome activity. J Neurosci 30, 11805–11814. doi:10.1523/JNEUROSCI.2862-09.2010

Vilchez, D., Morantte, I., Liu, Z., et al., 2012. RPN-6 determines C. elegans longevity under proteotoxic stress conditions. Nature 489, 263–268. doi:10.1038/nature11315

Walker, F.O., 2007. Huntington's disease. Lancet 369, 218–228. doi:10.1016/S0140-6736(07)60111-1

Walther, D.M., Kasturi, P., Zheng, M., et al., 2015. Widespread proteome remodeling and aggregation in aging C. elegans. Cell 161, 919–932. doi:10.1016/j.cell.2015.03.032

Yang, E.J., Kim, S.H., Yang, S.C., et al., 2011. Melittin restores proteasome function in an animal model of ALS. J Neuroinflammation 8, 69. doi:10.1186/1742-2094-8-69

Yuan, N.N., Cai, C.Z., Wu, M.Y., et al., 2019. Canthin-6-one accelerates alpha-synuclein degradation by enhancing ups activity: drug target identification by CRISPR-Cas9 whole genome-wide screening technology. Front Pharmacol 10, 16. doi:10.3389/fphar.2019.00016

Zaher, H.S., Green, R., 2009. Fidelity at the molecular level: lessons from protein synthesis. Cell 136, 746–762. doi:10.1016/j.cell.2009.01.036

Zhang, M., Cai, F., Zhang, S., et al., 2014. Overexpression of ubiquitin carboxyl-terminal hydrolase L1 (UCHL1) delays Alzheimer's progression in vivo. Sci Rep 4, 7298. doi:10.1038/srep07298

Zhang, Y., Chen, X., Zhao, Y., et al., 2017. The role of ubiquitin proteasomal system and autophagy-lysosome pathway in Alzheimer's disease. Rev Neurosci 28, 861–868. doi:10.1515/revneuro-2017-0013

Zhang, T., Periz, G., Lu, Y.N., et al., 2020. USP7 regulates ALS-associated proteotoxicity and quality control through the NEDD4L–SMAD pathway. Proc Natl Acad Sci U S A 117, 28114–28125. doi:10.1073/pnas.2014349117

CHAPTER THIRTEEN

The Proteasomal System in Cancer

Sema Arslan-Eseryel, Ulkugul Guven and Betul Karademir-Yilmaz

CONTENTS

13.1 INTRODUCTION

Cancer is described as the malignant transformation of normal cells to neoplastic cells due to disturbances in major metabolic pathways. These malignant cells grow in an uncontrolled manner and may spread around the body, resulting in metastasis. Several theories seek to clarify the cancer development process, and among others, disturbances in the proteolytic pathways have important roles in the stabilization of oncogenes and de-stabilization of tumor suppressor genes. In addition, the proteolytic network plays a central role in cancer, mainly in its progression, invasion and metastasis (Mason and Joyce, 2011).

Proteotoxic stress is one of the signs of cancer progression. Cancer relies on protein quality control for proliferation and survival. Unfolded protein response, heat shock proteins, autophagy and the ubiquitin-proteasome system (UPS) are the members of protein quality control network that orchestrate a cascade of events in carcinogenesis (Bastola et al., 2018). Protein damage due to impaired protein quality control leads to the accumulation of dysfunctional proteins that contributes to malignant cell formation in cancer (Li et al., 2011).

The UPS is composed of proteasome itself (different types such as 20S proteasome, the 26S proteasome and the immunoproteasome), ubiquitin conjugating complexes and deubiquitinating enzymes (Jung et al., 2009). This system has a crucial function in the degradation of abnormal, short-lived and partially unfolded cellular proteins and in that of oxidized proteins (Jung et al., 2009; Orlowski, 2005). The proteins regulated via degradation by the UPS have important roles in cell-cycle regulation, cell proliferation, apoptosis, cell growth, tumor suppression and metastasis, as will be detailed below. Moreover, the immunoproteasome has an important role in immune response

DOI: 10.1201/9781003048138-13

167

by preparing small peptides to be presented by major histocompatibility complex (MHC) class I molecules (Catalgol, 2012; Ciechanover and Schwartz, 1998).

13.2 THE UPS IN CANCER

The UPS is a multiprotein cellular complex located in both the cytoplasm and the nucleus (Manasanch and Orlowski, 2017). The UPS participates in cellular proteostasis by degrading damaged or short-lived proteins in an organized fashion (Brooks et al., 2000). Two lines of evidence highlight the importance of the UPS in cancer progression. First is the high activity and expression of proteasomes in cancer cells, which underlie resistance in standard chemo- and radiotherapies. Second is the wide range of proteasomal substrates related to cancer progression including the signaling molecules that play roles in the cell cycle, cell differentiation, metastasis, tumor suppression and apoptosis (Kaplan et al., 2017). Finally, the role of the UPS in cancer stem cells (CSCs) and mutations in proteasome subunits is a key focus in proteasome-related cancer studies (Voutsadakis, 2017).

As mentioned above, high proteasomal content in cancer cells makes the proteasomal system a relevant therapeutic target and is frequently used in cancer studies. Until now, studies have shown that high proteasome activity is generally accompanied by high expression of proteasome alpha and beta subunits in tumor types. In one study, the expression of PSMA1, PSMB5, PSMD2, PSMD1, PSMD8 and PSMD14 proteasome subunits increased more than threefold in breast cancer tissues compared to healthy ones in the tumor microenvironment (Deng et al., 2007). Upregulation of PSMA6, PSMB4, PSMC2 and PSMD12 was observed in hepatocellular p21-HBx transgenic mice with hepatocellular carcinoma (Cui et al., 2006). PSMA5 and PSMA4 levels were increased in colorectal cancer (Arlt et al., 2009) and mRNA expression levels of PSMA1–7 were significantly upregulated in breast, lung, stomach, bladder, head and neck cancer compared to healthy tissues (Li et al., 2017). In ovarian cancer, high expression of PSMB4 was related to clinically pathological conditions such as tumor grade and cancer stage (Liu et al., 2016). Hepatocellular cancer patients with relatively higher PSMD10 expression had larger tumors and higher vascular invasion as well as intrahepatic or distant metastasis, compared to patients with low PSMD10 expression (Fu et al., 2011). Similarly, increased PSMD9 expression in breast cancer was significantly associated with higher local recurrence rates after radiotherapy (Langlands et al., 2014). Another study reported a poor prognosis for lung adenocarcinoma patients with overexpression of PSMD2 (Matsuyama et al., 2011).

Besides high proteasome activity and increased expression of proteasome subunits in cancer cells, changes in the proteasomal system in CSCs have been also studied (Teicher and Tomaszewski, 2015; Voutsadakis, 2017). CSCs are a subpopulation of stem-like cells that are found in tumors, have differentiation capacities and unlike cancer cells they display reduced proteasome activity. Specifically, a study with glioblastoma and breast carcinoma cells showed that cells growing in sphere cultures enriched for CSCs and progenitor cells have lower proteasome activity compared to cells growing in monolayers (Vlashi et al., 2009). Some types of cancer cells, such as head and neck, colon, which grew in a stem cell-like manner displayed low proteasome activities and showed therapy resistance, supporting the relationship of decreased proteasome activity to increased tumorigenic potential (Lagadec et al., 2014; Munakata et al., 2016). The connection of reduced proteasomal activity and increased tumorigenic potential still needs more detailed studies but on the other side explains how CSCs contribute to the ineffectiveness of cancer treatments that target the proteasome.

13.3 ROLE OF PROTEASOME IN CANCER-RELATED PATHWAYS

Cancer progression and development mainly depend on cell proliferation, cell growth, apoptosis and angiogenesis. Cell proliferation is defined as an exponential increase in cell number at the end of cell growth. Cell proliferation is a tightly controlled and complex process that requires several growth factors and proteins. Growth factors activate cell proliferation by inducing resting cells to enter the cell cycle. The G1 phase, which is the first period of growth, prepares cells for DNA synthesis (the S phase). R is the restriction point between the G1 and S phases and is the commitment step for the rest of cell cycle. Until the R point, growth factor-mediated signaling pathways are on duty together with the cell cycle control system and growth factors retire from the scene when the R point is breached. The cell-cycle

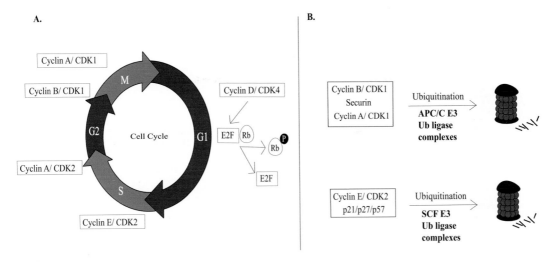

A.

Cyclin A/ CDK1

Cyclin B/ CDK1

M

G2

Cell Cycle

G1

Cyclin D/ CDK4

E2F Rb

Rb P

E2F

S

Cyclin A/ CDK2

Cyclin E/ CDK2

B.

| Cyclin B/ CDK1 |
| Securin |
| Cyclin A/ CDK1 |

Ubiquitination → **APC/C E3 Ub ligase complexes**

| Cyclin E/ CDK2 |
| p21/p27/p57 |

Ubiquitination → **SCF E3 Ub ligase complexes**

Figure 13.1 **The involvement of the proteasome in cell-cycle regulation.** (A) Schematic representation of the cell cycle. (B) Degradation of cell-cycle-related proteins by the proteasome. APC/C ubiquitin ligase complexes play a role in the proteolysis of cyclin B, cyclin A, securin and CDK1, whereas SCF ubiquitin ligase complexes play a role in the proteolysis of cyclin E, CDK2, p21, p27 and p57 by ubiquitin–proteasome system.

control system is mainly composed of cyclins and cyclin-dependent kinases (CDKs). A number of CDK inhibitors, retinoblastoma and E2F family members also regulate downstream events. The relationship of the UPS with cell proliferation and the cell cycle starts with these control system proteins, as cyclins, CDKs, retinoblastoma and E2F1 are all regulated by proteasomal degradation (Figure 13.1) (Koepp et al., 1999; Pagano et al., 1995; Hofmann et al., 1996).

Apoptosis is an important form of cell death that controls cell numbers. Apoptosis is usually induced either by ligation of tumor necrosis factor (TNF) receptors or by mitochondrial perturbation, resulting in cytochrome c release and the activation of the apoptosis complex containing Apaf1 (Bratton et al., 2000). Specific proteins play a role in the process, including the members of the Bcl-2 family and caspases (Nicholson, 1999). Proteasomal degradation is crucial to the regulation of apoptosis, and proteasome inhibitors appear to regulate both pro- and anti-apoptotic signals (Chen and Ping, 2010). As an example, proteasome inhibitors induce apoptosis in chronic lymphocytic leukemia cells and oral squamous cell carcinoma cells at doses that have no apoptotic effect on normal human lymphocytes and oral epithelial cells (Delic et al., 1998; Ikebe et al., 1998).

Angiogenesis is defined as the formation of new blood vessels under pathological conditions

such as inflammation and cancer (Folkman, 1995). Vascular endothelial growth factors (VEGFs), fibroblast growth factors and transforming growth factors (TGF) are the main proteins involved in angiogenesis (Yoshida et al., 1997). Hypoxia stimulates the angiogenic growth of vessels in tumor tissue by increasing VEGF-A expression (Shweiki et al., 1992). Bortezomib (PS341, Velcade®), the first approved proteasome inhibitor, was shown to inhibit tumor growth, prevent metastasis and inhibit VEGF expression and angiogenesis (Adams, 2002). In animal models, bortezomib showed 50% suppression of blood vessel development and inhibited angiogenesis (LeBlanc et al., 2002; Nawrocki et al., 2002; Sunwoo et al., 2001). In human studies, bortezomib adjuvant therapy showed antitumor activity in many metastatic cancers such as hormone receptor-positive metastatic breast cancer (Adelson et al., 2016), malignant melanoma (Markowitz et al., 2014) and advanced non-small-cell lung cancer (Piperdi et al., 2012) patients.

13.4 REGULATION OF CANCER-RELATED SIGNALING MOLECULES BY THE PROTEASOMAL SYSTEM

Cancer progression and development are complex processes that depend on many signaling molecules. The relationship of some of these molecules

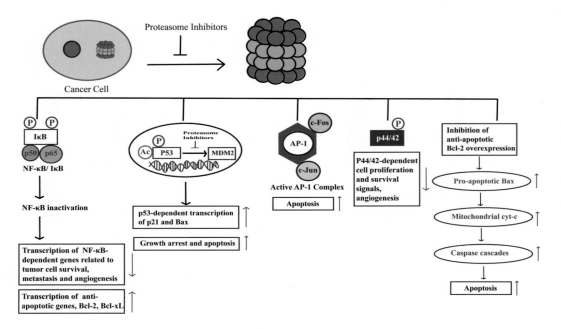

Figure 13.2 **Proteasome inhibitors act on cancer cells via regulation of various signaling molecules.** NF-κB inactivation, p53 stabilization, AP-1 and MAPK activation, and the regulation of pro- and anti-apoptotic proteins are all related to the effects of proteasome inhibitors. *Abbreviations:* Ac, acetylation; P, phosphorylation; U, ubiquitination.

with the proteasomal system, which play a role in cell proliferation, cell cycle, apoptosis and angiogenesis will be explained in the following sections (Figure 13.2).

13.4.1 Nuclear Factor-Kappa B

Nuclear factor-kappa B (NF-κB) is a well-known transcription factor responsible for the expression of chemokines, growth factors, cytokines, anti-apoptotic factors, cell adhesion molecules, surface receptors, some enzymes and cyclins in tumor development. Specifically, NF-κB activity increases in breast cancer, myeloma, prostate cancer, leukemia, Hodgkin lymphoma, B-cell lymphoma and T-cell lymphoma (Moorthy et al., 2006). NF-κB is a p65/p50 dimer and is bound to its inhibitor IκB in unstimulated cells. The UPS is involved in the release and activation of NF-κB (Ghosh et al., 1998). Activation of NF-κB can suppress apoptosis by increasing the expression of survival signals, including IAPs, TRAF1, TRAF2 and the Bcl-2 family homolog Bfl-1/A1 (Chu et al., 1997). Given the role of the proteasome in IκB degradation, proteasome inhibitors can induce apoptosis by stabilizing cytoplasmic IκB and blocking nuclear translocation of NF-κB

(Lin et al., 1995). Some anticancer agents and irradiation both decrease the activation of NF-κB. Therefore, a combination of proteasome inhibitors and certain chemotherapeutic agents should cause an increase in cytotoxicity on cancer cells.

Besides apoptosis, NF-κB plays a role in immune and inflammatory responses by controlling the expression of cytokines and cell adhesion molecules. NF-κB activity also controls tumor metastasis and angiogenesis. Anti-angiogenic agents prevent multiple myeloma (MM) cells from adhering to bone marrow stromal cells and suppress the production of NF-κB-dependent interleukin-6 in these cells (Hideshima et al., 2001). The ability of proteasome inhibitors to inhibit NF-κB is important in their activity against angiogenesis and metastasis (Sunwoo et al., 2001).

13.4.2 p53

The tumor suppressor p53 is a short-lived transcription factor regulated by the UPS. In response to cellular stress or DNA damage, wild-type p53 protein is activated by phosphorylation and other modifications and dissociates from its inhibitor MDM2, which is a RING-type E3 ubiquitin ligase. Once activated, p53 is no longer targeted

PROTEOSTASIS AND PROTEOLYSIS

for proteasomal degradation and therefore binds to its specific sequences in DNA to initiate growth arrest, DNA repair or the transcription of apoptosis-related genes (Maki et al., 1996).

P53 upregulates Bax, pro-apoptotic BH3-containing protein and death receptors CD95 (Fas/Apo-1) and DR5 (TRAIL-R2) (Vogelstein et al., 2000). In some cancer cells, proteasome inhibitors cause p53 stabilization and accumulation. For example, treatment of LNCaP-Pro5 prostate cancer cells with proteasome inhibitors resulted in the accumulation of transcriptionally active p53, which induces apoptosis (Williams and McConkey, 2003).

13.4.3 p44/42 Mitogen-Activated Protein Kinase

Mitogen-activated protein kinases (MAPKs) are widely conserved members of the serine/threonine-protein kinase family. p44/42 MAPK (Erk1/2) pathways can be activated in response to a diverse range of stimuli and play an important role in cell proliferation, suppression of apoptosis and angiogenesis (Voorhees et al., 2003). These kinases also regulate the activity of several important signaling molecules and transcription factors, including NF-κB. The p44/42 pathway has important roles in a variety of tumors, including hematological malignancies such as leukemias (Lee and McCubrey, 2002) and solid tumors such as breast cancer (Santen et al., 2002). Treatment of breast cancer cell lines with proteasome inhibitors resulted in apoptosis via the loss of the phosphorylated active p44/42 MAPK (Orlowski et al., 2002).

13.4.4 Pro-Apoptotic Bcl-2 Gene Family Members

Bcl-2 gene family members are divided into two main classes: anti-apoptotic and pro-apoptotic. Among these proteins, Bcl-2 is anti-apoptotic while Bax, Bik and Bid proteins are pro-apoptotic. In prostate carcinoma and Jurkat T-cell leukemia cells, proteasome inhibitors were able to overcome Bcl-2-mediated protection from apoptosis (An et al., 1998).

In Jurkat cells, proteasome inhibitors decrease the degradation of Bax, which causes increased apoptosis via the accumulation of this pro-apoptotic molecule. The same study also showed that increased degradation of Bax by the proteasome in aggressive human prostate cancer tissue samples is associated with decreased Bax protein levels (Li and Dou, 2000). Pro-apoptotic Bcl-2 family members, such as Bik and tBid, are also proteasomal substrates, and MG132, a peptide aldehyde proteasome inhibitor, was shown to stabilize Bik and tBid in head and neck squamous cell carcinoma cells (Sung et al., 2012). In another study, bortezomib induced accumulation of Bik and therefore apoptosis in human colon and lung cancer cell lines (Zhu et al., 2005).

13.4.5 AP-1

Activator protein-1 (AP-1) is a dimeric transcription factor involved in cell proliferation, differentiation and apoptosis. AP-1 is composed of a Jun subunit (c-Jun, Jun B and Jun D) and a Fos subunit (c-Fos, Fos B, Fra-1 or Fra-2). C-jun and c-fos proteins have short half-lives, and this rapid breakdown is important for normal cell growth (Jariel-Encontre et al., 1997). Studies have shown that the UPS plays a role mainly in the degradation of c-jun and c-fos (Stancovski et al., 1995). In some cell types, it has been found that the E1 ubiquitin-activating enzyme is inactive, indicating that the c-jun and c-fos proteins can be degraded in both ubiquitin-dependent and independent mechanisms (Basbous et al., 2008). Proteasome inhibition was shown to increase the accumulation of c-jun and phospho-c-jun in HepG2 hepatocellular carcinoma cells. In the same study, it was shown that silencing c-jun decreases the activity of AP-1 and the sensitivity of cells to apoptotic stimuli (Lauricella et al., 2006).

13.5 PROTEASOME INHIBITORS

Proteasome inhibitors act mainly on the catalytic subunits of the proteasome. They are investigated according to their structures and classified as first- and second-generation inhibitors which are used in clinical trials.

13.5.1 20S Proteasome Inhibitors

Proteasome inhibitors are divided into five main groups according to their chemical structure: peptide boronates, peptide aldehydes, peptide vinyl sulfones, peptide epoxyketones and (v)-lactone. For a peptide to inhibit proteasome catalytic activity, it is required to interact with

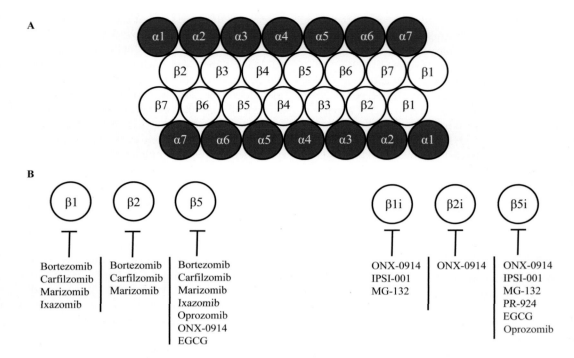

A

B

Figure 13.3 **Targeted proteasome subunits by proteasome inhibitors.** (A) The structure of proteasome 20S catalytic core. (B) The schematic representation for the targets of proteasome inhibitors. *Abbreviation:* βi, inducible beta-subunit).

substrate-binding sites in the 20S and i20S core particles (Figure 13.3A). Binding sites of different proteasome inhibitors are shown in Figure 13.3B.

Peptide aldehydes are structurally highly unstable and easily oxidized, so their ability to specifically inhibit the 20S proteasome is limited. MG132 and IPSI-001 are peptide aldehyde-based inhibitors. These compounds inhibit the proteasome by forming a covalent hemiacetal bond with the N-terminal Thr1 hydroxyl group in the catalytic subunits. This inhibition is reversible, and the recovery rate is relatively fast. IPSI-001 inhibits the β5i and β1i subunits of the immunoproteasome, strongly suppressing proliferation and inducing apoptosis in MM cell lines as well as in cells derived from MM patients (Kubiczkova et al., 2014).

Peptide boronates have been more successful as 20S proteasome inhibitors, especially due to their small molecular structures (Adams, 2003). Bortezomib is the first FDA-approved peptide boronate-structured proteasome inhibitor and has been used in clinics for many years. It was approved for the treatment of mantle cell lymphoma (MCL) in 2006 and of MM in 2008 (Kaplan et al., 2017). Bortezomib targets β5, β1

and β2 subunits and shows reversible action. A high number of clinical trials and research studies have demonstrated success in using bortezomib (Curran and McKeage, 2009). Ixazomib (MLN-9708) is another boronate-based reversible proteasome inhibitor and the first oral proteasome inhibitor to enter clinical trials in MM patients. Data from the phase I study suggest that ixazomib may have clinical activity in previously treated and relapsing MM patients (Richardson et al., 2014).

Carfilzomib (PR-171, Kyprolis®) is an approved second-generation proteasome inhibitor. It has an epoxyketone structure and is a highly specific, potent and irreversible proteasome inhibitor. High selectivity is mostly achieved through the interactions of β-epoxyketone with the hydroxyl group and free α-amino group of Thr1 in the catalytic β-subunits of the proteasome (Kaplan et al., 2017). Studies have shown that carfilzomib exhibits more specificity for proteasome than bortezomib and has little or no off-target activity (Wang 2011). Carfilzomib targets the β5 subunit in the proteasome as well as the β5i subunit in the immunoproteasome, which is highly expressed in MM (Kuhn et al., 2007). Oprozomib

PROTEOSTASIS AND PROTEOLYSIS

(ONX-0912 or PR-047) is another epoxyketone-based proteasome inhibitor. Orally administered oprozomib showed antitumor activity equivalent to that of intravenously administered carfilzomib in human tumor xenograft and mouse syngeneic models (Zhou et al., 2009). ONX-0914 and PR-924 are also peptide epoxyketone-based inhibitors, both targeting and irreversibly inhibiting the β5i subunit in the immunoproteasome. ONX-0914 shows a more specific and higher affinity against the β5i subunit than both the β5 and β1i subunits (Muchamuel et al., 2009).

Marizomib (salinosporamide A, NPI-0052) is a proteasome inhibitor with a β-lactone structure. The carbonyl group of β-lactone interacts with the hydroxyl group of Thr1 in the catalytic β-subunits, and the acylester bond causes a cyclic tetrahydrofuran ring with the proteasome, rendering the reaction irreversible (Chauhan et al., 2008).

Lactacystin is a synthetic inhibitor produced by *Streptomyces* and is mainly used in research studies focusing on proteasome function. Lactacystin is also converted into reactive β-lactone as marizomib and reacts with side-chain hydroxyl of the amino-terminal Thr1. It readily crosses the plasma membrane in cell culture medium. Normally, active β-lactone is a very unstable molecule under intracellular conditions and is rapidly inactivated. Despite this disadvantage, lactacystin is more selective than peptide aldehydes (Adams, 2003).

Besides the aforementioned proteasome inhibitors, natural dietary products such as epigallocatechin-3-gallate (EGCG), celastrol and curcumin have inhibitory effects on proteasome activity. EGCG is a flavonoid that can irreversibly inhibit the proteasomal β5 subunit and the β5i subunit of the immunoproteasome (Landis-Piwowar et al., 2006; Pang et al., 2010). Many studies, which are not detailed here, suggest the potential use of these natural proteasome inhibitors as supplements to conventional cancer therapies (Yang et al., 2008).

13.5.2 Clinical Studies

The potential of proteasome inhibitors has been demonstrated with the success of bortezomib in treating MM, which involves the carcinogenesis of antibody-producing plasma B cells (Chauhan et al., 2005). Bortezomib also inhibits angiogenesis and suppresses the interaction of MM cells with bone marrow stem cells, which may contribute to MM drug resistance (Richardson et al., 2003).

MCL is a highly aggressive type of lymphoma for which there is no standard patient care (Campo and Rule, 2015). The presence of additional mutations in oncogenes and tumor suppressor genes also contributes to the aggressive nature of MCL. Bortezomib has been approved by the FDA and EMA for use in the treatment of MCL (Jares et al., 2012).

In addition to MM and MCL, clinical trials of proteasome inhibitors have been conducted against other hematological malignancies such as non-Hodgkin lymphoma (Orlowski, 2005) and solid tumors (Papandreou et al., 2004) in combination with conventional anticancer drugs.

Despite the major advances in the clinic, many concerns remain about proteasome inhibitors, especially bortezomib. First of all, bortezomib has numerous side effects due to its off-target effects. The most common side effects associated with the use of bortezomib in clinical trials are classified as asthenic conditions, gastrointestinal effects, hematological toxicity, decreased sensation, paresthesia and high rate of peripheral neuropathy (Oakervee et al., 2005; Richardson et al., 2005). Carfilzomib, a second-generation proteasome inhibitor, is designed to overcome bortezomib's side effects. Importantly, carfilzomib has shown efficacy in patients with MM that are resistant to bortezomib (Steiner and Manasanch, 2017).

Secondly, bortezomib interacts with various natural compounds. The anticancer activity of boronic acid-derived proteasome inhibitors was shown to be blocked by green tea polyphenols (Golden et al., 2009). This is because the boric acid structure interacts with the catechol structure of EGCG to form a borate ester. It has also been found that ascorbic acid inhibits the anticancer effects of bortezomib (Perrone et al., 2009).

Finally, although bortezomib has some efficacy in hematological malignancies, it does not provide an effective treatment for solid tumors. In a study by Yang et al. (2006), no significant response was observed in a total of 12 metastatic breast cancer patients. This study suggested that the use of bortezomib as a single-agent demonstrates limited clinical activity against metastatic breast cancer. Similarly, a phase II study of bortezomib in patients with metastatic neuroendocrine tumors found that administering bortezomib to 16 patients did not cause any positive response

TABLE 13.1
Inhibitors targeting the ubiquitination process via E1, E2 and E3 enzymes

Enzyme	Inhibitor	Feature	References
Uba1 (E1)	PYR-41 PYZD-4409	Binds cysteine on the active site	Xu et al. (2010)
Cdc34 (E2)	CC0651	Allosteric inhibitor	Ceccarelli et al. (2011)
Ubc13-Uev1A (E2)	NSC697923	Significant effect on NF-κB activation	Cheng et al. (2014)
Mdm2 (RING-type E3)	Benzosulfonamide, urea analogs, imidazole derivatives, Nutlin derivatives	Significant effect on p53	Lai et al. (2002); Secchiero et al. (2011)
BCA2 (E3, zinc binding)	Disulfiram	Crucial in epidermal growth factor receptor trafficking	Chen et al. (2006)
E6-AP (E3)	4,4′-Dithiodimorpholine	Treatments limited to HPV-positive tumors	Traidej et al. (2000)

(Shah et al., 2004). In another study (Kondagunta et al., 2004), a partial response was observed in 4 of 37 patients when bortezomib used alone in advanced renal cell carcinoma. Results from laboratory and clinical studies showed the significant efficacy of bortezomib combined with pegylated liposomal doxorubicin (Orlowski et al., 2007), melphalan, dexamethasone, thalidomide (Terpos et al., 2008) and cyclophosphamide (Reece et al., 2008) in solid tumors.

13.6 TARGETING UBIQUITINATION IN CANCER TREATMENT

Inhibition of the proteasome first started by targeting catalytic subunits. As our knowledge progressed, the ubiquitination cascade has been also targeted in cancer studies. Ubiquitination is a complex process that requires three main steps, each catalyzed by specific enzymes. The process involves three types of enzymes: E1, E2 and E3, which are required for activation, conjugation and ligation, respectively (Catalgol, 2012). There are inhibitors in different structures designed to inhibit the enzymes involved in ubiquitination. These are summarized in Table 13.1. Among others, there are recruiting clinical trials using ring-type E3 inhibitors in salivary gland carcinoma, brain cancer, soft tissue carcinoma and acute myeloid leukemia. Milademetan, as a MDM2 inhibitor, has some promising data in acute myeloid leukemia patients (Daver et al., 2019; DiNardo et al., 2019).

13.7 CONCLUSIONS

Proteasome inhibitors are promising drugs in cancer treatment. However, proteasome inhibitors also have significant limitations due to their side effects, their low effectiveness in solid tumors and the resistance due to mutations in proteasome subunits. Although there have been many studies on proteasome inhibitors, more research is required to highlight the effects of second-generation inhibitors, effectiveness on solid tumors and inhibitors targeting ubiquitination.

REFERENCES

Adams, J., 2002. Preclinical and clinical evaluation of proteasome inhibitor PS-341 for the treatment of cancer. Curr. Opin. Chem. Biol. 6, 493–500. doi:10.1016/s1367-5931(02)00343-5.

Adams, J., 2003. Potential for proteasome inhibition in the treatment of cancer. Drug Discov. Today 8, 307–315. doi:10.1016/s1359-6446(03)02647-3

Adelson, K., Ramaswamy, B., Sparano, J. et al., 2016. Randomized phase II trial of fulvestrant alone or in combination with bortezomib in hormone receptor-positive metastatic breast cancer resistant to aromatase inhibitors: a New York Cancer Consortium trial. NPJ Breast Cancer 2, 16037. doi:10.1038/npjbcancer.2016.37.

An, B., Goldfarb, R.H., Siman, R., et al., 1998. Novel dipeptidyl proteasome inhibitors overcome Bcl-2 protective function and selectively accumulate the cyclin-dependent kinase inhibitor p27 and induce apoptosis in transformed, but not normal, human fibroblasts. Cell Death Differ. 12, 1062–1075. doi:10.1038/sj.cdd.4400436

Arlt, A., Bauer, I., Schafmayer, C., et al., 2009. Increased proteasome subunit protein expression and proteasome activity in colon cancer relate to an enhanced activation of nuclear factor E2-related factor 2 (Nrf2). Oncogene 28, 3983–3996. doi:10.1038/onc.2009.264

Basbous, J., Jariel-Encontre, I., Gomard, T., et al., 2008. Ubiquitin-independent-versus ubiquitin-dependent proteasomal degradation of the c-Fos and Fra-1 transcription factors: is there a unique answer?. Biochimie 90(2), 296–305. doi:10.1016/j.biochi.2007.07.016

Bastola, P., Oien, D.B., Cooley, M., et al., 2018. Emerging cancer therapeutic targets in protein homeostasis. AAPS J. 20(6), 94. doi:10.1208/s12248-018-0254-1

Bratton, S.B., MacFarlane, M., Cain, K., et al., 2000. Protein complexes activate distinct caspase cascades in death receptor and stress induced apoptosis. Exp. Cell Res. 256, 27–33. doi:10.1006/excr.2000.4835

Brooks, P., Fuertes, G., Murray, R.Z., et al., 2000. Subcellular localization of proteasomes and their regulatory complexes in mammalian cells. Biochem J. 346, 155–161. PMID: 10657252.

Campo, E., Rule, S., 2015. Mantle cell lymphoma: evolving management strategies. Blood 125, 48–55. doi:10.1182/blood-2014-05-521898

Catalgol, B., 2012. Proteasome and cancer. Prog. Mol. Biol. Transl. Sci. 109, 277–293. doi:10.1016/B978-0-12-397863-9.00008-0

Ceccarelli, D.F., Tang, X., Pelletier, B., et al., 2011. An allosteric inhibitor of the human Cdc34 ubiquitin-conjugating enzyme. Cell 145, 1075–1087. doi:10.1016/j.cell.2011.05.039

Chauhan, D., Hideshima, T., Mitsiades, C., et al., 2005. Proteasome inhibitor therapy in multiple myeloma. Mol. Cancer Ther. 4, 686–692. doi:10.1158/1535-7163.MCT-04-0338

Chauhan, D., Singh, D., Brahmandam, A., et al., 2008. Combination of proteasome inhibitors bortezomib and NPI-0052 trigger in vivo synergistic cytotoxicity in multiplemyeloma. Blood 111, 1654–1664. doi:10.1182/blood-2007-08-105601

Chen, D., Cui, Q.C., Yang, H., et al., 2006. Disulfiram, a clinically used anti-alcoholism drug and copper-binding agent, induces apoptotic cell death in breast cancer cultures and xenografts via inhibition of the proteasome activity. Cancer Res. 66, 10425–10433. doi:10.1158/0008-5472.CAN-06-2126

Chen, D., Ping, D.Q., 2010. The ubiquitin-proteasome system as a prospective molecular target for cancer treatment and prevention. Curr. Protein Pept. Sci. 11, 459–470. doi:10.2174/138920310791824057

Cheng, J., Fan, Y., Xu, X., et al., 2014. A small-molecule inhibitor of UBE2N induces neuroblastoma cell death via activation of p53 and JNK pathways. Cell Death Dis. 5(2), e1079. doi:10.1038/cddis.2014.54

Chu, Z.L., McKinsey, T.A., Liu, L., et al., 1997. Suppression of tumor necrosis factor-induced cell death by inhibitor of apoptosis c-IAP2 is under NF-kB control. Proc. Natl. Acad. Sci. U. S. A. 94(19), 10057–10062. doi:10.1073/pnas.94.19.10057

Ciechanover, A., Schwartz, A.L., 1998. The ubiquitin-proteasome pathway: the complexity and myriad functions of proteins death. Proc. Natl. Acad. Sci. U. S. A. 95(6), 2727–2730. doi:10.1073/pnas.95.6.2727

Cui, F., Wang, Y., Wang, J., et al., 2006. The up-regulation of proteasome subunits and lysosomal proteases in hepatocellular carcinomas of the HBx gene knockin transgenic mice. Proteomics 6, 498–504. doi:10.1002/pmic.200500218

Curran, M.P., McKeage, K., 2009. Bortezomib: a review of its use in patients with multiple myeloma. Drugs 69, 859–888. doi:10.2165/00003495-200969070-00006

Daver, N. G., Zhang, W., Graydon, R., et al., 2019. A phase 1 study of milademetan in combination with quizartinib in patients with newly diagnosed (ND) or relapsed/refractory (R/R) FLT3-ITD acute myeloid leukemia (AML). Blood 134 (Suppl. 1), 1389. doi:10.1182/blood-2019-122585

DiNardo, C. D., Olin, R., Ishizawa, J., 2019. A phase 1 dose escalation study of milademetan in combination with 5-azacitidine (AZA) in patients with acute myeloid leukemia (AML) or high-risk myelodysplastic syndrome (MDS). Blood 134 (Suppl. 1), 3932. doi:10.1182/blood-2019-122241.

Delic, J., Masdehors, P., Omura, S., et al., 1998. The proteasome inhibitor lactacystin induces apoptosis and sensitizes chemo- and radioresistant human chronic lymphocytic leukaemia lymphocytes to TNF-alpha-initiated apoptosis. Br. J. Cancer 77(7), 1103–1107. doi:10.1038/bjc.1998.183

Deng, S., Zhou, H., Xiong, R., et al., 2007. Over-expression of genes and proteins of ubiquitin specific peptidases (USPs) and proteasome subunits (PSs) in breast cancer tissue observed by the methods of RFDD-PCR and proteomics. Breast Cancer Res. Treat. 104, 21–30. doi:10.1007/s10549-006-9393-7

Folkman, J., 1995. Angiogenesis in cancer, vascular, rheumatoid and other disease. Nat. Med. 1(1), 27–30. doi:10.1038/nm0195-27

Fu, J., Chen, Y., Cao, J., et al., 2011. p28GANK overexpression accelerates hepatocellular carcinoma invasiveness and metastasis via phosphoinositol 3-kinase/AKT/hypoxia-inducible factor-1α pathways. Hepatology 53(1), 181–192. doi:10.1002/hep.24015

Ghosh, S., May, M.J., Kopp, E.B., 1998. NF-kappa B and Rel proteins: evolutionarily conserved mediators of immune responses. Annu. Rev. Immunol. 16, 225–260. doi:10.1146/annurev.immunol.16.1.225

Golden, E.B., Lam, P.Y., Kardosh, A., et al., 2009. Green tea polyphenols block the anticancer effects of bortezomib and other boronic acid-based proteasome inhibitors. Blood 113, 5927–5937. doi:10.1182/blood-2008-07-171389

Hideshima, T., Richardson, P., Chauhan, D., et al., 2001. The proteasome inhibitor PS-341 inhibits growth, induces apoptosis, and overcomes drug resistance in human multiple myeloma cells. Cancer Res. 61(7), 3071–3076. PMID: 11306489.

Hofmann, F., Martelli, F., Livingston, D.M., et al., 1996. The retinoblastoma gene product protects E2F-1 from degradation by the ubiquitin-proteasome pathway. Genes Dev. 10, 2949–2959. doi:10.1101/gad.10.23.2949

Ikebe, T., Takeuchi, H., Jimi, E., et al., 1998. Involvement of proteasomes in migration and matrix metalloproteinase-9 production of oral squamous cell carcinoma. Int. J. Cancer 77, 578–585. doi:10.1002/(sici)1097-0215(19980812)77:4<578::aid-ijc18>3.0.co;2-2

Jares, P., Colomer, D., Campo, E., et al., 2012. Molecular pathogenesis of mantle cell lymphoma. J. Clin. Invest. 122(10), 3416–3423. doi:10.1172/JCI61272

Jariel-Encontre, I., Salvat, C., Steff, A.M., et al., 1997. Complex mechanisms for c-fos and c-jun degradation. Mol. Biol. Rep. 24, 51–56. doi:10.1023/a:1006804723722

Jung, T., Catalgol, B., Grune, T., 2009. The proteasomal system. Mol. Aspects Med. 30(4), 191–296. doi:10.1016/j.mam.2009.04.001

Kaplan, G.S., Torcun, C.C., Grune, T., et al., 2017. Proteasome inhibitors in cancer therapy: treatment regimen and peripheral neuropathy as a side effect. Free Radic. Biol. Med. 103, 1–13. doi:10.1016/j.freeradbiomed.2016.12.007

Koepp, D.M., Harper, J.W., Elledge, S.J., 1999. How the cyclin became a cyclin: regulated proteolysis in the cell cycle. Cell 97, 431–434. doi:10.1016/s0092-8674(00)80753-9

Kondagunta, G.V., Drucker, B., Schwartz, L., et al., 2004. Phase II trial of bortezomib for patients with advanced renal cell carcinoma. J. Clin. Oncol. 22, 3720–3425. doi:10.1200/JCO.2004.10.155

Kubiczkova, L., Pour, L., Sedlarikova, L., et al., 2014. Proteasome inhibitors: molecular basis and current perspectives in multiple myeloma. J. Cell Mol. Med. 18, 947–961. doi:10.1111/jcmm.12279

Kuhn, D. J., Chen, Q., Voorhees, P. M., et al., 2007. Potent activity of carfilzomib, a novel, irreversible inhibitor of the ubiquitin-proteasome pathway, against preclinical models of multiple myeloma. Blood 110(9), 3281–3290.

Lagadec, C., Vlashi, E., Bhuta, S., et al., 2014. Tumor cells with low proteasome subunit expression predict overall survival in head and neck cancer patients. BMC Cancer 14, 152–159. doi:10.1186/1471-2407-14-152

Lai, Z., Yang, T., Kim, Y.B., et al., 2002. Differentiation of Hdm2-mediated p53 ubiquitination and Hdm2 autoubiquitination activity by small molecular weight inhibitors. Proc. Natl. Acad. Sci. U. S. A. 99(23), 14734–14739. doi:10.1073/pnas.212428599

Landis-Piwowar, K.R., Milacic, V., Chen, D., et al., 2006. The proteasome as a potential target for novel anticancer drugs and chemosensitizers. Drug Resist. Update 9, 263–273. doi:10.1016/j.drup.2006.11.001

Langlands, F.E., Dodwell, D., Hanby, A.M., et al., 2014. PSMD9 expression predicts radiotherapy response in breast cancer. Mol. Cancer 13(1), 73. doi:10.1186/1476-4598-13-73

Lauricella, M., Emanuele, S., D'Anneo, A., 2006. JNK and AP-1 mediate apoptosis induced by bortezomib in HepG2 cells via FasL/caspase-8 and mitochondria-dependent pathways. Apoptosis 11(4), 607–625. doi:10.1007/s10495-006-4689-y

LeBlanc, R., Catley, L.P., Hideshima, T., et al., 2002. Proteasome inhibitor PS-341 inhibits human myeloma cell growth in vivo and prolongs survival in a murine model. Cancer Res. 62, 4996–5000. PMID: 12208752.

Lee, J.T.J., McCubrey, J.A., 2002. The Raf/MEK/ERK signal transduction cascade as a target for chemotherapeutic intervention in leukemia. Leukemia 16, 486–507. doi:10.1038/sj.leu.2402460

Li, B., Dou, Q. P., 2000. Bax degradation by the ubiquitin/proteasome-dependent pathway: involvement in tumor survival and progression. Proc. Natl. Acad. Sci. U. S. A. 97(8), 3850–3855. doi:10.1073/pnas.070047997

Li, X., Zhang, K., Li, Z., 2011. Unfolded protein response in cancer: the physician's perspective. J. Hematol. Oncol. 4, 8. doi.org/10.1186/1756-8722-4-8

Li, Y., Huang, J., Sun, J., et al., 2017. The transcription levels and prognostic values of seven proteasome alpha subunits in human cancers. Oncotarget 8(3), 4501–4519. doi:10.18632/oncotarget.13885

Lin, Y.C., Brown, K., Siebenlist, U., 1995. Activation of NF-kB requires proteolysis of the inhibitor IkB: signal-induced phosphorylation of IkB alone does not release active NF-kB. Proc. Natl. Acad. Sci. U. S. A. 92(2), 552–556. doi:10.1073/pnas.92.2.552

Liu, R., Lu, S., Deng, Y., et al., 2016. PSMB4 expression associates with epithelial ovarian cancer growth and poor prognosis. Arch. Gynecol. Obstet. 293, 1–11. doi:10.1007/s00404-015-3904-x

Maki, C.G., Huibregtse, J.M., Howley, P.M., 1996. In vivo ubiquitination and proteasome-mediated degradation of p53. Cancer Res. 56(11), 2649–2654. PMID: 8653711.

Manasanch, E.E., Orlowski, R.Z., 2017. Proteasome inhibitors in cancer therapy. Nat. Rev. Clin. Oncol. 14(7), 417. doi:10.1038/nrclinonc.2016.206.

Markowitz, J., Luedke, E. A., Grignol, V. P., et al., 2014. A phase I trial of bortezomib and interferon-α-2b in metastatic melanoma. J. Immunother. 37(1), 55–62. doi:10.1097/CJI.0000000000000009.

Mason, S.D., Joyce, J.A., 2011. Proteolytic networks in cancer. Trends Cell Biol. 21(4), 228–237. doi:10.1016/j.tcb.2010.12.002

Matsuyama, Y., Suzuki, M., Arima, C., et al., 2011. Proteasomal non-catalytic subunit PSMD2 as a potential. Mol. Carcinog. 50, 301–309. doi:10.1002/mc.20632.

Moorthy, A.K., Savinova, O.V., Ho, J.Q., et al., 2006. The 20S proteasome processes NF-kappaB1 p105 into p50 in a translation-independent manner. EMBO J. 25(9), 1945–1956. doi:10.1038/sj.emboj.7601081

Muchamuel, T., Basler, M., Aujay, M. A., et al., 2009. A selective inhibitor of the immunoproteasome subunit LMP7 blocks cytokine production and attenuates progression of experimental arthritis. Nat. Med. 15(7), 781–787.

Munakata, K., Uemura, M., Tanaka, S., et al., 2016. Cancer stem-like properties in colorectal cancer cells with low proteasome activity. Clin. Cancer Res. 22, 5277–5286. doi:10.1158/1078-0432.CCR-15-1945

Nawrocki, S.T., Bruns, C.J., Harbison, M.T., et al., 2002. Effects of the proteasome inhibitor PS-341 on apoptosis and angiogenesis in orthotopic human pancreatic tumor xenografts. Mol. Cancer Ther. 1(14), 1243–1253. PMID: 12516957.

Nicholson, D.W., 1999. Caspase structure, proteolytic substrates, and function during apoptotic cell death. Cell Death Differ. 6, 1028–1042. doi:10.1038/sj.cdd.4400598

Oakervee, H.E., Popat, R., Curry, N., et al., 2005. PAD combination therapy (PS-341/bortezomib, doxorubicin and dexamethasone) for previously untreated patients with multiple myeloma. Br. J. Haematol. 129, 755–762. doi:10.1111/j.1365-2141.2005.05519.x

Orlowski, R.Z., 2005. The ubiquitin proteasome pathway from bench to bedside. Hematology 1, 220–225. doi:10.1182/asheducation-2005.1.220

Orlowski, R.Z., Small, G.W., Shi, Y.Y., 2002. Evidence that inhibition of p44/42 mitogen-activated protein kinase signaling is a factor in proteasome inhibitor-mediated apoptosis. J. Biol. Chem. 277, 27864–27871. doi:10.1074/jbc.M201519200

Orlowski, R.Z., Nagler, A., Sonneveld, P., et al., 2007. Randomized phase III study of pegylated liposomal doxorubicin plus bortezomib compared with bortezomib alone in relapsed or refractory multiple myeloma: combination therapy improves time to progression. J. Clin. Oncol. 25, 3892–3901. doi:10.1200/JCO.2006.10.5460

Pagano, M., Tam, S.W., Theodoras, A.M., et al., 1995. Role of the ubiquitin-proteasome pathway in regulating abundance of the cyclin-dependent kinase inhibitor P27. Science 269, 682–685. doi:10.1126/science.7624798.

Pang, X., Yi, Z., Zhang, J., et al., 2010. Celastrol suppresses angiogenesis-mediated tumor growth through inhibition of AKT/mammalian target of rapamycin pathway. Cancer Res. 70(5), 1951–1959. doi:10.1158/0008-5472.CAN-09-3201

Papandreou, C.N., Daliani, D.D., Nix, D., et al., 2004. Phase I trial of the proteasome inhibitor bortezomib in patients with advanced solid tumors with observations in androgen-independent prostate cancer. J. Clin. Oncol. 22, 2108–2121. doi:10.1200/JCO.2004.02.106

Perrone, G., Hideshima, T., Ikeda, H., et al., 2009. Ascorbic acid inhibits antitumor activity of bortezomib in vivo. Leukemia 23, 1679–1686. doi:10.1038/leu.2009.83

Piperdi, B., Walsh, W. V., Bradley, K., et al., 2012. Phase I/II study of bortezomib in combination with carboplatin and bevacizumab as first-line therapy in patients with advanced non–small-cell lung cancer. J. Thor. Oncol. 7 (6), 1032–1040. doi:10.1097/JTO.0b013e31824de2fa.

Reece, D.E., Rodriguez, G.P., Chen, C., et al., 2008. Phase I–II trial of bortezomib plus oral cyclophosphamide and prednisone in relapsed and refractory multiple myeloma. J. Clin. Oncol. 26, 4777–4783. doi:10.1200/JCO.2007.14.2372

Richardson, P.G., Hideshima, T., Anderson, K.C., 2003. Bortezomib (PS-341): a novel, first-in-class proteasome inhibitor for the treatment of multiple myeloma and other cancers. Cancer Control 10, 361–369. doi:10.1177/107327480301000502

Richardson, P.G., Sonneveld, P., Schuster, M.W., et al., 2005. Bortezomib or high-dose dexamethasone for relapsed multiple myeloma. N. Engl. J. Med. 352(24), 2487–2498. doi:10.1056/NEJMoa043445

Richardson, P.G., Baz, R., Wang, M., et al., 2014. Phase 1 study of twice-weekly ixazomib, an oral proteasome inhibitor, in relapsed/refractory multiple myeloma patients. Blood 124(7), 1038–1046. doi:10.1182/blood-2014-01-548826.

Santen, R.J., Song, R.X., McPherson, R., et al., 2002. The role of mitogen-activated protein (MAP) kinase in breast cancer. J. Steroid Biochem. Mol. Biol. 80(2), 239–256. doi:10.1016/s0960-0760(01)00189-3

Secchiero, P., Bosco, R., Celeghini, C., et al., 2011. Recent advances in the therapeutic perspectives of Nutlin-3. Curr. Pharm. Des. 17(6), 569–577. doi:10.2174/138161211795222586

Shah, M.H., Young, D., Kindler, H.L., et al., 2004. Phase II study of the proteasome inhibitor bortezomib (PS-341) in patients with metastatic neuroendocrine tumors. Clin. Cancer Res. 10, 6111–6118. doi:10.1158/1078-0432.CCR-04-0422

Shweiki, D., Itin, A., Soffer, D., et al., 1992. Vascular endothelial growth factor induced by hypoxia may mediate hypoxia-initiated angiogenesis. Nature 359, 843–845. doi:10.1038/359843a0

Stancovski, I., Gonen, H., Orian, A., et al., 1995. Degradation of the proto-oncogene product c-Fos by the ubiquitin proteolytic system in vivo and in vitro: identification and characterization of the conjugating enzymes. Mol. Cell Biol. 15, 7106–7116. doi:10.1128/mcb.15.12.7106.

Steiner, R.E., Manasanch, E.E., 2017. Carfilzomib boosted combination therapy for relapsed multiple myeloma. Onco Targets Ther. 10, 895–907. doi:10.2147/OTT.S102756

Sung, E.S., Park, K.J., Choi, H.J., et al., 2012. The proteasome inhibitor MG132 potentiates TRAIL receptor agonist-induced apoptosis by stabilizing tBid and Bik in human head and neck squamous cell carcinoma cells. Exp. Cell Res. 318(13), 1564–1576. doi:10.1016/j.yexcr.2012.04.003

Sunwoo, J.B., Chen, Z., Dong, G., et al., 2001. Novel proteasome inhibitor PS-341 inhibits activation of nuclear factor-kappa B, cell survival, tumor growth, and angiogenesis in squamous cell carcinoma. Clin. Cancer Res. 7, 1419–1428. PMID: 11350913.

Teicher, B.A., Tomaszewski, J.E., 2015. Proteasome inhibitors. Biochem. Pharmacol. 96, 1–9. doi:10.1016/j.bcp.2015.04.008

Terpos, E., Kastritis, E., Roussou, M., et al., 2008. The combination of bortezomib, melphalan, dexamethasone and intermittent thalidomide is an effective regimen for relapsed/refractory myeloma and is associated with improvement of abnormal bone metabolism and angiogenesis. Leukemia 22, 2247–2256. doi:10.1038/leu.2008.235

Traidej, M., Chen, L., Yu, D., et al., 2000. The roles of E6-AP and MDM2 in p53 regulation in human papillomavirus-positive cervical cancer cells. Antisense Nucleic Acid Drug Dev. 10(1), 17–27. doi:10.1089/oli.1.2000.10.17

Vlashi, E., Kim, K., Lagadec, C., et al., 2009. In vivo imaging, tracking, and targeting of cancer stem cells. J. Natl. Cancer Inst. 101(5), 350–359. doi:10.1093/jnci/djn509

Vogelstein, B., Lane, D., Levine, A.J., 2000. Surfing the p53 network. Nature 408, 307–310. doi:10.1038/35042675

Voorhees, P.M., Dees, E.C., O'Neil, B., et al., 2003. The proteasome as a target for cancer therapy. Clin. Cancer Res. 9(17), 6316–6325. PMID: 14695130.

Voutsadakis, I.A., 2017. Proteasome expression and activity in cancer and cancer stem cells. Tumour Biol. 39(3). doi:10.1177/1010428317692248

Wang, M., 2011. Comparative mechanisms of action of proteasome inhibitors. Oncology (Williston Park) 2, 19–24. PMID: 25188479.

Williams, S.A., McConkey, D.J., 2003. The proteasome inhibitor bortezomib stabilizes a novel active form of p53 in human LNCaP-Pro5 prostate cancer cells. Cancer Res. 63(21), 7338–7344. PMID: 14612532.

Xu, G.W., Ali, M., Wood T.E., et al., 2010. The ubiquitin-activating enzyme E1 as a therapeutic target for the treatment of leukemia and multiple myeloma. Blood 115(11), 2251–2259. doi:10.1182/blood-2009-07-231191

Yang, C.H., Gonzalez-Angulo, A.M., Reuben, J.M., et al., 2006. Bortezomib (VELCADE) in metastatic breast cancer: pharmacodynamics, biological effects, and prediction of clinical benefits. Ann. Oncol. 17, 813–817. doi:10.1093/annonc/mdj131

Yang, H., Landis-Piwowar, K. R., Chen, D., et al., 2008. Natural compounds with proteasome inhibitory activity for cancer prevention and treatment. Curr. Protein Peptide Sci. 9(3), 227–239. doi:10.2174/138920308784533998

Yoshida, S., Ono, M., Shono, T., et al., 1997. Involvement of interleukin-8, vascular endothelial growth factor, and basic fibroblast growth factor in tumor necrosis factor alpha-dependent angiogenesis. Mol. Cell Biol. 17, 4015–4023. doi:10.1128/mcb.17.7.4015

Zhou, H.J., Aujay, M.A, Bennett, M.K., et al., 2009. Design and synthesis of an orally bioavailable and selective peptide epoxyketone proteasome inhibitor (PR-047). J. Med. Chem. 52(9), 3028–3038. doi:10.1021/jm801329v

Zhu, H., Zhang, L., Dong, F., et al., 2005. Bik/NBK accumulation correlates with apoptosis-induction by bortezomib (PS-341, Velcade) and other proteasome inhibitors. Oncogene 24, 4993–4999. doi:10.1038/sj.onc.1208683

CHAPTER FOURTEEN

Proteostasis and Skin Aging

Anne-Laure Bulteau and Bertrand Friguet

CONTENTS

14.1 INTRODUCTION

The skin is located at the body's interface interacting with the external environment. It is a complex structure made up of several layers of different tissues, each with a distinct cellular composition. It has crucial functions like thermal protection and hydration and acts as a physical barrier against pathogens. The top part of the skin is the epidermis, a layered structure containing hair follicles. The outer epidermal surface consists of corneocytes which form a dense protective barrier. The basal layer of the epidermis contains stem cells, which constantly regenerate the epidermis. The dermis is located under the epidermis and is composed of dermal fibroblasts which produce collagen and elastin that form the extracellular matrix (ECM), as well as melanocytes which produce the photo-protective pigment melanin (Rittie and Fisher, 2015). Upon chronological aging, the epidermis and dermis become thinner and lose their regenerative capacity, resulting in formation of wrinkles and skin dryness. Extrinsic aging is characterized by deep wrinkles, sagging skin and hyperpigmentation and is primarily caused by chronic sun exposure and pollution (D'Orazio et al., 2013). Regardless of the type of aging, wrinkles and reduction in elasticity are typical phenomena of skin aging and result from progressive

atrophy of the dermis. One of the main mechanisms of skin atrophy is believed to be the reduction in the amount of ECM, especially collagen. The elastic fiber system undergoes age-associated structural changes and fibrillin-rich microfibrils are selectively degraded in the papillary dermis during intrinsic skin aging (Langton et al., 2012). In aged skin, collagen production decreases while its degradation increases, which leads to an overall reduction in the amount of collagen. Skin aging is a natural process that would occur over time due to the different hallmarks of aging as previously reviewed (Lopez-Otin et al., 2013), even in the absence of sun exposure. These hallmarks are genomic instability, telomere attrition, epigenetic alterations, loss of proteostasis, deregulated nutrient-sensing, mitochondrial dysfunction, cellular senescence, stem-cell exhaustion and altered intercellular communication. While the precise mechanisms of photoaging are still under investigation, collagen alteration, mitochondrial mtDNA damage, increased reactive oxygen species (ROS) production and inflammation are all key features (Cavinato and Jansen-Durr, 2017; Fisher et al., 2002) (Figure 14.1). The structural and functional phenotypes of skin aging appear prematurely and are more severe in photo-aged skin compared to chronologically aged skin. In addition, as photoaging is a cumulative process, it is more pronounced

DOI: 10.1201/9781003048138-14

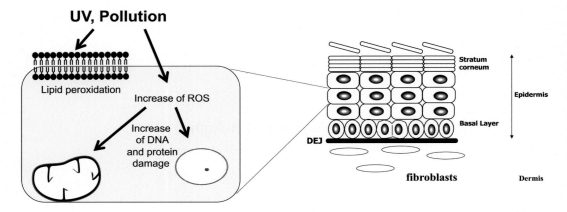

Figure 14.1 Impact of UV and pollution on skin aging. UV and pollution through enhanced lipid peroxidation and ROS production are targeting cellular components such as mitochondrial and nuclear DNA but also proteins that need to be degraded. *Abbreviation:* DEJ, dermis-epidermis junction.

in older people who have been regularly exposed to the sun for long periods (Ortonne, 1990). Skin changes constitute the first obvious evidence of aging, a complex and multifactorial biological process affecting the whole body. Skin aging is influenced by several factors including hormonal changes and metabolic processes. Indeed, skin aging is affected by growth factor modifications and hormone activity that decline with age. The best-known decline is that of sex steroids such as estrogen, testosterone, dehydroepiandrosterone (DHEA) and its sulfate ester (DHEAS). Other hormones including melatonin, insulin, cortisol, thyroxine and growth hormone also decline (Phillips et al., 2001). Regarding metabolic processes, skin aging is marked by a senescence-related decline in lipid and water content, which ultimately impairs epidermal barrier function. It is well documented that aged skin is prone to develop altered drug permeability, increased susceptibility to irritants, contact dermatitis and severe xerosis. Thus, the *stratum corneum* barrier function appears more "fragile" in elderly. Lipids and more particularly ceramide deficiency may account, at least in part, for the age-associated dysfunction of the *stratum corneum* (Rogers et al., 1996).

The skin is an organ with high renewal rates and the epidermis is in continuous regeneration. The epidermal stem cells are therefore metabolically active and depend on adenosine triphosphate (ATP) for their energy needs. ATP is primarily produced by oxidative phosphorylation (OXPHOS) in the mitochondria. The mitochondrial electron transport chain (ETC) contains a series of protein complexes located in the inner membrane level. These complexes sequentially transport electrons and translocate protons to the intermembrane space creating a proton gradient, which is used to generate ATP. The natural by-products of OXPHOS include ROS such as superoxide anion, which can disrupt macromolecular and cellular structures. Due to its physical proximity to ETC and therefore ROS, mtDNA is very sensitive to mutations leading to dysfunctional ETC complexes, thus triggering a vicious cycle of increased ROS production and damaged mtDNA (Neiman and Taylor, 2009). Several studies have directly or indirectly linked mitochondrial dysfunction to both skin chronological aging and photoaging (Krutmann and Schroeder, 2009). At the molecular level, aged skin is characterized by damaged mitochondria, mtDNA deletions, high levels of ROS and oxidative stress which is accompanied by a constant decline in mitochondrial function in both dermal and epidermal layers (Krutmann and Schroeder, 2009). An extensive 4,977 bases' pair region of mtDNA, including genes encoding respiratory chain complexes I, IV and V, is frequently deleted in aged human skin (J. H. Yang et al., 1995). In addition to UV, exposure to pollution may be a cause of skin aging. Diesel particles and ozone have been shown to increase oxidative stress, to promote DNA and protein damage by ROS (Y. K. Kim et al., 2017; Piao et al., 2018) resulting in the formation of pigment spots and skin wrinkles (Vierkotter et al., 2010). Although epidemiological and clinical studies highlight the harmful effects of pollution on human health, scarce biological research is available to date on its effects on the skin (Krutmann and Schroeder, 2009). Altered protein quality control systems (proteasome and Lon protease activity) in keratinocytes exposed to a volatile organic compounds (VOC) cocktail have been reported, suggesting that the proteolytic systems in the skin may be affected by this type of pollutants (Dezest et al., 2017). In addition to this, mtDNA damage is generally seen at higher levels in photoexposed skin, leading to an accumulation of cellular damage and decreased mitochondrial activity (Berneburg et al., 1997; Bowman and Birch-Machin, 2016; Krishnan et al., 2004). Mitochondrial proteins modified by oxidation that accumulate during aging are eliminated by specific proteolytic systems such as the Lon protease. Because mitochondria are the main source of ATP in many types of cells, while the ubiquitin–proteasome system (UPS) is consuming ATP, it is conceivable that the functional link between mitochondria and proteasomes may be related with alterations in intracellular ATP concentration during skin aging. Indeed, ATP levels increased upon inhibition of the proteasome while inhibition of mitochondrial function in turn reduced the activity of the proteasome, suggesting that these two systems are interconnected (Koziel et al., 2011). During stress such as the accumulation of misfolded proteins in the mitochondria, mitochondrial unfolded protein response (mtUPR) induces the synthesis of mitochondrial proteases to degrade aberrant proteins. However, when stress promotes excessive damage in individual mitochondria, the entire organelle can be degraded in a process known as mitophagy (mitochondrial specific autophagy). An increased focus on mitochondrial proteostasis, autophagy and proteasome activity should provide a more informed perspective on the etiology and treatment of skin aging. This chapter focuses on this topic. Firstly, proteases that are implicated in mitochondrial protein homeostasis in the different compartments of the mitochondrion are described with a special emphasis on the mitochondrial matrix Lon and Clp proteases. Then, the implication of autophagy and proteasome dysfunction in the age-related impairment of skin protein quality control, as well as its potential prevention, is addressed. Overall, this chapter highlights the important role of mitochondrial proteases, proteasome and autophagy in both mitochondrial and cellular protein maintenance and intracellular redox homeostasis that

are altered in the skin upon aging, resulting in the less efficient degradation of oxidatively modified proteins and their subsequent accumulation (Figure 14.2).

14.2 ROLE OF MITOCHONDRIAL PROTEASES IN SKIN AGING

Mitochondria are a major source of reactive intracellular ROS, the production of which increases with aging. These organelles are also preferential targets of oxidative damage (oxidation of mtDNA, lipids, proteins). The deleterious effects of ROS may be responsible for altered mitochondrial function observed in various pathophysiological states associated with oxidative stress and aging. An important factor for the maintenance of proteins in the presence of oxidative stress is the enzymatic reversal of the oxidative damage and the degradation of oxidized proteins. The failure of these processes is most likely to represent a critical part of the aging process in the skin. Mitochondrial proteases degrade misfolded/damaged polypeptides, thus performing protein quality control in the organelle. Mammalian mitochondria contain four major ATP-dependent proteases; Lon, Clp, as well as m-AAA and i-AAA proteases (Figure 14.3). Clp and Lon proteases are hetero-oligomeric complexes located in the matrix, whereas m-AAA and i-AAA types are found in the inner mitochondrial membrane (Van Dyck and Langer, 1999). As evidenced by various genetic studies, these proteases contribute to the degradation of misfolded and damaged proteins and/or to the maintenance of the mitochondrial genome stability. In addition, both proteolytic systems appear to exert chaperone activity. Currently, information regarding the regulation of each of the ATP-dependent proteases and/or the identities of specific protein substrates is limited. The two matrix soluble proteases are mainly degrading substrate proteins in the matrix, whereas the inner membrane AAA proteases degrade membrane-integrated targets. Nevertheless, the exposure of hydrophobic residues is probably a common recognition element, and the chaperone and proteolytic functions participate in the prevention of the accumulation of aggregated material (Friguet et al., 1994; Voos, 2009). We will mainly focus on the two matrix proteases in this chapter.

The Lon protease plays a critical role in the removal of oxidized and damaged proteins (Voos,

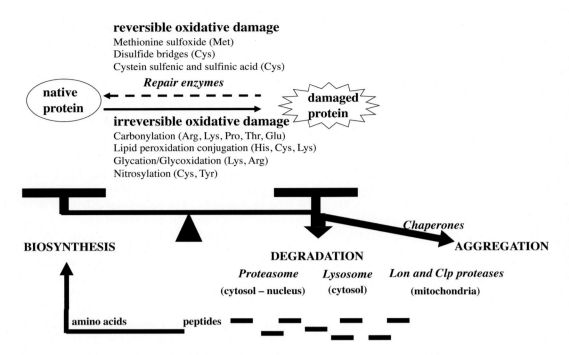

reversible oxidative damage
Methionine sulfoxide (Met)
Disulfide bridges (Cys)
Cystein sulfenic and sulfinic acid (Cys)

Repair enzymes

native protein

damaged protein

irreversible oxidative damage
Carbonylation (Arg, Lys, Pro, Thr, Glu)
Lipid peroxidation conjugation (His, Cys, Lys)
Glycation/Glycoxidation (Lys, Arg)
Nitrosylation (Cys, Tyr)

Chaperones

BIOSYNTHESIS

DEGRADATION

AGGREGATION

Proteasome
(cytosol – nucleus)

Lysosome
(cytosol)

Lon and Clp proteases
(mitochondria)

amino acids   peptides

Figure 14.2 **Protein oxidative damage, proteostasis and aging.** During aging and upon oxidative stress, proteins represent critical targets for oxidative modifications. Depending on whether the oxidative damage is reversible or not, oxidatively modified proteins can be either repaired or eliminated by dedicated proteolytic pathways (previously reviewed in Friguet, 2006). Oxidized protein repair is restricted to a few sulfur-containing amino-acid residues (cystein and methionine) oxidation products that are reduced by such antioxidant systems as thioredoxin/thioredoxin reductase, glutaredoxin/glutathione/glutathione reductase and methionine sulfoxide reductases. Irreversible oxidative damage to proteins targets them to degradation by intracellular proteolytic systems such as the proteasome, the lysosome and the mitochondrial Lon and Clp proteases. Loss of proteostasis has been described as an important contributor of the age-associated accumulation of oxidized proteins, aggregation of misfolded proteins due to chaperone overload and subsequent alteration of cellular function, including mitochondrial dysfunction.

2009). Like many mitochondrial proteins, the Lon protease is encoded by the nuclear genome. This protease is active in the form of a homo-oligomeric complex composed of six monomers in *Escherichia coli* and of seven monomers in eukaryotes such as yeast with a molecular weight of approximately 106 kDa. The Lon protease is made up of three domains that are conserved within different species (Garcia-Nafria et al., 2003). The N-terminal domain (N domain) is able to interact with protein substrates together with the second domain, also called the AAA+ module. The AAA+ module is composed of two segments, one involved in the binding of ATP and the other in the hydrolysis of ATP. A third domain (P domain) carries the active site residues serine and lysine which form the catalytic dyad for proteolytic activity (Amerik et al., 1990). An important characteristic of Lon proteolytic activity is its stimulation by ATP. In contrast, ADP acts as a protease inhibitor

indicating a regulation in response to changes in energy load (Watabe et al., 2001). In addition to its proteolytic activity, the mammalian Lon has been also shown to exhibit chaperone properties and specifically to bind to human mitochondrial DNA and RNA sequences, as well as to interact with the mitochondrial DNA polymerase gamma and the Twinkle helicase. Consequently, Lon also regulates mitochondrial DNA replication (Lee and Suzuki, 2008; Liu et al., 2004; Lu et al., 2003, 2007).

The Lon protease has been involved in the degradation of oxidized proteins (Bayot et al., 2010). The enzymes with oxidation-sensitive prosthetic groups are prone to oxidative damage and are the primary targets of Lon protease (Bender et al., 2010). As a key player in the elimination of oxidized mitochondrial proteins, Lon protease plays a critical role in maintaining the homeostasis of mitochondrial matrix proteins and therefore

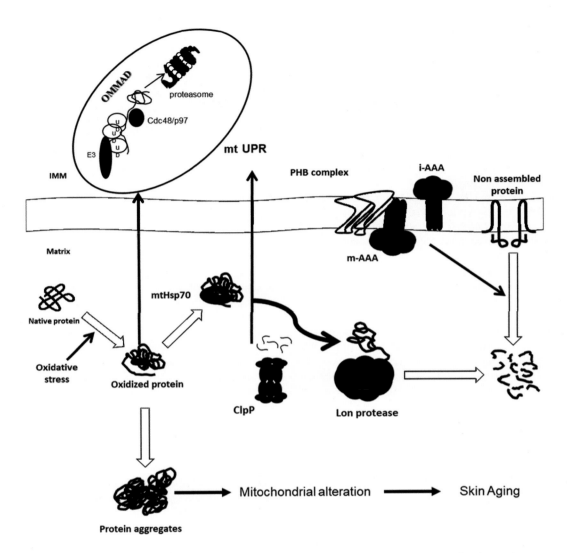

Figure 14.3 **Mitochondrial protein quality control in the matrix compartment.** Upon aging, mitochondria tend to produce more ROS that may damage proteins. Due to oxidative stress conditions, proteins become partially misfolded and more prone to aggregation. By degrading oxidized proteins, Lon and Clp proteases prevent their accumulation and subsequent mitochondria alteration. *Abbreviations:* AAA, ATPase associated with diverse cellular activities; Clp, caseinolytic protease; IM, inner membrane; IMS, intermembrane space; M, matrix; OM, outer membrane; OMMAD, outer mitochondrial membrane-associated degradation; mtUPR, mitochondrial unfolded protein response.

in the functional and structural integrity of the mitochondria (Bulteau, Szweda, et al., 2006). Thus, Bota et al. (2005) demonstrated that the reduction of Lon expression in human pulmonary fibroblasts induces alterations in mitochondrial morphology and function and activates apoptosis. In addition, Ngo and Davies (2009) proposed that Lon is a human stress protein given the significant increase in Lon levels as well as the preserved mitochondrial function and the improvement in

the survival of cells exposed to different stresses. The Lon gene promoter carries HIF-1 (hypoxia-inducible factor 1)-binding sites and its transcription is activated by HIF-1 in mammalian hypoxic cells (Fukuda et al., 2007), suggesting that in the skin it can be upregulated in the dermis where the oxygen level is low. This involvement of the Lon protease in mitochondrial homeostasis upon aging has been observed in aged tissues where the levels and/or the activity of Lon decrease

(Bota and Davies, 2002). This deleterious effect of aging on Lon has been exacerbated by increased chronic oxidative stress. Lon inducibility was also abolished after oxidative treatment in senescent WI-38 fibroblasts (Ngo et al., 2011). As mtDNA mutations accumulate throughout lifespan, a role of Lon in mtDNA renewal and maintenance represents another possibility for its involvement in the aging processes. Due to its multiple functions affecting both mitochondrial activities and integrity, the role of Lon in aging and longevity can therefore take different forms (Hamon et al., 2015). Recent work has claimed that in nonfunctional mitochondria, characterized by a low inner membrane potential (Δψ), Lon (together with ClpP) might even have a direct protective role against oxidative stress by degrading respiratory complex I subunits, which is a major source of ROS under these conditions (Pryde et al., 2016) suggesting a collaboration between the two proteases. Clp proteases are a protein family encompassing ClpAP, ClpCP, ClpEP, ClpXP and ClpYQ, which are found in a wide variety of prokaryotes and in the mitochondria of eukaryotic cells as well as in the chloroplast of algae and plant cells. They are active in the form of complexes comprising two functional elements: a proteolytic subunit (ClpP or ClpQ) sandwiched between two ATPase-active chaperone rings (ClpA, ClpC, ClpE, ClpX or ClpY) depending on the species (Kress et al., 2009). ClpX and ClpA have loops located in their central channel that recognize terminal sequences, or "tags", in substrate proteins (Hinnerwisch et al., 2005).

The best-known implication of the ClpP protease in the quality control of mitochondrial proteins is probably its role in the mtUPR. This function places ClpP at the center of age-related processes since aging is associated with the accumulation of misfolded and unfolded protein aggregates, leading to adverse effects on cellular function (Sherman and Goldberg, 2001). Like the Lon protease, ClpP can be considered as a stress protein since its expression is induced under stress conditions, in eukaryotes and in prokaryotes (Gispert et al., 2013). Since loss of mitochondrial protein homeostasis has been linked to disease and aging, the alteration of mtUPR, and therefore ClpP, appears to play a central role in the aging process (Hamon et al., 2015). Durieux et al. (2011) showed that mtUPR is necessary to mediate longevity in *Caenorhabditis elegans*. Silencing

the mtUPR components of long-lived ETC mutants abolished their extended lifespan. Mice presenting decreased cytochrome *c* oxidase activity demonstrated increased longevity and induction of mtUPR (Pulliam et al., 2014). ClpP activity has also been linked to mtDNA maintenance since ClpP deficiency has been associated with increased mutations (Gispert et al., 2013). Mitochondrial protein homeostasis is vital for various physiological processes and its dysregulation results in severe mitochondrial dysfunction.

14.3 AGE-ASSOCIATED IMPAIRMENT OF AUTOPHAGY IN SKIN

When mitochondria are damaged and their proteases are overwhelmed, mitochondria are targeted to autophagy also referred as to mitophagy.

Autophagy is a fundamental eukaryotic cellular process conserved from yeast to humans that controls protein and organelle degradation and has essential roles in survival, development, cellular homeostasis and aging. Three types of autophagy are currently described: chaperone-mediated autophagy, microautophagy and macroautophagy (Mizushima and Komatsu, 2011). Autophagy, which usually refers to macroautophagy, is a process related to the inclusion of cellular components into vesicles called autophagosomes that fuse with lysosomes and their acid hydrolases can degrade nucleic acids, carbohydrates, lipids and proteins (Levine and Klionsky, 2004). Different points of control are under the influence of numerous factors that regulate the autophagic flux. The regulatory process of autophagy is divided into two distinct forms: selective that is regulated under homeostatic conditions and nonselective autophagy that is induced upon starvation or in response to stress-related conditions (Klionsky and Schulman, 2014). The core autophagic machinery is the same in both forms and is structured by a number of autophagy-related genes (ATG) firstly identified in yeast almost three decades ago (Mehrpour et al., 2010). Moreover, there is remarkable conservation of the proteins and pathways that carry out and regulate autophagy from yeast to mammals (Klionsky and Schulman, 2014; Mehrpour et al., 2010; Z. Yang and Klionsky, 2010). Autophagy is one of the main cytosolic proteolytic pathways responsible for the elimination of damaged proteins. Indeed, autophagy has been implicated in the removal of damaged proteins and organelles that are accumulating during aging (Mizushima and Komatsu, 2011). An age-associated decline of the autophagic flux has been previously reported in many organs while stimulation of autophagy has been shown to result in the extension of cellular replicative lifespan and/or in the inhibition of stress-induced cellular senescence (Rubinsztein et al., 2011). Moreover, pharmacological or genetic interventions stimulating autophagy were found to extend the lifespan of model organisms (Madeo et al., 2015). Conversely, inhibition of autophagy has been reported to promote premature aging. Interestingly, using radioisotopic pulse-chase labeling for assessing the degradation of long-lived proteins and a macroautophagy-specific inhibitor, autophagy was measured in dermal fibroblasts from three phylogenetically distinct pairs of short- and long-lived mammals (rodents, marsupials and bats) and was found to be significantly enhanced in all three long-lived species (Pride et al., 2015). Nevertheless, whether autophagy is impaired or not in aged human skin fibroblasts remains controversial since autophagic flux, measured by monitoring the level and turnover of intracellular microtubule-associated protein-1 light chain-3 (LC3-II), was found to either decline in primary fibroblasts derived from surgical breast skin samples of female human donors aged 20–67 years (Kalfalah et al., 2016) or remain unchanged when comparing young males foreskin fibroblasts (mean age 12.7) with old males truncal dermal fibroblasts (mean age 67) (H. S. Kim et al., 2018). Autophagy vesicles were found increased in both fibroblasts and keratinocytes upon replicative and stress-induced replicative senescence, respectively (Gerland et al., 2003; Gosselin et al., 2009). In addition, basal levels of LC3 were found increased in aging human dermal fibroblasts upon replicative senescence in vitro but no increase in LC3 fluorescence was detected in full-thickness skin sections from the biopsies obtained from 10 healthy young (age 25–30 years) and 10 old (age 60–65 years) donors. No difference was also observed in the basal level of LC3 in the skin sections from photoprotected and photo-exposed areas of the arm (Demirovic et al., 2015). Nevertheless, the age-pigment lipofuscin, which is composed of oxidized and partially degraded proteins, lipids and carbohydrates as well as transition metals, has been reported to accumulate within lysosomes of aged cells, including skin dermal fibroblasts and keratinocytes (Tonolli et al., 2017; von Zglinicki

et al., 1995). Such an accumulation of lipofuscin has been suggested to result from the combined dysfunction of mitochondria and lysosomes (Tonolli et al., 2017; von Zglinicki et al., 1995). In addition, the formation of lipofuscin can be generated upon stress-induced premature senescence by cell exposure to UV light or oxidants such as hydrogen peroxide or paraquat in both human fibroblasts and keratinocytes (Hohn et al., 2012; Tonolli et al., 2020). Interestingly, lipofuscin has been reported to inhibit the proteasome (Hohn et al., 2010; Sitte et al., 2001) and to induce cell cycle arrest and cell death (von Zglinicki et al., 1995) in human fibroblasts while lipofuscin-bound iron was found to be a major intracellular source of oxidants in senescent human dermal fibroblasts (Hohn et al., 2010). Moreover, due to its chromophore properties with reference to visible light, lipofuscin has been also described as a fluorophore and a photosensitizer which, when excited by blue light, can generate a variety of oxidant species such as singlet oxygen, hence promoting oxidative damage and its auto-catalytic formation (Crouch et al., 2015; Sparrow and Duncker, 2014).

Using UVB-induced senescent human dermal fibroblasts as an experimental model of skin photoaging (Cavinato and Jansen-Durr, 2017), increased ROS production followed by inhibition of the proteasome and activation of autophagosome formation was shown to represent early events in this process. The establishment of senescence was dependent on autophagy since its inhibition resulted in a phenotypic switch from senescence to apoptosis (Cavinato and Jansen-Durr, 2017). Moreover, treatment with the anti-oxidant N-acetyl-cysteine was found to prevent autophagy activation and to promote apoptotic-mediated cell death, indicating that ROS acts as signaling molecules that are implicated in autophagy activation (Cavinato et al., 2017). Other ROS-related signaling pathways, such as those relying on mTOR or AMPK that have been implicated in aging and lifespan regulation, have been also involved in the modulation of autophagy upon UV radiation (Wang et al., 2019). Indeed, UVB-induced AMPK activation has been reported to increase LC3-II levels and autophagic flux while PI3K/Akt activation, which is also induced by UVB, has been shown to inhibit autophagy through mTOR activation (Qiang et al., 2013; B. Zhang et al., 2018). Interestingly, upon UVB exposure AMPK-dependent activation of autophagy was found to protect human keratinocytes from apoptosis (Martinez-Lopez et al., 2015).

Several studies have demonstrated that ROS activate autophagy by regulating the PI3K/Akt/mTORC1-signaling pathway (Byun et al., 2009; H. Zhang et al., 2009) while AMPK activity has been also shown to be upregulated by ROS and to consequently activate autophagy (Rodney et al., 2016). Furthermore, the existence of regulation loops between ROS and the circadian clock has been previously demonstrated suggesting that the endogenous rhythmicity of ROS homeostasis could be implicated in the control of the circadian modulation of autophagy (O'Neill and Reddy, 2011). Hence, the age-associated deregulation of the circadian clock might explain, at least in part, the decreased autophagy activity observed during aging.

Recent studies on skin fibroblasts have shown a direct link between age-related changes of clock gene expression and the decline of autophagy levels (Kalfalah et al., 2016). More specifically, the expression levels of transcriptional repressor components of the circadian oscillator PER2 were strongly reduced in primary dermal fibroblasts from aged humans, as well as the autophagic flux. Therefore, an altered core clock gene expression may participate in the age-related reduction of autophagic flux and promote cellular aging of human skin fibroblasts. Interestingly, the dysfunctional autophagy mechanism of aged cells could be partially restored by elevating the expression of a single-core clock gene, namely PER2 (Kalfalah et al., 2016). Hence, the age-related changes of the clock gene expression may promote the decline of autophagy levels.

14.4 IMPAIRMENT OF PROTEASOME IN SKIN AGING AND PHOTOAGING

Accumulation of oxidized proteins is a hallmark of aging in different organs and tissues, including skin (Levine & Stadtman, 2001; Petropoulos et al., 2000). In addition, damage to macromolecules resulting from chronic exposure of human skin to UV irradiation represents one of the main features of photoaging. Indeed, upon UVA irradiation of the skin, which represents most of the terrestrial UV solar energy and penetrates more efficiently in the basal layer of the epidermis than UVB, ROS are formed through the absorption of photons by endogenous photosensitizers.

The high concentration of ROS generated after UV irradiation of the skin has damaging effects (Cunningham et al., 1985; Tyrrell, 1995; Van der Zee et al., 1993). ROS such as singlet oxygen, superoxide anion and hydroxyl radical are produced and react with various intracellular targets including lipids, nucleic acids and proteins (Vile and Tyrrell, 1995). We have previously shown that UV irradiation of human keratinocytes induces increased levels of oxidized proteins (Bulteau, Moreau, et al., 2002).

Protein degradation is the most efficient mechanism to prevent toxicity associated with the accumulation of damaged proteins. The proteasome is the main intracellular proteolytic system implicated in the degradation of oxidized proteins. The 20S proteasome is made of 28 subunits arranged as 4 stacked rings. The two outer rings are formed by α-subunits and the two inner rings are formed by β-subunits that carry the proteolytic activities. The core 20S proteasome can combine with the 19S regulator complex to form a 26S-proteolytic complex which is responsible for the ATP- and ubiquitin-dependent proteolysis (Glickman and Ciechanover, 2002). Earlier studies of our laboratory on the effect of aging in human keratinocytes and epidermis have evidenced a decrease in proteasome activity and content and age-related alterations of proteasome subunits (Bulteau et al., 2000). Indeed, in collaboration with Yoram Milner laboratory, a decrease of proteasome peptidase activities was found in epidermis extracts from old donors as compared to young ones, which, as expected, was associated with an age-related increase of damaged proteins. A similar decrease of proteasome activities was also found in senescent keratinocyte cultures, which could be explained, at least in part, by the decrease of proteasome content found in old cells (Petropoulos et al., 2000). Interestingly, in cell cultures, as well as in skin, an inverse relationship was found between the senescence marker SA-β-galactosidase activity and the proteasome content. These results suggested that proteasome is downregulated during replicative senescence of keratinocytes in vitro as well as in aged epidermis in vivo, possibly resulting in the accumulation of modified proteins. To gain further insight in other mechanisms that may be implicated in decreased proteasome activities with age, 20S proteasome was purified from the epidermis of donors of different ages: young, middle-aged and old. The patterns of proteasome subunits were analyzed by 2D gel electrophoresis to determine whether proteasome structure is also affected with age. The 2D gel pattern of proteasome subunits was found to be modified for four subunits identified as α3, α4, α5 and β4, indicating that the observed decline in proteasome activity with age may also be related to alterations of its subunits (Bulteau et al., 2000). Moreover, in vitro studies in human lung and also dermal fibroblasts have shown a decreased proteasomal activity or expression of specific proteasome subunits during cellular senescence (Chondrogianni et al., 2003; Sitte, Huber, et al., 2000; Sitte, Merker, et al., 2000a, 2000b, 2000c). By comparing the proteasome composition of in vitro cultured dermal fibroblasts from healthy subjects of different ages (i.e. 15, 41 and 82 years old), a reduction of the proteasome subunit α3 and the ubiquitin carboxyl-terminal hydrolase isozyme L1 has been observed with aging (Boraldi et al., 2003). This is consistent with the hypothesis of a less efficient proteasomal system during senescence. Taken together, these results may explain, at least in part, the accumulation of oxidized proteins in the cytosol of cells during skin aging (Gu et al., 2020; Ma et al., 2001).

In addition, ex vivo studies of proteasomal peptidase activities in human dermal fibroblasts from individuals ranging from 20 to 78 years old also showed an age-associated decrease of proteasome activities. All three proteasome activities were found dramatically decreased in human dermal fibroblasts from 20- to 50-year-old donors but remained stable in donors aged 50–78 years (Hwang et al., 2007). This age-associated loss of proteasome activities could be explained, at least in part, by a decrease in the expression of proteasome subunits and was accompanied by the accumulation of oxidized and ubiquitinated proteins. By using a system to determine proteasome activity in living cells based on a destabilized version of green fluorescent protein (GFP), this age-associated decreased proteasome activity in skin fibroblasts from middle-aged donors compared with young donors was further confirmed by Jansen-Dürr and colleagues (Koziel et al., 2011). Conversely, fibroblasts from healthy centenarians were reported to exhibit proteasome activity and proteasome subunit expression levels similar to those of younger individuals than older ones, suggesting that sustained proteasome activity could have contributed to their successful aging (Chondrogianni et al., 2000). In

accordance, improved protein homeostasis has been reported for long-lived species compared to phylogenetically related shorter lived species (Pride et al., 2015), while earlier studies had previously shown that the long-lived rodent naked-mole-rat exhibited increased liver proteasome activity and decreased accumulation of ubiquitinated proteins as compared to mice (Perez et al., 2009). Interestingly, using skin fibroblasts from three phylogenetically distinct pairs of short- and long-lived mammals, proteasome activity was found to be increased in the long-lived rodent and marsupial but not in bats while autophagy was significantly enhanced in all three long-lived species (Pride et al., 2015).

The trend for a decline in proteasome function observed during aging may be related to different reasons, including tissue specificity. Downregulated expression of proteasomal subunits in aging heart, epidermal keratinocytes and dermal fibroblasts (Bulteau, Szweda, et al., 2002; Hwang et al., 2007; Petropoulos et al., 2000), unbalanced levels of catalytic subunits in senescent WI-38 fibroblasts (Chondrogianni et al., 2003), defective expression of regulatory subunits in aged muscle (Ferrington et al., 2005), damaging modifications in critical proteasome subunits in aged epidermis and peripheral blood mononuclear cells (Carrard et al., 2003), extensive protein oxidation and cross-linking of proteasome substrates in senescent MRC-5 fibroblasts (Sitte, Merker, et al., 2000b) have all been reported as possible causes of proteasome dysfunction. Recent studies carried out in our laboratory on senescent dermal fibroblasts as a cellular model of aging, suggested that the decreased proteasome activity observed during cellular aging may be related to the circadian clock perturbation (Desvergne et al., 2016). Indeed, following dexamethasone synchronization, a functional circadian clock machinery has been previously described in primary cultures of human dermal fibroblasts, keratinocytes and melanocytes (Sandu et al., 2012). Using human dermal fibroblasts synchronized upon serum shock, we have shown that the levels of carbonylated protein and proteasome activity vary rhythmically following a 24-h period (Desvergne et al., 2016). Such modulation of proteasome activity is explained, at least in part, by the circadian expression of both Nuclear factor (erythroid-derived 2)-like 2 (Nrf2) transcription factor and the proteasome activator PA28$\alpha\beta$, which have been previously implicated in the efficient proteasomal degradation of oxidized proteins (Pickering et al., 2012). Indeed, Pickering et al. (2012) have shown that adaptation to oxidative stress includes Nrf2-dependent increase in cellular capacity to degrade oxidized proteins through increased expression of 20S proteasome and immunoproteasome subunits and of PA28αβ. Interestingly, stimulation of the proteasome upon repeated mild heat stress in human dermal fibroblasts was previously shown to be due to the induction and enhanced binding of PA28αβ activator (Beedholm et al., 2004). Moreover, the circadian rhythmicity of the Nrf2 protein was reported to be essential for the regulation of the rhythmic expression of antioxidant genes involved in glutathione redox homeostasis in the mouse lung (Pekovic-Vaughan et al., 2014). Alteration of the circadian rhythmicity characterized by a decrease of the amplitude as well as an increase of the period's length of the gene expression of the clock genes bmal1 and per2 was observed in synchronized senescent dermal fibroblasts (Desvergne et al., 2016). Furthermore, in contrast to the synchronized young fibroblasts, proteasome activities, protein carbonyls, Nrf2, ROS and oxidized glutathione levels did not show circadian rhythmicity in senescent fibroblasts. The link between the biological clock and protein redox homeostasis, including proteasome-mediated oxidized protein degradation, could thus explain, at least in part, the previously observed changes during the aging process.

There is evidence that aging and photoaging have, at least in part, overlapping biochemical mechanisms and we have previously reported inhibition of proteasome activity during UV irradiation of cultured human keratinocytes (Bulteau, Moreau, et al., 2002). This inhibition of the proteasome peptidase activities is most likely due to the presence of highly oxidatively damaged proteins (Bulteau, Moreau, et al., 2002). Indeed, such highly damaged proteins as proteins cross-linked by the lipid peroxidation product 4-hydroxy-2-nonenal (HNE) can act as endogenous inhibitors as previously demonstrated in in vitro reconstitution studies (Friguet and Szweda, 1997). Impairment of the proteasome together with autophagy activation has been also implicated in the UVB-induced senescence of human dermal fibroblasts (Cavinato et al., 2017). Pioneering studies of our laboratory were undertaken to determine whether natural plant or algae extracts could have a beneficial effect by preventing the UV irradiation-induced inhibition of the proteasome. Several extracts were analyzed for their ability to stimulate proteasome activity in vitro and to preserve proteasome function after UV irradiation of cultured human keratinocytes. We have reported that an extract prepared from the algae Phaeodactylum tricornutum can stimulate and protect proteasome peptidase activities from the detrimental effects of UV exposure (Bulteau, Moreau, et al., 2006). In addition, as evidenced by assaying the carbonyl content of total protein load, oxidized protein levels were reduced in the extract-treated cells both before and, to an even bigger extent, after UV irradiation (Bulteau, Moreau, et al., 2006). Another study performed by Katsiki et al. (2007) has shown a stimulatory impact on proteasome activities in vitro by oleuropein, the most abundant of the phenolic compounds in Olea europaea leaf extract, olive oil and olives. The increased activities promoted cellular resistance to oxidants and conferred extension of human fibroblast lifespan. Oleuropein most likely acts through structural changes of the 20S α-gated channel conformation in a similar manner to Sodium dodecyl sulfate (SDS), although its effects are considerably stronger. Moreover, 18α-glycyrrhetinic acid (18α-GA) was identified as a novel Nrf2 inducer, which by stimulating the Nrf2-mediated proteasome activation resulted in enhanced survival of HFL-1 primary human fibroblasts against oxidants and in lifespan extension (Kapeta et al., 2010). Quercetin caprylate (QU-CAP) was also identified as a proteasome activator with antioxidant properties that consequently influenced cellular lifespan, survival and viability of HFL-1 fibroblasts (Chondrogianni and Gonos, 2010). Other attempts to activate the proteasome have been achieved through overexpression of the proteasome β5 subunit in human lung IMR-90 fibroblasts which resulted in a 15–20% increase in cellular lifespan (Chondrogianni and Gonos, 2007). Additionally, the restoration of the normal levels of proteasome catalytic subunits, using a lentiviral gene-delivery system for the β1 and β5 proteasome subunits, decreased the severity of the aging markers P21 and SA-β-galactosidase activity in dermal fibroblasts from elderly donors (Hwang et al., 2007), suggesting that the maintenance of normal proteasome activities could delay skin aging.

14.5 CONCLUSIONS

In conclusion, mitochondrial proteases, proteasome and autophagy emerge as major controllers of skin health, through the general maintenance of intracellular protein homeostasis and mitochondrial integrity and the selective degradation of damaged proteins. Proteasomal and lysosomal proteolytic pathways have been already validated as valuable targets for preserving and even improving proteostasis despite the complexity of both systems while mitochondrial proteases are only emerging as potential ones. Only few specific substrates of Lon protease have been identified in mammalian systems. Thus, further studies in cellular systems and animal models are needed to decipher the specific pathways controlled by Lon and Clp and to develop an advanced understanding of their involvement in skin aging. Moreover, the increasing exposure of skin to UV and pollutants due to environmental changes underscore the importance of intracellular proteases in skin aging. Indeed, among other deleterious effects, UV and pollutants damage various proteins, including mitochondrial proteins. Identification of substrates and development of specific activators for mitochondrial proteases should hasten skincare. In addition, conditions that adversely affect the efficient removal of oxidatively modified proteins must be further characterized.

REFERENCES

Amerik, A., Antonov, V. K., Ostroumova, N. I., et al. 1990. Cloning, structure and expression of the full-size Lon gene in Escherichia coli coding for ATP-dependent La-proteinase. *Bioorg Khim*, 16(7), 869–880.

Bayot, A., Gareil, M., Rogowska-Wrzesinska, A., et al. 2010. Identification of novel oxidized protein substrates and physiological partners of the mitochondrial ATP-dependent Lon-like protease Pim1. *J Biol Chem*, 285(15), 11445–11457. doi:10.1074/jbc.M109.065425

Beedholm, R., Clark, B. F., & Rattan, S. I. 2004. Mild heat stress stimulates 20S proteasome and its 11S activator in human fibroblasts undergoing aging in vitro. *Cell Stress Chaperones*, 9(1), 49–57. doi:10.1379/475.1

Bender, T., Leidhold, C., Ruppert, T., et al. 2010. The role of protein quality control in mitochondrial protein homeostasis under oxidative stress. *Proteomics*, 10(7), 1426–1443. doi:10.1002/pmic.200800619.

Berneburg, M., Gattermann, N., Stege, H., et al. 1997. Chronically ultraviolet-exposed human skin shows a higher mutation frequency of mitochondrial DNA as compared to unexposed skin and the hematopoietic system. *Photochem Photobiol*, 66(2), 271–275. doi:10.1111/j.1751-1097.1997.tb08654.x

Boraldi, F., Bini, L., Liberatori, S., et al. 2003. Proteome analysis of dermal fibroblasts cultured in vitro from human healthy subjects of different ages. *Proteomics*, 3(6), 917–929. doi:10.1002/pmic.200300386

Bota, D. A., & Davies, K. J. 2002. Lon protease preferentially degrades oxidized mitochondrial aconitase by an ATP-stimulated mechanism. *Nat Cell Biol*, 4(9), 674–680.

Bota, D. A., Ngo, J. K., & Davies, K. J. 2005. Downregulation of the human Lon protease impairs mitochondrial structure and function and causes cell death. *Free Radic Biol Med*, 38(5), 665–677. doi:10.1016/j.freeradbiomed.2004.11.017

Bowman, A., & Birch-Machin, M. A. 2016. Age-dependent decrease of mitochondrial complex II activity in human skin fibroblasts. *J Invest Dermatol*, 136(5), 912–919. doi:10.1016/j.jid.2016.01.017

Bulteau, A. L., Petropoulos, I., & Friguet, B. 2000. Age-related alterations of proteasome structure and function in aging epidermis. *Exp Gerontol*, 35(6–7), 767–777. doi:10.1016/s0531-5565(00)00136-4

Bulteau, A. L., Moreau, M., Nizard, C., et al. 2002. Impairment of proteasome function upon UVA- and UVB-irradiation of human keratinocytes. *Free Radic Biol Med*, 32(11), 1157–1170. doi:10.1016/s0891-5849(02)00816-x

Bulteau, A. L., Szweda, L. I., & Friguet, B. 2002. Age-dependent declines in proteasome activity in the heart. *Arch Biochem Biophys*, 397(2), 298–304. doi:10.1006/abbi.2001.2663

Bulteau, A. L., Moreau, M., Saunois, A., et al. 2006. Algae extract-mediated stimulation and protection of proteasome activity within human keratinocytes exposed to UVA and UVB irradiation. *Antioxid Redox Signal*, 8(1–2), 136–143. doi:10.1089/ars.2006.8.136

Bulteau, A. L., Szweda, L. I., & Friguet, B. 2006. Mitochondrial protein oxidation and degradation in response to oxidative stress and aging. *Exp Gerontol*, 41(7), 653–657.

Byun, Y. J., Kim, S. K., Kim, Y. M., et al. 2009. Hydrogen peroxide induces autophagic cell death in C6 glioma cells via BNIP3-mediated suppression of the mTOR pathway. *Neurosci Lett*, 461(2), 131–135. doi:10.1016/j.neulet.2009.06.011

Carrard, G., Dieu, M., Raes, M., et al. 2003. Impact of ageing on proteasome structure and function in human lymphocytes. *Int J Biochem Cell Biol*, 35(5), 728–739. doi:10.1016/s1357-2725(02)00356-4

Cavinato, M., & Jansen-Durr, P. 2017. Molecular mechanisms of UVB-induced senescence of dermal fibroblasts and its relevance for photoaging of the human skin. *Exp Gerontol*, 94, 78–82. doi:10.1016/j.exger.2017.01.009

Cavinato, M., Koziel, R., Romani, N., et al. 2017. UVB-induced senescence of human dermal fibroblasts involves impairment of proteasome and enhanced autophagic activity. *J Gerontol A Biol Sci Med Sci*, 72(5), 632–639. doi:10.1093/gerona/glw150

Chondrogianni, N., & Gonos, E. S. 2007. Overexpression of hUMP1/POMP proteasome accessory protein enhances proteasome-mediated antioxidant defence. *Exp Gerontol*, 42(9), 899–903. doi:10.1016/j.exger.2007.01.012

Chondrogianni, N., & Gonos, E. S. 2010. Proteasome function determines cellular homeostasis and the rate of aging. *Adv Exp Med Biol*, 694, 38–46. doi:10.1007/978-1-4419-7002-2_4

Chondrogianni, N., Petropoulos, I., Franceschi, C., et al. 2000. Fibroblast cultures from healthy centenarians have an active proteasome. *Exp Gerontol*, 35(6-7), 721–728. doi:10.1016/s0531-5565(00)00137-6

Chondrogianni, N., Stratford, F. L., Trougakos, I. P., et al. 2003. Central role of the proteasome in senescence and survival of human fibroblasts: induction of a senescence-like phenotype upon its inhibition and resistance to stress upon its activation. *J Biol Chem*, 278(30), 28026–28037. doi:10.1074/jbc.M301048200

Crouch, R. K., Koutalos, Y., Kono, M., et al. 2015. A2E and lipofuscin. *Prog Mol Biol Transl Sci*, 134, 449–463. doi:10.1016/bs.pmbts.2015.06.005

Cunningham, M. L., Johnson, J. S., Giovanazzi, S. M., et al. 1985. Photosensitized production of superoxide anion by monochromatic (290-405 nm) ultraviolet irradiation of NADH and NADPH coenzymes. *Photochem Photobiol*, 42(2), 125–128. doi:10.1111/j.1751-1097.1985.tb01549.x

D'Orazio, J., Jarrett, S., Amaro-Ortiz, A., et al. 2013. UV radiation and the skin. *Int J Mol Sci*, 14(6), 12222–12248. doi:10.3390/ijms140612222

Demirovic, D., Nizard, C., & Rattan, S. I. 2015. Basal level of autophagy is increased in aging human skin fibroblasts in vitro, but not in old skin. *PLoS One*, 10(5), e0126546. doi:10.1371/journal.pone.0126546

Desvergne, A., Ugarte, N., Radjei, S., et al. 2016. Circadian modulation of proteasome activity and accumulation of oxidized protein in human embryonic kidney HEK 293 cells and primary dermal fibroblasts. *Free Radic Biol Med*, 94, 195–207. doi:10.1016/j.freeradbiomed.2016.02.037

Dezest, M., Le Bechec, M., Chavatte, L., et al. 2017. Oxidative damage and impairment of protein quality control systems in keratinocytes exposed to a volatile organic compounds cocktail. *Sci Rep*, 7(1), 10707. doi:10.1038/s41598-017-11088-1

Durieux, J., Wolff, S., & Dillin, A. 2011. The cell-non-autonomous nature of electron transport chain-mediated longevity. *Cell*, 144(1), 79–91. doi:10.1016/j.cell.2010.12.016

Ferrington, D. A., Husom, A. D., & Thompson, L. V. 2005. Altered proteasome structure, function, and oxidation in aged muscle. *FASEB J*, 19(6), 644–646. doi:10.1096/fj.04-2578fje

Fisher, G. J., Kang, S., Varani, J., et al. 2002. Mechanisms of photoaging and chronological skin aging. *Arch Dermatol*, 138(11), 1462–1470. doi:10.1001/archderm.138.11.1462

Friguet B., 2006. Oxidized protein degradation and repair in ageing and oxidative stress. *FEBS Lett* 580:2910–6. 1 doi:0.1016/j.febslet.2006.03.028

Friguet, B., & Szweda, L. I. 1997. Inhibition of the multicatalytic proteinase (proteasome) by 4-hydroxy-2-nonenal cross-linked protein. *FEBS Lett*, 405(1), 21–25. doi:10.1016/s0014-5793(97)00148-8

Friguet, B., Szweda, L. I., & Stadtman, E. R. 1994. Susceptibility of glucose-6-phosphate dehydrogenase modified by 4-hydroxy-2-nonenal and metal-catalyzed oxidation to proteolysis by the multicatalytic protease. *Arch Biochem Biophys*, 311(1), 168–173.

Fukuda, R., Zhang, H., Kim, J. W., et al. 2007. HIF-1 regulates cytochrome oxidase subunits to optimize efficiency of respiration in hypoxic cells. *Cell*, 129(1), 111–122.

Garcia-Nafria, J., Ondrovicova, G., Blagova, E., et al. 2003. Structure of the catalytic domain of the human mitochondrial Lon protease: proposed relation of oligomer formation and activity. *Protein Sci*, 19(5), 987–999.

Gerland, L. M., Peyrol, S., Lallemand, C., et al. 2003. Association of increased autophagic inclusions labeled for beta-galactosidase with fibroblastic aging. *Exp Gerontol*, 38(8), 887–895. doi:10.1016/s0531-5565(03)00132-3

Gispert, S., Parganlija, D., Klinkenberg, M., et al. 2013. Loss of mitochondrial peptidase ClpP leads to infertility, hearing loss plus growth retardation via accumulation of CLPX, mtDNA and inflammatory factors. *Hum Mol Genet*, 22(24), 4871–4887. doi:10.1093/hmg/ddt338

Glickman, M. H., & Ciechanover, A. 2002. The ubiquitin-proteasome proteolytic pathway: destruction for the sake of construction. *Physiol Rev*, 82(2), 373–428. doi:10.1152/physrev.00027.2001

Gosselin, K., Deruy, E., Martien, S., et al. 2009. Senescent keratinocytes die by autophagic programmed cell death. *Am J Pathol*, 174(2), 423–435. doi:10.2353/ajpath.2009.080332

Gu, Y., Han, J., Jiang, C., et al. 2020. Biomarkers, oxidative stress and autophagy in skin aging. *Ageing Res Rev*, 59, 101036. doi:10.1016/j.arr.2020.101036

Hamon MP, Bulteau AL, Friguet B., 2015. Mitochondrial proteases and protein quality control in ageing and longevity. *Ageing Res Rev* 23:56–66. doi:10.1016/j.arr.2014.12.010.

Hinnerwisch, J., Fenton, W. A., Furtak, K. J., et al. 2005. Loops in the central channel of ClpA chaperone mediate protein binding, unfolding, and translocation. *Cell*, 121(7), 1029–1041. doi:10.1016/j.cell.2005.04.012

Hohn, A., Jung, T., Grimm, S., et al. 2010. Lipofuscin-bound iron is a major intracellular source of oxidants: role in senescent cells. *Free Radic Biol Med*, 48(8), 1100–1108. doi:10.1016/j.freeradbiomed.2010.01.030

Hohn, A., Sittig, A., Jung, T., et al. 2012. Lipofuscin is formed independently of macroautophagy and lysosomal activity in stress-induced prematurely senescent human fibroblasts. *Free Radic Biol Med*, 53(9), 1760–1769. doi:10.1016/j.freeradbiomed.2012.08.591

Hwang, J. S., Hwang, J. S., Chang, I., et al. 2007. Age-associated decrease in proteasome content and activities in human dermal fibroblasts: restoration of normal level of proteasome subunits reduces aging markers in fibroblasts from elderly persons. *J Gerontol A Biol Sci Med Sci*, 62(5), 490–499. doi:10.1093/gerona/62.5.490

Kalfalah, F., Janke, L., Schiavi, A., et al. 2016. Crosstalk of clock gene expression and autophagy in aging. *Aging (Albany NY)*, 8(9), 1876–1895. doi:10.18632/aging.101018

Kapeta, S., Chondrogianni, N., & Gonos, E. S. 2010. Nuclear erythroid factor 2-mediated proteasome activation delays senescence in human fibroblasts. *J Biol Chem*, 285(11), 8171–8184. doi:10.1074/jbc.M109.031575

Katsiki, M., Chondrogianni, N., Chinou, I., et al. 2007. The olive constituent oleuropein exhibits proteasome stimulatory properties in vitro and confers life span extension of human embryonic fibroblasts. *Rejuv Res*, 10(2), 157–172. doi:10.1089/rej.2006.0513

Kim, Y. K., Koo, S. M., Kim, K., et al. 2017. Increased antioxidant activity after exposure of ozone in murine asthma model. *Asia Pac Allergy*, 7(3), 163–170. doi:10.5415/apallergy.2017.7.3.163

Kim, H. S., Park, S. Y., Moon, S. H., et al. 2018. Autophagy in human skin fibroblasts: impact of age. *Int J Mol Sci*, 19(8). doi:10.3390/ijms19082254

Klionsky, D. J., & Schulman, B. A. 2014. Dynamic regulation of macroautophagy by distinctive ubiquitin-like proteins. *Nat Struct Mol Biol*, 21(4), 336–345. doi:10.1038/nsmb.2787

Koziel, R., Greussing, R., Maier, A. B., et al. 2011. Functional interplay between mitochondrial and proteasome activity in skin aging. *J Invest Dermatol*, 131(3), 594–603. doi:10.1038/jid.2010.383

Kress, W., Maglica, Z., & Weber-Ban, E. 2009. Clp chaperone-proteases: structure and function. *Res Microbiol*, 160(9), 618–628. doi:10.1016/j.resmic.2009.08.006

Krishnan, K. J., Harbottle, A., & Birch-Machin, M. A. 2004. The use of a 3895 bp mitochondrial DNA deletion as a marker for sunlight exposure in human skin. *J Invest Dermatol*, 123(6), 1020–1024. doi:10.1111/j.0022-202X.2004.23457.x

Krutmann, J., & Schroeder, P. 2009. Role of mitochondria in photoaging of human skin: the defective powerhouse model. *J Investig Dermatol Symp Proc*, 14(1), 44–49. doi:10.1038/jidsymp.2009.1

Langton AK, Sherratt MJ, Griths CE, et al., 2012. Diferential expression of elastic fibre components in intrinsically aged skin. *Biogerontology* 13, 37–48. doi:10.1007/Fs10522-011-9332-9

Lee, I., & Suzuki, C. K. 2008. Functional mechanics of the ATP-dependent Lon protease: lessons from endogenous protein and synthetic peptide substrates. *Biochim Biophys Acta*, 1784(5), 727–735.

Levine, B., & Klionsky, D. J. 2004. Development by self-digestion: molecular mechanisms and biological functions of autophagy. *Dev Cell*, 6(4), 463–477. doi:10.1016/s1534-5807(04)00099-1

Levine RL, Stadtman ER., 2001. Oxidative modification of proteins during aging. *Exp Gerontol* 36: 1495–1502. doi:10.1016/s0531-5565(01)00135-8

Liu, T., Lu, B., Lee, I., et al. 2004. DNA and RNA binding by the mitochondrial Lon protease is regulated by nucleotide and protein substrate. *J Biol Chem*, 279(14), 13902–13910.

Lopez-Otin, C., Blasco, M. A., Partridge, L., et al. 2013. The hallmarks of aging. *Cell*, 153(6), 1194–1217. doi:10.1016/j.cell.2013.05.039

Lu, B., Liu, T., Crosby, J. A., et al. 2003. The ATP-dependent Lon protease of Mus musculus is a DNA-binding protein that is functionally conserved between yeast and mammals. *Gene*, 306, 45–55.

Lu, B., Yadav, S., Shah, P. G., et al. 2007. Roles for the human ATP-dependent Lon protease in mitochondrial DNA maintenance. *J Biol Chem*, 282(24), 17363–17374.

Ma, W., Wlaschek, M., Tantcheva-Poor, I., et al. 2001. Chronological ageing and photoageing of the fibroblasts and the dermal connective tissue. *Clin Exp Dermatol*, 26(7), 592–599. doi:10.1046/j.1365-2230.2001.00905.x

Madeo, F., Zimmermann, A., Maiuri, M. C., et al. 2015. Essential role for autophagy in life span extension. *J Clin Invest*, 125(1), 85–93. doi:10.1172/JCI73946

Martinez-Lopez, N., Athonvarangkul, D., & Singh, R. 2015. Autophagy and aging. *Adv Exp Med Biol*, 847, 73–87. doi:10.1007/978-1-4939-2404-2_3

Mehrpour, M., Esclatine, A., Beau, I., et al. 2010. Overview of macroautophagy regulation in mammalian cells. *Cell Res*, 20(7), 748–762. doi:10.1038/cr.2010.82

Mizushima, N., & Komatsu, M. 2011. Autophagy: renovation of cells and tissues. *Cell*, 147(4), 728–741. doi:10.1016/j.cell.2011.10.026

Neiman, M., & Taylor, D. R. 2009. The causes of mutation accumulation in mitochondrial genomes. *Proc Biol Sci*, 276(1660), 1201–1209. doi:10.1098/rspb.2008.1758

Ngo, J. K., & Davies, K. J. 2009. Mitochondrial Lon protease is a human stress protein. *Free Radic Biol Med*, 46(8), 1042–1048. doi:10.1016/j.freeradbiomed.2008.12.024

Ngo, J. K., Pomatto, L. C., Bota, D. A., et al. 2011. Impairment of lon-induced protection against the accumulation of oxidized proteins in senescent WI-38 fibroblasts. *J Gerontol A Biol Sci Med Sci*, 66(11), 1178–1185. doi:10.1093/gerona/glr145

O'Neill, J. S., & Reddy, A. B. 2011. Circadian clocks in human red blood cells. *Nature*, 469(7331), 498–503. doi:10.1038/nature09702

Ortonne, J. P. 1990. Pigmentary changes of the ageing skin. *Br J Dermatol*, 122(Suppl. 35), 21–28. doi:10.1111/j.1365-2133.1990.tb16121.x

Pekovic-Vaughan, V., Gibbs, J., Yoshitane, H., et al. 2014. The circadian clock regulates rhythmic activation of the NRF2/glutathione-mediated antioxidant defense pathway to modulate pulmonary fibrosis. *Genes Dev*, 28(6), 548–560. doi:10.1101/gad.237081.113

Perez, V. I., Buffenstein, R., Masamsetti, V., et al. 2009. Protein stability and resistance to oxidative stress are determinants of longevity in the longest-living rodent, the naked mole-rat. *Proc Natl Acad Sci U S A*, 106(9), 3059–3064. doi:10.1073/pnas.0809620106

Petropoulos, I., Conconi, M., Wang, X., et al. 2000. Increase of oxidatively modified protein is associated with a decrease of proteasome activity and content in aging epidermal cells. *J Gerontol A Biol Sci Med Sci*, 55(5), B220–227. doi:10.1093/gerona/55.5.b220

Phillips TJ, Demircay Z, Sahu M., 2001. Hormonal effects on skin aging. *Clin Geriatr Med*. 17:661–72. doi:10.1016/s0749-0690(05)70092-

Piao, M. J., Ahn, M. J., Kang, K. A., et al. 2018. Particulate matter 2.5 damages skin cells by inducing oxidative stress, subcellular organelle dysfunction, and apoptosis. *Arch Toxicol*, 92(6), 2077–2091. doi:10.1007/s00204-018-2197-9

Pickering, A. M., Linder, R. A., Zhang, H., et al. 2012. Nrf2-dependent induction of proteasome and Pa28alphabeta regulator are required for adaptation to oxidative stress. *J Biol Chem*, 287(13), 10021–10031. doi:10.1074/jbc.M111.277145

Pride, H., Yu, Z., Sunchu, B., et al. 2015. Long-lived species have improved proteostasis compared to phylogenetically-related shorter-lived species. *Biochem Biophys Res Commun*, 457(4), 669–675. doi:10.1016/j.bbrc.2015.01.046

Pryde, K. R., Taanman, J. W., & Schapira, A. H. 2016. A Lon-ClpP proteolytic axis degrades complex I to extinguish ROS production in depolarized mitochondria. *Cell Rep*, 17(10), 2522–2531. doi:10.1016/j.celrep.2016.11.027

Pulliam, D. A., Deepa, S. S., Liu, Y., et al. 2014. Complex IV-deficient Surf1(-/-) mice initiate mitochondrial stress responses. *Biochem J*, 462(2), 359–371. doi:10.1042/BJ20140291

Qiang, L., Wu, C., Ming, M., et al. 2013. Autophagy controls p38 activation to promote cell survival under genotoxic stress. *J Biol Chem*, 288(3), 1603–1611. doi:10.1074/jbc.M112.415224

Rittie, L., & Fisher, G. J. 2015. Natural and sun-induced aging of human skin. *Cold Spring Harb Perspect Med*, 5(1), a015370. doi:10.1101/cshperspect.a015370

Rodney, G. G., Pal, R., & Abo-Zahrah, R. 2016. Redox regulation of autophagy in skeletal muscle. *Free Radic Biol Med*, 98, 103–112. doi:10.1016/j.freeradbiomed.2016.05.010

Rogers J, Harding C, Mayo A, et al., 1996. Stratum corneum lipids: The effect of ageing and the seasons. *Arch Dermatol Res* 288(12): 765–70. doi: 10.1007/BF02505294

Rubinsztein, D. C., Marino, G., & Kroemer, G. 2011. Autophagy and aging. *Cell*, 146(5), 682–695. doi:10.1016/j.cell.2011.07.030

PROTEOSTASIS AND PROTEOLYSIS

Sandu, C., Dumas, M., Malan, A., et al. 2012. Human skin keratinocytes, melanocytes, and fibroblasts contain distinct circadian clock machineries. *Cell Mol Life Sci*, 69(19), 3329–3339. doi:10.1007/s00018-012-1026-1

Sherman, M. Y., & Goldberg, A. L. 2001. Cellular defenses against unfolded proteins: a cell biologist thinks about neurodegenerative diseases. *Neuron*, 29(1), 15–32. doi:10.1016/s0896-6273(01)00177-5

Sitte, N., Huber, M., Grune, T., et al. 2000. Proteasome inhibition by lipofuscin/ceroid during postmitotic aging of fibroblasts. *FASEB J*, 14(11), 1490–1498. doi:10.1096/fj.14.11.1490

Sitte, N., Merker, K., Grune, T., et al. 2001. Lipofuscin accumulation in proliferating fibroblasts in vitro: an indicator of oxidative stress. *Exp Gerontol*, 36(3), 475–486. doi:10.1016/s0531-5565(00)00253-9

Sitte, N., Merker, K., Von Zglinicki, T., et al. 2000a. Protein oxidation and degradation during cellular senescence of human BJ fibroblasts: part II—aging of nondividing cells. *FASEB J*, 14(15), 2503–2510. doi:10.1096/fj.00-0210com

Sitte, N., Merker, K., Von Zglinicki, T., et al. 2000b. Protein oxidation and degradation during proliferative senescence of human MRC-5 fibroblasts. *Free Radic Biol Med*, 28(5), 701–708. doi:10.1016/s0891-5849(99)00279-8

Sitte, N., Merker, K., Von Zglinicki, T., et al. 2000c. Protein oxidation and degradation during cellular senescence of human BJ fibroblasts: part I—effects of proliferative senescence. *FASEB J*, 14(15), 2495–2502. doi:10.1096/fj.00-0209com

Sparrow, J. R., & Duncker, T. 2014. Fundus autofluorescence and RPE lipofuscin in age-related macular degeneration. *J Clin Med*, 3(4), 1302–1321. doi:10.3390/jcm3041302

Tonolli, P. N., Chiarelli-Neto, O., Santacruz-Perez, C., et al. 2017. Lipofuscin generated by UVA turns keratinocytes photosensitive to visible light. *J Invest Dermatol*, 137(11), 2447–2450. doi:10.1016/j.jid.2017.06.018

Tonolli, P. N., Martins, W. K., Junqueira, H. C., et al. 2020. Lipofuscin in keratinocytes: production, properties, and consequences of the photosensitization with visible light. *Free Radic Biol Med*, 160, 277–292. doi:10.1016/j.freeradbiomed.2020.08.002

Tyrrell, R. M. 1995. Ultraviolet radiation and free radical damage to skin. *Biochem Soc Symp*, 61, 47–53. doi:10.1042/bss0610047

Van der Zee, J., Krootjes, B. B., Chignell, C. F., et al. 1993. Hydroxyl radical generation by a light-dependent Fenton reaction. *Free Radic Biol Med*, 14(2), 105–113. doi:10.1016/0891-5849(93)90001-b

Van Dyck, L., & Langer, T. 1999. ATP-dependent proteases controlling mitochondrial function in the yeast Saccharomyces cerevisiae. *Cell Mol Life Sci*, 56(9–10), 825–842.

Vierkotter, A., Schikowski, T., Ranft, U., et al. 2010. Airborne particle exposure and extrinsic skin aging. *J Invest Dermatol*, 130(12), 2719–2726. doi:10.1038/jid.2010.204

Vile, G. F., & Tyrrell, R. M. 1995. UVA radiation-induced oxidative damage to lipids and proteins in vitro and in human skin fibroblasts is dependent on iron and singlet oxygen. *Free Radic Biol Med*, 18(4), 721–730. doi:10.1016/0891-5849(94)00192-m

von Zglinicki, T., Nilsson, E., Docke, W. D., et al. 1995. Lipofuscin accumulation and ageing of fibroblasts. *Gerontology*, 41(Suppl. 2), 95–108. doi:10.1159/000213728

Voos, W. 2009. Mitochondrial protein homeostasis: the cooperative roles of chaperones and proteases. *Res Microbiol*, 160(9), 718–725.

Wang, M., Charareh, P., Lei, X., et al. 2019. Autophagy: multiple mechanisms to protect skin from ultraviolet radiation-driven photoaging. *Oxid Med Cell Longev*, 2019, 8135985. doi:10.1155/2019/8135985

Watabe, S., Hara, M., Yamamoto, M., et al. 2001. Activation of mitochondrial ATP-dependent protease by peptides and proteins. *Tohoku J Exp Med*, 195(3), 153–161.

Yang, Z., & Klionsky, D. J. 2010. Eaten alive: a history of macroautophagy. *Nat Cell Biol*, 12(9), 814–822. doi:10.1038/ncb0910-814

Yang, J. H., Lee, H. C., & Wei, Y. H. 1995. Photoageing-associated mitochondrial DNA length mutations in human skin. *Arch Dermatol Res*, 287(7), 641–648. doi:10.1007/BF00371736

Zhang, H., Kong, X., Kang, J., et al. 2009. Oxidative stress induces parallel autophagy and mitochondria dysfunction in human glioma U251 cells. *Toxicol Sci*, 110(2), 376–388. doi:10.1093/toxsci/kfp101

Zhang, B., Zhao, Z., Meng, X., et al. 2018. Hydrogen ameliorates oxidative stress via PI3K-Akt signaling pathway in UVB-induced HaCaT cells. *Int J Mol Med*, 41(6), 3653–3661. doi:10.3892/ijmm.2018.3550

CHAPTER FIFTEEN

Reactive Oxygen Species and Protein Homeostasis in Skeletal Muscle Regeneration

Sofia Lourenço dos Santos, Aurore L'honoré and Isabelle Petropoulos

CONTENTS

15.1 INTRODUCTION

Adult stem cells are among the longest living cells in an organism, and their dysfunction is therefore a highly relevant driver of organismal aging, health and longevity. Recent advances implicate energy metabolism, mitochondrial dynamics and proteostasis networks in the preservation of stemness and the regulation of adult stem cell function (García-Prat et al., 2017). Greater understanding of the metabolic and proteostatic changes that occur during aging is mandatory to identify interventions that could be relevant to the treatment of age-related diseases and to regenerative medicine.

Understanding how tissue homeostasis is maintained throughout life is a fundamental question in biology, with critical implications for aging and disease. The homeostatic requirements for cellular replacements are tissue and context specific. In contrast to skin, intestine and blood, the skeletal muscle undergoes relatively little tissue turnover under homeostatic conditions, being mostly made up of long-term multinucleated and post-mitotic myofibers that possess the contractile machinery. It is therefore characterized by a remarkable capacity to regenerate in situations where acute or chronic injuries, such as intense physical activity, genetic degenerative diseases or aging, lead to loss of muscle mass.

Muscle stem cells, also termed satellite cells, are a prerequisite for regeneration of skeletal muscle (Relaix and Zammit, 2012). Like other stem cells, they are able to both regenerate the injured muscle and self-renew. Muscle regeneration starts with the activation and migration of satellite cells along the damaged fiber. After several cycles of proliferation, the resulting myogenic precursor cells differentiate and fuse to form new fibers, while the others will re-enter into quiescence to replenish the stem cell pool (Schmidt et al., 2019). Recently, the redox state of satellite cells and the mechanisms permitting the proteome quality control have emerged as key regulators of the satellite cell's fate (García-Prat et al., 2017; Le Moal et al., 2017; L'Honoré et al., 2018).

In this chapter, we describe the contributions of the redox state and proteostasis in both quiescence maintenance and regulation of satellite cells during regeneration, with particular emphasis on the mechanisms that control the satellite cell redox state. Moreover, we discuss how the alterations of these mechanisms can affect the muscle integrity during aging and in pathologies.

15.2 SKELETAL MUSCLE REGENERATION

Skeletal muscle regeneration is a highly orchestrated process that involves the activation of various cellular and molecular responses in order to reform a complete innervated and vascularized contractile muscle apparatus (Schmidt et al., 2019). As adult skeletal muscle stem cells, satellite cells play an indispensable role in this process (Lepper

et al., 2011; Sambasivan et al., 2011; Murphy et al., 2011). Located in a specialized niche between the myofiber sarcolemma and the surrounding extracellular matrix called basal lamina (Mauro, 1961), these cells remain mostly quiescent through life and under homeostatic conditions (Montarras et al., 2013). When stimulated by environmental signals (e.g. damage or stress) or in pathological conditions (e.g. degenerative muscle diseases), these cells are activated and migrate to the regeneration site. There, activated satellite cells, called myoblasts, enter the cell cycle and proliferate massively to generate a large number of myogenic progenitors. Thereafter, these progenitors exit the cell cycle and either undergo differentiation and fusion to form newly regenerating fibers or self-renew to replenish the satellite cell pool (Schmidt et al., 2019) (Figure 15.1). Each step of this regenerative process is regulated through the coordinated expression of specific transcription factors including Pax3/7 homeodomain transcription factors, and the myogenic regulatory factors (MRFs); MyoD, Myf5, Myogenin and MRF4 (Schmidt et al., 2019).

In the adult, quiescence of muscle stem cells is an actively regulated process (Montarras et al., 2013). This state exists in a balance of opposing pathways that maintain high expression of cell

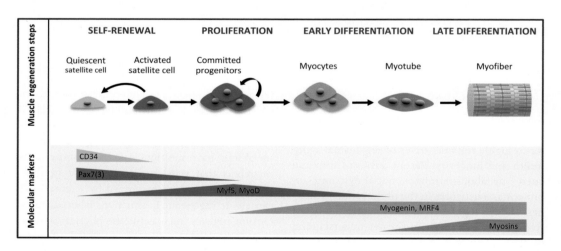

Figure 15.1 **Skeletal muscle regeneration process.** In adult muscle, satellite cells, which reside between the basal lamina and the sarcolemma, are in a quiescent state and marked by the expression of the membrane protein CD34 and the transcription factors Pax7(3). Following muscle injury, satellite cells are activated, inducing expression of Myf5 and MyoD transcription factors. They proliferate as myoblasts and either self-renew to reconstitute their reservoir or undergo differentiation. Differentiation is accompanied by down-regulation of *Pax* genes and upregulation of *Myogenin*. At this stage, early-differentiating myocytes will fuse to form myotubes, which then mature into myofibers. These later stages of differentiation are marked by the assembly of sarcomeres and the expression of Mrf4 transcription factor and structural proteins such as myosins.

PROTEOSTASIS AND PROTEOLYSIS

cycle inhibitors (McCroskery et al., 2003) while priming the cell for rapid activation in case of injury (Crist et al., 2012). Quiescent satellite cells are marked by the expression of Pax7 homeodomain transcription factor in all skeletal muscles and the co-expression of Pax3 in a subset of muscles, including diaphragm, abdominal and *pectoralis* muscles (Relaix et al., 2006). Upon activation, commitment and differentiation of satellite cells require the progressive expression of MRFs. Committed myogenic progenitors marked by the expression of Myf5 (Gayraud-Morel et al., 2007; Ustanina et al., 2007) and MyoD (Cornelison and Wold, 1997; Sabourin et al., 1999; Yablonka-Reuveni et al., 1999) will undergo several rounds of cell division, before they differentiate. Early and terminal differentiation of myogenic progenitor cells are marked by the expression of Myogenin and MRF4, and the subsequent expression of myofibrillar proteins, such as myosin heavy chain, just prior to fusion into myotubes (Figure 15.1).

Gene regulatory networks and signaling pathways involved in the regulation of the different muscle regeneration steps have been intensively studied (Schmidt et al., 2019) thanks to a large array of *in vivo* and *ex vivo* models. Muscle regeneration can be experimentally induced *in vivo* by two main injury-induced protocols: intramuscular injections of myotoxins or chemical agent barium chloride, and mechanical damage induced by freeze injury, crush, denervation or contraction. While these models slightly differ in the kinetic of the regenerative process, they share a similar experimental paradigm involving a first phase of acute muscle fiber necrosis and inflammation followed by satellite cell activation, proliferation and differentiation, leading to complete muscle regeneration after 3–4 weeks (Hardy et al., 2016). Although they are not considered to recapitulate the whole muscle regeneration process, primary myogenic cells are extensively used to study *ex vivo* satellite cell regulation. Quiescent satellite cells, freshly purified from human or mouse muscles by flow cytometry, spontaneously activate, proliferate and differentiate in culture (Montarras et al., 2005; Bosnakovski et al., 2008; Sambasivan et al., 2011; Jean et al., 2011; Almada and Wagers, 2016). Use of these different models has led to a better understanding of gene regulatory networks and signaling pathways involved in satellite cell regulation, including redox regulation.

15.3 REDOX HOMEOSTASIS: ROS PRODUCTION, SCAVENGING AND SIGNALING

The cellular redox state can be defined as the balance between ROS production and their scavenging by the antioxidant systems within a cell. Originally thought to be harmful by-products of oxidative metabolism implicated in diseases and aging, there is increasing evidence demonstrating a role for ROS in the regulation of crucial biological processes from early development to adulthood (Hernández-García et al., 2010; Holmström and Finkel, 2014; Shadel and Horvath, 2015), including regulation of stem cell fate and behavior (Bigarella et al., 2014; Perales-Clemente et al., 2014).

Reactive oxygen species, which include superoxide anions ($O2^{\bullet-}$), hydroxyl radicals ($^{\bullet}OH$) and hydrogen peroxide (H_2O_2), consist of radical and non-radical oxygen species formed by the partial reduction of oxygen (O_2) (Figure 15.2A). The three major cellular sources of ROS are the mitochondrial respiratory chain, the NADPH oxidase (NOX) complexes (Brandes et al., 2014) and the cytochrome P450 (Seliskar and Rozman, 2007). Mitochondrial reactive oxygen species (mROS) are considered the principal source of ROS in mammalian cells (Murphy, 2009; Willems et al., 2015). In muscle, mROS formation has also been shown to be critically dependent on phospholipase A2 (PLA2) activation, by interaction with complex I of the electron transport chain (ETC) (Nethery et al., 2000; Zhou et al., 2019). In fact, usually, mROS are formed during ATP production as by-products of electron leakage of ETC. In addition to these three main sources, the activity of various cytosolic enzymes such as xanthine oxidases (XO) or nitric oxide synthases (NOS) can also be responsible for the formation of ROS (Cantu-Medellin and Kelley, 2013; Gomez-Cabrera et al., 2005).

The maintenance of intracellular redox homeostasis is dependent on a large array of antioxidant molecules. These antioxidants include both non-enzymatic molecules such as the tripeptide glutathione (GSH) and enzymatic molecules defined as scavenger systems. ROS scavengers are antioxidant enzymatic defenses that can neutralize ROS by directly reacting with them. Cells contain multiple types of ROS scavengers characterized by specific subcellular localizations and chemical reactivity (Figure 15.2A).

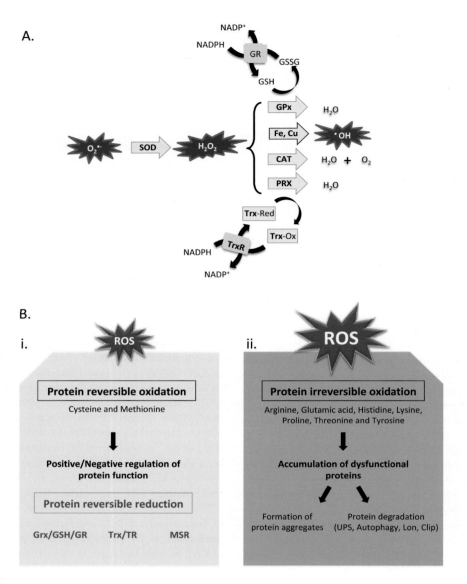

Figure 15.2 (A) **Reactive oxygen species and cellular scavenger enzymatic systems.** Superoxide dismutases, MnSOD (or SOD2) localized in the mitochondrial matrix and CuZnSOD localized mainly in cytosol (SOD1) but also present in the intermembrane space of mitochondria (SOD3), catalyze the rapid dismutation of anion superoxide ($O_2^{\cdot-}$) into hydrogen peroxide (H_2O_2). H_2O_2 can then be further reduced to H_2O by several enzymes, including catalase (CAT), glutathione peroxidases (GPx) and peroxiredoxins (PRX). While catalase activity leads directly to both molecular oxygen (O_2) and H_2O production, the two other cellular systems, namely the glutathione peroxiredoxin (GPx)/glutathione (GSH)/glutathione reductase (GR) pathway and the peroxiredoxin (PRX)/thioredoxin (Trx)/thioredoxin reductase (TrxR) system, use NADPH as ultimate reducing power. When H_2O_2 levels overcome the enzymatic antioxidants activity, these non-radical ROS can interact with reduced transition metals (Fe^{2+}, Cu^+) through the Fenton reaction to form hydroxyl radicals ($\cdot$OH), leading to a well-known condition called oxidative stress. (B) **Protein oxidation and subsequent reduction or degradation.** (i) In homeostatic conditions, ROS can reversibly oxidize target proteins on their Cys and/or Met residues, leading to positive or negative regulation of their function. While oxidized Cys can be reduced by the glutaredoxin (Grx)/glutathione (GSH)/ glutathione reductase (GR) system or by the thioredoxin (Trx)/thioredoxin reductase (TR) system, methionine oxidation in its turn can only be reversed by the methionine sulfoxide reductase (MSR). (ii) In the presence of excessive ROS levels, proteins accumulate irreversible oxidative damages. To avoid their accumulation into toxic aggregates, damaged proteins can be degraded by proteolytic systems including the ubiquitin–proteasome system (UPS), the autophagy-lysosomal system as well as mitochondrial proteases (Lon, Clip).

Under physiological conditions, both ROS production and intracellular antioxidant defense systems are finely modulated. A subset of molecules has been identified as critical regulators of this redox homeostasis. These include transcription factors such as Nrf2 (Dai et al., 2020), NF-κB (Lingappan, 2018), p53 (Maillet and Pervaiz, 2012), Meis1 (Kocabas et al., 2012), Pitx2/3 (L'honoré et al., 2018), FoxO (de Keizer et al., 2011), AP-1 (Yin et al., 2017) and HIF transcription factor families (Simsek et al., 2010; Takubo et al., 2010) and enzymes such as ATM (Zhang et al., 2018), mTOR (Dodson et al., 2013), AKT (Juntilla et al., 2010), APE1/Ref-1 (Tell et al., 2009) and Sirtuins (Singh et al., 2018). These molecules, called redox regulators, act mainly by regulating the antioxidant systems or by modulating ROS production through the regulation of mitochondrial biogenesis and/or function (Wang et al., 2013).

It is now well established that in physiological conditions ROS act as signaling molecules and that redox homeostasis is an important modulator of a wide variety of cellular processes such as proliferation, differentiation or migration (Schieber and Chandel, 2014; Shadel and Horvath, 2015; Sies and Jones, 2020). To serve as second messengers and trigger signaling cascades, ROS act via protein oxidation (Roos and Messens, 2011; Di Marzo et al., 2018). While non-specific oxidation in case of oxidative stress can affect almost all amino acids, physiological redox regulation is based on reversible modification of specific reactive cysteine (Cys) and methionine (Met) residues on proteins (Figure 15.2B). Such oxidation that requires substrate specificity cannot be achieved by all ROS (Bigelow and Squier, 2011). Among the different ROS species, H_2O_2 is thought to function in vivo as a second messenger and to trigger signaling cascades. This is due to the substrate specificity of H_2O_2, which reacts preferentially with Met and Cys residues, but also due to its long half-life, which allows its diffusion within the cell and through cellular membranes (Holmström and Finkel, 2014).

Cys oxidation generates reactive sulfenic acid that can either form disulfide bonds with nearby Cys or undergo further oxidation to sulfinic or sulfonic acid. In the case of Met, incorporation of an extra oxygen atom can lead to the formation of two diastereoisomers of Met sulfoxide, Met-S-sulfoxide and Met-R-sulfoxide, which can both undergo further oxidation resulting in the production of Metsulfone. With the exception of sulfonic acid, Metsulfone, and to a lesser degree sulfinic acid, the other oxidative modifications can be reduced by thioredoxin and glutaredoxin for Cys and Metsulfoxide reductase for Met (Lourenço Dos Santos et al., 2018) (Figure 15.2B). Therefore, by regulating reversible protein oxidation, these proteins are not only critical actors of redox regulation, but also new potential regulators of stem cell function and behavior.

Technological advances such as redox proteomics have identified more than 500 redox reactive Cys and Met in a wide range of proteins that includes transcription factors, kinases, phosphatases and metabolic enzymes (Jones and Go, 2011). Oxidation of these residues results in conformational changes of the respective proteins that affect their stability, their function, their subcellular localization and their interactions with other proteins. Interestingly, among these ROS sensitive molecules also called redox sensors, a growing number has been implicated in stem cell regulation, including the p38 MAP kinase family, mTOR, AKT, AMP kinase, p53, NF-κB, FoxO and Hif1α transcription factors (Bigarella et al., 2014; Moldogazieva et al., 2018).

The interest in ROS regarding cell fate has long been focused on their damaging effects and the antioxidant mechanism cells have established in order to protect themselves. These conclusions have come mostly from in vitro models in which excessive ROS levels impair cell function. However, increasing evidence is now supporting the notion that ROS are required for normal cell function in many systems and especially in stem cells (Le Moal et al., 2017). In the following sections, we will focus on redox regulation of muscle stem cells.

15.4 REDOX REGULATION OF SATELLITE CELLS IN QUIESCENCE, REGENERATION, AGING AND PATHOLOGICAL CONDITIONS

Growing evidence suggests that excessive ROS, i.e. oxidative stress, and inflammation have been involved in a subset of disorders such as in aging, muscle dystrophies or cachexia that affect muscle integrity and function, and more specifically satellite cell regenerative capacity.

15.4.1 Redox Regulation of Satellite Cell Quiescence

Like most long-lived tissue-resident stem cells, adult satellite cells are maintained in a quiescent state during healthy resting periods (Cheung and Rando, 2013). Quiescence is required for satellite cell long-term maintenance, as its disruption leads to their spontaneous activation in absence of injury and to the progressive depletion of their reservoir (Mourikis et al., 2012; Bjornson et al., 2012). The quiescent state is marked by low fatty acids oxidation (FAO) (Ryall et al., 2015) and by reduced mitochondrial mass leading to low ROS production (Latil et al., 2012; Montarras et al., 2013; Tang and Rando 2014) (Figure 15.3). Quiescent satellite cells are also characterized by a high enzymatic and non-enzymatic antioxidant capacity (Pallafacchina et al., 2010). This capacity, as suggested by the increased SOD activity and GSH level (Urish et al., 2009), is higher than that of myoblasts, their activated descendants and confers them greater survival capacity both *ex vivo* (Pallafacchina et al., 2010) and *in vivo* (Urish et al., 2009). Aldehyde dehydrogenase 1a1 (Aldh1a1), an enzyme that detoxifies aldehydic products of ROS-mediated lipid peroxidation, has been identified as a marker of most tissue resident stem cells including hematopoietic (Hess et al., 2006), mammary (Ginestier et al., 2007) and neural (Corti et al., 2006) stem cells. This enzyme is also highly expressed in quiescent satellite cells and chemical inhibition of its activity by diethylaminobenzaldehyde (DEAB) treatment leads to a decrease of their viability (Pallafacchina et al., 2010). The above clearly demonstrate the necessity of low ROS content for the high viability of quiescent satellite cells.

In addition to modulating satellite cell survival, low redox state may also be involved in the maintenance of their quiescence. While this has been clearly demonstrated in other models such as hematopoietic stem cells in which increased ROS production is sufficient to induce activation (Kocabas et al., 2012), a direct functional link between ROS levels and the transition from satellite cell quiescence to activation is still missing. Analysis of selenoprotein N deficient mice that are characterized by increased oxidized proteins in skeletal muscle has revealed a progressive decline in the satellite cell pool (Castets et al., 2011), suggesting a spontaneous activation of satellite cells. While further

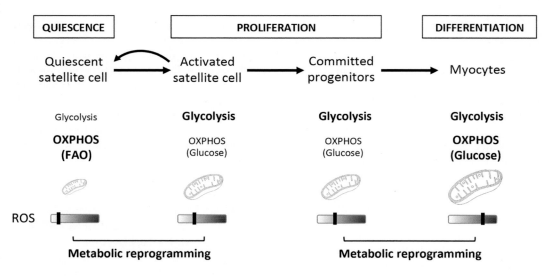

Figure 15.3 Changes in energy metabolism, mitochondrial mass and redox status of satellite cells during muscle regeneration. Quiescent state is marked by low fatty acids oxidation (FAO) metabolic activity, characterized by a reduced mitochondrial mass leading to low ROS production. Activated and proliferating satellite cells mostly rely on glycolysis, with a reduced OXPHOS activity. The onset of differentiation is accompanied by intense mitochondrial biogenesis along with an increased reliance on respiration leading to increased ROS production.

PROTEOSTASIS AND PROTEOLYSIS

experiments involving conditional mutants specifically targeted to the quiescent satellite cells are required, this result suggests a critical role of redox state in the regulation of satellite cell quiescence.

15.4.2 Redox Regulation of Satellite Cell Function during Regeneration

In contrast to quiescence, which is marked by a low metabolic state, satellite cell activation and differentiation are accompanied by metabolic reprograming, required for the cell to support its high energetic demands. While activated and proliferating satellite cells mostly rely on glycolysis, cell cycle exit and differentiation are accompanied by intense mitochondrial biogenesis along with increased respiration (Remels et al., 2010; Malinska et al., 2012; L'honoré et al., 2018; Yucel et al., 2019), leading to both increased ATP and ROS production (Figure 15.3). Together with mitochondrial respiration, the activity of NOX increases with differentiation (Piao et al., 2005), resulting into a global increase in ROS content.

A number of studies have investigated the role of ROS during myogenic differentiation, *in vitro* and *ex vivo* in cultured cells, and more recently, *in vivo* in satellite cells during regeneration. Although conflicting results have been published, mainly due to a large heterogeneity of cellular models and approaches, a general consensus has been reached on the positive role of ROS in satellite cell differentiation. Decrease in mROS using mitochondria targeted antioxidants (MitoQ and MitoTEMPOL) and mitochondria targeted catalase leads to suppression of muscle differentiation without affecting the amount of available ATP (Lee et al., 2011). Similar results were obtained by the inhibition of mitochondrial complex I through small interfering RNAs that led to the reduction of ROS production (Lee et al., 2011). In the same line, the use of the antioxidant N-acetyl-cysteine (NAC) (L'honoré et al., 2018), the overexpression of the transcription factor Nrf2 (Al-Sawaf et al., 2014) or of the antioxidant enzymes heme oxygenase 1 (HO-1) (Kozakowska et al., 2012), Prx2 (Won et al., 2012) or MnSOD (Togliatto et al., 2013) promotes satellite cell proliferation and delays their differentiation, as revealed by the increase and decrease in Pax7 and Myogenin expression, respectively. In contrast,

increasing total or mROS content in proliferating satellite cells in mice by *Gpx1* or *Pitx3* inactivation, respectively, leads to decreased proliferation, premature differentiation and subsequently to the formation of hypotrophic myofibers (Lee et al., 2006; L'honoré et al., 2018).

Among the different signaling pathways acting in satellite cells, a recent study has identified the p38α MAP kinase as a critical mediator of the positive role of ROS on differentiation (L'honoré et al., 2018). Using a mutant mouse for the redox regulator Pitx3, it was shown that increased ROS content in myogenic progenitors leads to premature differentiation by activation of p38α MAP kinase. In addition to being a well-known redox sensor activated either directly by oxidation of Cys 119 and 162 residues or indirectly by oxidation of the upstream kinase Ask (Bigarella et al., 2014), p38α MAP kinase is also a well-described positive regulator of myogenic differentiation (Perdiguero et al., 2007; Brien et al., 2013).

Most of the above studies examine the role of ROS produced in satellite cells. However, non-cell autonomous sources of ROS can also contribute to satellite cell regulation, *in vivo*. For example, during the first three days after injury, high ROS levels are produced and released both by injured muscle fibers (Kerkweg et al., 2007) and by cells of the immune system such as neutrophils and M1/M2 macrophages (Barbieri and Sestili, 2012; Chazaud, 2014). While immune cells are indispensable for muscle regeneration, one cannot exclude that ROS generated in proximity of satellite cells can also contribute to the activation of signaling regulatory pathways, such as the p38α MAP kinase (Le Moal et al., 2017).

15.4.3 Redox Regulation of Satellite Cell Function during Aging

Aging is a process in which the role of ROS has been often implicated but never definitively established. There is a large body of literature suggesting that aging is accompanied by an increase in ROS levels and in the levels of oxidized proteins, lipids and DNA (López-Otín et al., 2013). Accordingly, up-regulation of antioxidant genes in some lower organisms (Dai et al., 2014) or reduction of mitochondrial ROS production by deletion of the $p66shc^{-/-}$ gene in mice (Trinei et al., 2009) both result in lifespan extension. While the role of ROS in organismal aging is still a matter of debate, accumulating evidence suggests that ROS might regulate the age-related regenerative capacity by being an important determinant of stem cell function.

Aged skeletal muscle is characterized by progressive loss of mass, a decline of functional properties and the accumulation of fibrosis and intramuscular fat, traits generally referred to as sarcopenia (Cruz-Jentoft and Sayer, 2019). Sarcopenia is accompanied by a decline in regenerative potential due to a decrease in both satellite cell number and function (see below, Cosgrove et al., 2014; Sousa-Victor et al., 2014; Hwang and Brack, 2018). Although it is not clear whether sarcopenia progresses independently of satellite cells (Fry et al., 2015; Snijders and Parise, 2017), several studies including heterochronic parabiosis (Conboy et al., 2005) and the use of mutant mice (Chakkalakal et al., 2012; Schwörer et al., 2016; Liu et al., 2018) have revealed an important contribution of the satellite cell niche and the systemic environment to the aging of satellite cells. Thus, the aged niche becomes stimulatory, driving satellite cells out of quiescence, and leading to their spontaneous activation and to progressive exhaustion of their pool (Chakkalakal et al., 2012). During regeneration, while the young systemic environment stimulates satellite cell proliferation, the aged one has been shown to be inhibitory (Conboy et al., 2005). Cell-autonomous modifications of geriatric satellite cells have also been associated with loss of their function. Satellite cells from young and geriatric mice were isolated and their regenerative capacities were compared by serial transplantation into young secondary recipients. Satellite cells from geriatric donors exhibited marked deficits in self-renewal, expansion and differentiation (Bernet et al., 2014; Cosgrove et al., 2014; Sousa-Victor et al., 2014). Further analyses revealed that with advanced age, quiescent satellite cells become pre-senescent, a state that leads to senescence in response to regenerative pressure (Cosgrove et al., 2014; Sousa-Victor et al., 2014). The link between mitochondrial dysfunction, high ROS levels and satellite cell senescence has been confirmed by several mouse mutants in which satellite cells exhibit reduced NAD+ levels or defective mitochondrial respiration (Zhang et al., 2016; Sahu et al., 2018; L'honoré et al., 2018). Taken all together, the above clearly illustrate the critical role of ROS levels in satellite cell maintenance and function.

15.4.4 Redox Regulation of Satellite Cell Function in Muscular Dystrophies

Muscular dystrophies are a group of genetic, hereditary muscle diseases characterized by defective expression or activity of structural and metabolic muscle proteins (Mercuri and Muntoni, 2013). These different pathologies share common features including progressive skeletal muscle damage marked by muscle wasting and myofiber degeneration. Duchenne muscular dystrophy (DMD) is a lethal X-linked recessive muscle disease and is one of the most common and severe progressive muscular dystrophies. The causes of this disease are loss-of-function mutations of the *dystrophin* gene (Koenig et al., 1987). Absence of the membrane-bound protein dystrophin makes the skeletal myofibers more sensitive to sarcolemma damage after each muscle contraction. Repeated cycles of damage and regeneration lead to chronic inflammation and progressive replacement of myofibers by fat and fibrosis (Terrill et al., 2013). Concomitantly, the satellite cell pool becomes exhausted with recent works showing that dystrophic satellite cells are also impaired in various aspects. For instance, these cells demonstrate prolonged mitosis, an inability to establish cell polarity and to be activated, defects that are associated with defective amplification and differentiation (Dumont et al., 2015). These results clearly suggest that the severity of DMD can be influenced not only by myofiber defects but also by the regenerative functions of endogenous muscle stem cells.

As mentioned above, dystrophic muscle is characterized by chronic inflammation that can lead to overproduction of ROS by inflammatory cells. Analyses of biopsies from DMD patients and *mdx* mice, a mouse model of DMD marked by absence of dystrophin (Bulfield et al., 1984; Bockhold et al., 1998), have revealed an increased level of protein oxidation while NAC antioxidant treatment of *mdx* mice slows down the progression of the dystrophic phenotype (Terrill et al., 2013). While the beneficial effects of NAC cannot be solely attributed to its putative action on satellite cells, these results clearly suggest that dystrophic muscle constitutes a particularly oxidative environment for satellite cells, making them vulnerable to cellular damage. Although further experimentation is needed, it is tempting to speculate that repeated accumulation of oxidative damage within satellite cells may profoundly affect their function and may be involved in their defective activation, amplification and differentiation.

15.4.5 Redox Regulation of Satellite Cell Function in Cachexia

Cachexia is a devastating muscle-wasting syndrome that occurs in patients with chronic diseases, like renal disease, chronic obstructive pulmonary disease (COPD), chronic heart failure, diabetes and advanced cancer. In particular, being observed in 80% of individuals with advanced cancer, cachexia is one of the primary causes of morbidity and mortality associated with this disease. Cachexia is considered to be a metabolic disease during which muscle fibers switch to a catabolic state and break down their own essential structural proteins. Production of pro-inflammatory cytokines by tumor cells is believed to trigger this process. Fukawa et al. have shown that pro-inflammatory factors, such as tumor necrosis factor, trigger excessive fatty acid metabolism in muscles, leading to excessive ROS production and p38α MAPK activation. Remarkably, pharmacological inhibition of FAO counteracted p38α MAPK activation and prevented muscle wasting in cancer-induced cachexia in mice (Fukawa et al., 2016).

In addition to atrophy, muscles undergoing cachexia become damaged, showing structural alterations of the sarcolemma, which lead to satellite cell activation. In contrast to normal muscle, activated satellite cells in cachexic muscles fail to downregulate Pax7 and to express MyoD or Myogenin, leading to defective fusion into muscle fibers and absence of regeneration (He et al., 2013). Interestingly, satellite cells isolated from cachexic muscles have normal differentiation and fusion *ex vivo*, leading to the conclusion that the muscle microenvironment and more specifically circulating tumor-derived factors may be responsible for the defective regenerative capacity of satellite cells observed *in vivo* in cachexic muscle (He et al., 2013).

Among these tumor-derived factors, TNF-α has been previously reported to inhibit skeletal muscle regeneration *in vivo* after injury through excessive ROS production (Coletti et al., 2005). Accordingly, *in vitro* treatment of myoblasts with TNF-α leads to impaired differentiation as a result of increased ROS levels and consequently the depletion of the antioxidant GSH (Langen et al., 2002). Pretreatment with the antioxidant NAC before exposure to TNF-α completely restored myoblast differentiation. The inhibition of muscle differentiation by TNF-α was shown to act through IκB oxidation that leads to the degradation of the latter, the release of NF-κB and the subsequent activation of NF-κB pathway. Notably, NF-κB activity has been reported highly induced in satellite cells from cachexic muscle and inhibition of this signaling pathway is sufficient to rescue their regenerative capacity (He et al., 2013).

15.5 PROTEIN HOMEOSTASIS: A NEW PLAYER IN THE REGULATION OF SKELETAL MUSCLE REGENERATION

Protein maintenance, also referred to as protein quality control or proteostasis, is required for the proteome integrity, thus being crucial for many physiological and pathological processes. It includes the unfolded protein response (UPR) in charge of protein misfolding and the elimination of damaged proteins through degradation and repair. The UPS and macroautophagy are both proteolytic systems critical for maintaining functional satellite cells in adult muscle but also for preventing accumulation and aggregation of damaged proteins, especially under oxidative stress conditions and aging.

15.5.1 The Role of the Unfolded Protein Response in Satellite Cell Function

Accumulation of misfolded proteins triggers endoplasmic reticulum (ER) stress, which initiates an evolutionarily conserved mechanism known as UPR^{ER}. A recent study, investigating UPR^{ER} function during muscle regeneration, has revealed that the kinase PERK, one of the three arms of the UPR^{ER}, is critical for satellite cell survival and function (Xiong et al., 2017). Conditional inactivation of PERK in satellite cells leads to defective expansion and to apoptosis of myogenic progenitors at the onset of differentiation through hyperactivation of p38α MAP kinase. Interestingly, PERK has been shown to promote cell antioxidant capacity by direct phosphorylation of NF-E2-related factor 2 transcription factor (Xiong et al., 2017). Taken together, these findings indicate that proteostatic stress responses regulate stem cell function and that PERK-mediated signaling may fine-tune the survival of myogenic progenitors at the onset of differentiation.

15.5.2 The Role of the Ubiquitin–Proteasome System in Satellite Cell Function

The UPS is responsible for most protein degradation in mammalian cells (Collins and Goldberg, 2017). This pathway functions by catalyzing the selective degradation of short-lived as well as abnormal proteins and has emerged as a central player in the regulation of several diverse cellular processes.

In vertebrate cells, the proteasome is found in the nucleus and cytoplasm in different forms. The 20S proteasome core, which functions in an ATP-independent manner, is able to recognize and rapidly degrade oxidized, unfolded or partially unfolded proteins. The 26S proteasome, also called constitutive proteasome, is the most common form and it degrades proteins that have been tagged by ubiquitin. Ubiquitin transfer occurs in a sequential ATP-dependent reaction involving three different enzymes: Ubiquitin-activating enzymes (E1), ubiquitin-conjugating enzymes (E2) and ubiquitin-ligases (E3) (Pickart, 2001). Once the protein target has been ubiquitinated, proteolysis is mediated by the 26S proteasome in an ATP-dependent manner (Hershko and Ciechanover, 1998). By removing ubiquitin moieties, deubiquitination enzymes are also required to regulate the ubiquitin status of substrates, and like ubiquitination enzymes, they play a major role in the ubiquitin–proteasome pathway (Eletr and Wilkinson, 2014).

Maintenance of satellite cell quiescence is essential for its self-renewal but also for its role in muscle regeneration (Mourikis et al., 2012; Bjornson et al., 2012) and requires surveillance mechanisms to preserve the quality of the proteome and to maintain long-term cellular homeostasis. Therefore, proteasome inactivation in satellite cells leads to apoptosis through activation of the p53 pathway and to progressive depletion of their reservoir (Kitajima et al., 2018).

After injury, the activation of satellite cells requires rapid changes in protein composition, elimination of proteins involved in quiescence maintenance and expression of new proteins involved in cell-cycle regulation and differentiation. Notably, the sequential expression of members of the MRF family together with the downregulation of Pax3/7 are key steps that fine-tune activation and proliferation, while preventing premature terminal differentiation (Schmidt et al., 2019). Like many transcription factors, Pax3/7 and MRFs are short-lived regulatory proteins, which all have been shown to be degraded by the UPS (Boutet et al., 2010; Bustos et al., 2015), allowing a spatiotemporal control of protein degradation during myogenic progression. Soon after activation, both Pax3 and Pax7 levels are downregulated, a step necessary for differentiation, as satellite cells become myogenic progenitors (Boutet et al., 2007; Bustos et al., 2015). The reduction of Pax3 and Pax7 proteins occurs through proteasomal degradation after selective ubiquitination by Taf1 (Boutet et al., 2010) and Nedd4 (Bustos et al., 2015) E3 ligases, respectively.

In contrast to ubiquitination enzymes, deubiquitinase activity protects some proteins from degradation during the muscle regeneration process. The deubiquitinase USP7 (ubiquitin-specific protease 7) has been shown to directly interact with Myogenin (de la Vega et al., 2020). While USP7 is transiently expressed in myogenic progenitors, correlating with the onset of Myogenin expression, inhibition of its activity during regeneration results in persistent expression of early regeneration markers, decrease of Myogenin content and significant reduction of the diameter of regenerating myofibers (de la Vega et al., 2020). These results suggest that USP7 regulates the onset of differentiation during regeneration by enhancing Myogenin stability.

15.5.3 The Role of Autophagy in Satellite Cell Function

Macroautophagy, also called autophagy, is a self-catabolic process by which cellular components are sequestered into double-membrane vacuoles, called auto-phagosomes, and then degraded after being delivered to the lysosomal machinery (Mizushima, 2007). Products of this degradation such as amino acids can then be recycled to maintain nutrient and energy levels during periods of metabolic starvation or stress. In addition to this recycling function, autophagy is also described as a cellular quality control mechanism that eliminates damaged molecules and organelles, such as

PROTEOSTASIS AND PROTEOLYSIS

mitochondria, or intracellular pathogens such as viruses and bacteria.

Basal autophagy is a characteristic of young and adult quiescent satellite cells where this process is essential for both establishment and maintenance of the stemness feature (García-Prat et al., 2016). Autophagy impairment by *Atg7* deletion in young mice leads to a reduction of the satellite cell pool. Moreover, these satellite cells are unable to undergo myogenic differentiation and become senescent, a phenotype that recapitulates the one observed in satellite cells from mice over 28 months old. In addition, geriatric satellite cells have been shown to exhibit defective autophagosome clearance, resulting in the accumulation of damaged mitochondria and high ROS levels, and to enter a senescent state. Autophagy dysfunction in these cells was linked to the age-related regenerative function impairment as pharmacological activation of autophagy in geriatric satellite cells prevents their senescence and completely restores their regenerative capacity (García-Prat et al., 2016).

The autophagic activity has been reported to increase during activation, proliferation and early differentiation of satellite cells (Tang and Rando, 2014; Fiacco et al., 2016). Autophagy is critical for the efficient activation of satellite cells and myogenic progenitor cells to meet their bioenergetic demands through the different stages of regeneration (Tang and Rando, 2014). In fact, while inhibition of autophagy by 3-methyladenine delays the muscle regenerative response, its pharmacological activation with the use of rapamycin enhances satellite cell activation and proliferation (Fiacco et al., 2016).

Interestingly, dystrophic muscle, in which satellite cells exhibit defective amplification and differentiation, is characterized by a premature decline of the autophagic activity. Reactivation of this process by low protein diet has been shown to prevent the decline in muscle regeneration and to delay disease progression (Fiacco et al., 2016). Collectively, these data render autophagy an essential process for the efficient activation of satellite cells throughout the whole regeneration process.

15.6 CONCLUSIONS AND PERSPECTIVES

Adult stem cells are crucial for the preservation of healthy and functional organs throughout life. In skeletal muscle, both maintenance of satellite cell quiescence and activation after injury are regulated by specific signaling pathways and are associated with many transcriptional and metabolic changes. As discussed in this chapter, recent compelling evidences point out the importance of redox regulation and protein quality control in muscle stem cell function. In particular, the role of autophagy and UPS in stemness maintenance and in muscle regeneration have been recently highlighted, although their interconnections with the metabolic changes driving adult stem cell behavior remain to be explored. Both systems coordinate the precise elimination of specific proteins during muscle regeneration and may ensure the appropriate metabolic pathway changes. Proteostasis decline has been described as a hallmark of aging (López-Otín et al., 2013), causing accumulation of aggregated proteins, mitochondrial dysfunction and senescence. Likewise, less effective degradation systems impair the quiescence/activation balance of satellite cells. Preserving proteostasis could therefore prevent stem cell exhaustion and defects in muscle regeneration and could potentially improve the quality of life during normal aging or in pathological conditions.

REFERENCES

Almada A.E., Wagers A.J., 2016. Molecular circuitry of stem cell fate in skeletal muscle regeneration, ageing and disease. Nat Rev Mol Cell Biol. 17: 267–79. doi:10.1038/nrm.2016.7

Al-Sawaf O., Fragoulis A., Rosen C. et al., 2014. Nrf2 augments skeletal muscle regeneration after ischaemia-reperfusion injury. J Pathol. 234: 538–47. doi:10.1002/path.4418

Barbieri E., Sestili P., 2012. Reactive oxygen species in skeletal muscle signaling. J Signal Transduct. 2012: 982794. doi:10.1155/2012/982794

Bernet J.D., Doles J.D., Hall J.K., et al., 2014. p38 MAPK signaling underlies a cell-autonomous loss of stem cell self-renewal in skeletal muscle of aged mice. Nat Med. 20: 265–71. doi:10.1038/nm.3465

Bigarella C.L., Liang R., Ghaffari S., 2014. Stem cells and the impact of ROS signaling. Development. 141: 4206–18. doi:10.1242/dev.107086

Bigelow D.J., Squier T.C., 2011. Thioredoxin-dependent redox regulation of cellular signaling and stress response through reversible oxidation of methionines. Mol Biosyst. 7: 2101–9. doi:10.1039/c1mb05081h

Bjornson C.R., Cheung T.H., Liu L., et al., 2012. Notch signaling is necessary to maintain quiescence in adult muscle stem cells. Stem Cells. 30: 232–42. doi:10.1002/stem.773

Bockhold K.J., Rosenblatt J.D., Partridge T.A., 1998. Aging normal and dystrophic mouse muscle: analysis of myogenicity in cultures of living single fibers. Muscle Nerve. 21: 173–83. doi:10.1002/(sici)1097-4598(199802)21:2<173::aid-mus4>3.0.co;2-8

Bosnakovski D., Xu Z., Li W. et al., 2008. Prospective isolation of skeletal muscle stem cells with a Pax7 reporter. Stem Cells. 26: 3194–204. doi:10.1634/stemcells.2007-1017

Boutet S.C., Disatnik M.H., Chan L.S., et al., 2007. Regulation of Pax3 by proteasomal degradation of monoubiquitinated protein in skeletal muscle progenitors. Cell. 130: 349–62. doi:10.1016/j.cell.2007.05.044

Boutet S.C., Biressi S., Iori K., et al., 2010. Taf1 regulates Pax3 protein by monoubiquitination in skeletal muscle progenitors. Mol Cell. 40: 749–61. doi:10.1016/j.molcel.2010.09.029

Brandes R.P., Weissmann N., Schröder K., 2014. Nox family NADPH oxidases: molecular mechanisms of activation. Free Radic Biol Med. 76: 208–26. doi:10.1016/j.freeradbiomed.2014.07.046

Brien P., Pugazhendhi D., Woodhouse S., et al., 2013. p38alpha MAPK regulates adult muscle stem cell fate by restricting progenitor proliferation during postnatal growth and repair. Stem Cells. 31: 1597–610. doi:10.1002/stem.1399

Bulfield G., Siller W.G., Wight P.A., et al., 1984. X chromosome-linked muscular dystrophy (mdx) in the mouse. Proc Natl Acad Sci U S A. 81: 1189–92. doi:10.1073/pnas.81.4.1189

Bustos F., de la Vega E., Cabezas F., et al., 2015. NEDD4 regulates PAX7 levels promoting activation of the differentiation program in skeletal muscle precursors. Stem Cells. 33: 3138–51. doi:10.1002/stem.2125

Cantu-Medellin N., Kelley E.E., 2013. Xanthine oxidoreductase-catalyzed reactive species generation: a process in critical need of reevaluation. Redox Biol. 1: 353–8. doi:10.1016/j.redox.2013.05.002

Castets P., Bertrand A.T., Beuvin M. et al., 2011. Satellite cell loss and impaired muscle regeneration in selenoprotein N deficiency. Hum Mol Genet. 20: 694–704. doi:10.1093/hmg/ddq515

Chakkalakal J.V., Jones K.M., Basson M.A., et al., 2012. The aged niche disrupts muscle stem cell quiescence. Nature. 490: 355–60. doi:10.1038/nature11438

Chazaud B., 2014. Macrophages: supportive cells for tissue repair and regeneration. Immunobiology. 219: 172–8. doi:10.1016/j.imbio.2013.09.001

Cheung T.H., Rando T.A., 2013. Molecular regulation of stem cell quiescence. Nat Rev Mol Cell Biol. 14: 329–40. doi:10.1038/nrm3591

Coletti D., Moresi V., Adamo S., et al., 2005. Tumor necrosis factor-alpha gene transfer induces cachexia and inhibits muscle regeneration. Genesis. 43: 120–8. doi:10.1002/gene.20160

Collins G.A., Goldberg A.L., 2017. The Logic of the 26S Proteasome. Cell. 169: 792–806. doi:10.1016/j.cell.2017.04.023

Conboy I.M., Conboy M.J., Wagers A.J., et al., 2005. Rejuvenation of aged progenitor cells by exposure to a young systemic environment. Nature. 433: 760–4. doi:10.1038/nature03260

Cornelison D.D., Wold B.J., 1997. Single-cell analysis of regulatory gene expression in quiescent and activated mouse skeletal muscle satellite cells. Dev Biol. 191: 270–83. doi:10.1006/dbio.1997.8721

Corti S., Locatelli F., Papadimitriou D. et al., 2006. Identification of a primitive brain-derived neural stem cell population based on aldehyde dehydrogenase activity. Stem Cells. 24: 975–85. doi:10.1634/stemcells.2005-0217

Cosgrove B.D., Gilbert P.M., Porpiglia E. et al., 2014. Rejuvenation of the muscle stem cell population restores strength to injured aged muscles. Nat Med. 20: 255–64. doi:10.1038/nm.3464

Crist C.G., Montarras D., Buckingham M., 2012. Muscle satellite cells are primed for myogenesis but maintain quiescence with sequestration of Myf5 mRNA targeted by microRNA-31 in mRNP granules. Cell Stem Cell. 11: 118–26. doi:10.1016/j.stem.2012.03.011

Cruz-Jentoft A.J., Sayer A.A., 2019. Sarcopenia. Lancet. 393: 2636–2646. doi:10.1016/S0140-6736(19)31138-9

Dai D.F., Chiao Y.A., Marcinek D.J., et al., 2014. Mitochondrial oxidative stress in aging and healthspan. Longev Healthspan. 3: 6. doi:10.1186/2046-2395-3-6

Dai X., Yan X., Wintergerst K.A., et al., 2020. Nrf2: redox and metabolic regulator of stem cell state and function. Trends Mol Med. 26: 185–200. doi:10.1016/j.molmed.2019.09.007

de Keizer P.L., Burgering B.M., Dansen T.B., 2011. Forkhead box o as a sensor, mediator, and regulator of redox signaling. Antioxid Redox Signal. 14: 1093–106. doi:10.1089/ars.2010.3403

de la Vega E., González N., Cabezas F., et al., 2020. USP7-dependent control of Myogenin stability is required for terminal differentiation in skeletal muscle progenitors. FEBS J. doi:10.1111/febs.15269.

Di Marzo N., Chisci E., Giovannoni R., 2018. The role of hydrogen peroxide in redox-dependent signaling: homeostatic and pathological responses in mammalian cells. Cells. 7(10): 156. doi:10.3390/cells7100156

Dodson M., Darley-Usmar V., Zhang J., 2013. Cellular metabolic and autophagic pathways: traffic control by redox signaling. Free Radic Biol Med. 63: 207–21. doi:10.1016/j.freeradbiomed.2013.05.014

Dumont N.A., Wang Y.X., von Maltzahn J. et al., 2015. Dystrophin expression in muscle stem cells regulates their polarity and asymmetric division. Nat Med. 21: 1455–63. doi:10.1038/nm.3990

Eletr Z.M., Wilkinson K.D., 2014. Regulation of proteolysis by human deubiquitinating enzymes. Biochim Biophys Acta Mol Cell. 1843: 114–28. doi:10.1016/j.bbamcr.2013.06.027

Fiacco E., Castagnetti F., Bianconi V. et al., 2016. Autophagy regulates satellite cell ability to regenerate normal and dystrophic muscles. Cell Death Differ. 23: 1839–49. doi:10.1038/cdd.2016.70

Fry C.S., Lee J.D., Mula J., et al., 2015. Inducible depletion of satellite cells in adult, sedentary mice impairs muscle regenerative capacity without affecting sarcopenia. Nat Med. 21: 76–80. doi:10.1038/nm.3710

Fukawa T., Yan-Jiang B.C., Min-Wen J.C., et al., 2016. Excessive fatty acid oxidation induces muscle atrophy in cancer cachexia. Nat Med. 22: 666–71. doi:10.1038/nm.4093

García-Prat L., Martinez-Vicente M., Perdiguero E., et al., 2016. Autophagy maintains stemness by preventing senescence. Nature. 529: 37–42. doi:10.1038/nature16187

García-Prat L., Sousa-Victor P., Muñoz-Cánoves P., 2017. Proteostatic and metabolic control of stemness. Cell Stem Cell. 20: 593–608. doi:10.1016/j.stem.2017.04.011

Gayraud-Morel B., Chretien F., Flamant P., et al., 2007. A role for the myogenic determination gene Myf5 in adult regenerative myogenesis. Dev Biol. 312: 13–28. doi:10.1016/j.ydbio.2007.08.059

Ginestier C., Hur M.H., Charafe-Jauffret E., et al., 2007. ALDH1 is a marker of normal and malignant human mammary stem cells and a predictor of poor clinical outcome. Cell Stem Cell. 1: 555–67. doi:10.1016/j.stem.2007.08.014

Gomez-Cabrera M.C., Borrás C., Pallardó F.V., et al., 2005. Decreasing xanthine oxidase-mediated oxidative stress prevents useful cellular adaptations to exercise in rats. J Physiol. 567: 113–20. doi:10.1113/jphysiol.2004.080564

Hardy D., Besnard A., Latil M., et al., 2016. Comparative study of injury models for studying muscle regeneration in mice. PLoS One. 11: e0147198. doi:10.1371/journal.pone.0147198

He W.A., Berardi E., Cardillo V.M., et al., 2013. NF-kappaB-mediated Pax7 dysregulation in the muscle microenvironment promotes cancer cachexia. J Clin Invest. 123: 4821–35. doi:10.1172/JCI68523

Hernández-García D., Wood C.D., Castro-Obregón S., et al., 2010. Reactive oxygen species: a radical role in development? Free Radic Biol Med. 49: 130–43. doi:10.1016/j.freeradbiomed.2010.03.020

Hershko A., Ciechanover A., 1998. The ubiquitin system. Annu. Rev. Biochem. 67(1): 425–79. doi:10.1146/annurev.biochem.67.1.425

Hess D.A., Wirthlin L., Craft T.P., et al., 2006. Selection based on CD133 and high aldehyde dehydrogenase activity isolates long-term reconstituting human hematopoietic stem cells. Blood. 107: 2162–9. doi:10.1182/blood-2005-06-2284

Holmström K.M., Finkel T., 2014. Cellular mechanisms and physiological consequences of redox-dependent signalling. Nat Rev Mol Cell Biol. 15: 411–21. doi:10.1038/nrm3801

Hwang A.B., Brack A.S., 2018. Muscle stem cells and aging. Curr Top Dev Biol. 126: 299–322. doi:10.1016/bs.ctdb.2017.08.008

Jean E., Laoudj-Chenivesse D., Notarnicola C., et al., 2011. Aldehyde dehydrogenase activity promotes survival of human muscle precursor cells. J Cell Mol Med. 15: 119–33. doi:10.1111/j.1582-4934.2009.00942.x

Jones D.P., Go Y.M., 2011. Mapping the cysteine proteome: analysis of redox-sensing thiols. Curr Opin Chem Biol. 15: 103–12. doi:10.1016/j.cbpa.2010.12.014

Juntilla M.M., Patil V.D., Calamito M., et al., 2010. AKT1 and AKT2 maintain hematopoietic stem cell function by regulating reactive oxygen species. Blood. 115: 4030–8. doi:10.1182/blood-2009-09-241000

Kerkweg U., Petrat F., Korth H.G., et al., 2007. Disruption of skeletal myocytes initiates superoxide release: contribution of NADPH oxidase. Shock. 27: 552–8. doi:10.1097/01.shk.0000245027.39483.e4

Kitajima Y., Suzuki N., Nunomiya A., et al., 2018. The ubiquitin-proteasome system is indispensable for the maintenance of muscle stem cells. Stem Cell Rep. 11: 1523–38. doi:10.1016/j.stemcr.2018.10.009

Koenig M., Hoffman E.P., Bertelson C.J., et al., 1987. Complete cloning of the Duchenne muscular dystrophy (DMD) cDNA and preliminary genomic organization of the DMD gene in normal and affected individuals. Cell. 50: 509–17. doi:10.1016/0092-8674(87)90504-6

Kocabas F., Zheng J., Thet S., et al., 2012. Meis1 regulates the metabolic phenotype and oxidant defense of hematopoietic stem cells. Blood. 120: 4963–72. doi:10.1182/blood-2012-05-432260

Kozakowska M., Ciesla M., Stefanska A., et al., 2012. Heme oxygenase-1 inhibits myoblast differentiation by targeting myomirs. Antioxid Redox Signal. 16: 113–27. doi:10.1089/ars.2011.3964

Le Moal E., Pialoux V., Juban G., et al., 2017. Redox control of skeletal muscle regeneration. Antioxid Redox Signal. 27: 276–310. doi:10.1089/ars.2016.6782

L'Honoré A., Commère P.H., Negroni E., et al., 2018. The role of Pitx2 and Pitx3 in muscle stem cells gives new insights into P38α MAP kinase and redox regulation of muscle regeneration. Elife. 7: e32991. doi:10.7554/eLife.32991

Lingappan K., 2018. NF-κB in oxidative stress. Curr Opin Toxicol. 7: 81–86. doi:10.1016/j.cotox.2017.11.002

Langen R.C., Schols A.M., Kelders M.C., et al., 2002. Tumor necrosis factor-alpha inhibits myogenesis through redox-dependent and -independent pathways. Am J Physiol Cell Physiol. 283: C714–21. doi:10.1096/fj.03-0251com

Latil M., Rocheteau P., Châtre L., et al., 2012. Skeletal muscle stem cells adopt a dormant cell state post mortem and retain regenerative capacity. Nat Commun. 3: 903. doi:10.1038/ncomms1890

Lee S., Shin H.S., Shireman P.K., et al., 2006. Glutathione-peroxidase-1 null muscle progenitor cells are globally defective. Free Radic Biol Med. 41: 1174–84. doi:10.1016/j.freeradbiomed.2006.07.005

Lee S., Tak E., Lee J., et al., 2011. Mitochondrial H_2O_2 generated from electron transport chain complex I stimulates muscle differentiation. Cell Res. 21: 817–34. doi:10.1038/cr.2011.55

Lepper C., Partridge T.A., Fan C.M., 2011. An absolute requirement for Pax7-positive satellite cells in acute injury-induced skeletal muscle regeneration. Development. 138: 3639–46. doi:10.1242/dev.067595

Liu L., Charville G.W., Cheung T.H., et al., 2018. Impaired notch signaling leads to a decrease in p53 activity and mitotic catastrophe in aged muscle stem cells. Cell Stem Cell. 2: 544–556.e4. doi:10.1016/j.stem.2018.08.019

López-Otín C., Blasco M.A., Partridge L., et al., 2013. The hallmarks of aging. Cell. 153: 1194–217. doi:10.1016/j.cell.2013.05.039

Lourenço Dos Santos S., Petropoulos I., Friguet B., 2018. The oxidized protein repair enzymes methionine sulfoxide reductases and their roles in protecting against oxidative stress, in ageing and in regulating protein function. Antioxidants (Basel) 7(12): 191. doi:10.3390/antiox7120191

Maillet A., Pervaiz S., 2012. Redox regulation of p53, redox effectors regulated by p53: a subtle balance. Antioxid Redox Signal. 16: 1285–94. doi:10.1089/ars.2011.4434

Malinska D., Kudin A.P., Bejtka M., et al., 2012. Changes in mitochondrial reactive oxygen species synthesis during differentiation of skeletal muscle cells. Mitochondrion. 12: 144–8. doi:10.1016/j.mito.2011.06.015

Mauro A., 1961. Satellite cell of skeletal muscle fibers. J Biophys Biochem Cytol. 9: 493–5. doi:10.1083/jcb.9.2.493

McCroskery S., Thomas M., Maxwell L., et al., 2003. Myostatin negatively regulates satellite cell activation and self-renewal. J Cell Biol. 162: 1135–47. doi:10.1083/jcb.200207056

Mercuri E., Muntoni F., 2013. Muscular dystrophies. Lancet. 381: 845–60. doi:10.1016/S0140-6736(12)61897-2

Mizushima N., 2007. Autophagy: process and function. Genes Dev. 21: 2861–73. doi:10.1101/gad.1599207

Moldogazieva N.T., Mokhosoev I.M., Feldman N.B., et al., 2018. ROS and RNS signalling: adaptive redox switches through oxidative/nitrosative protein modifications. Free Radic Res. 52: 507–43. doi:10.1080/10715762.2018.1457217

Montarras D., Morgan J., Collins C., et al., 2005. Direct isolation of satellite cells for skeletal muscle regeneration. Science. 309: 2064–7. doi:10.1126/science.1114758

Montarras D., L'Honoré A., Buckingham M., 2013. Lying low but ready for action: the quiescent muscle satellite cell. FEBS J. 280: 4036–50. doi:10.1111/febs.12372

Mourikis P., Sambasivan R., Castel D., et al., 2012. A critical requirement for notch signaling in maintenance of the quiescent skeletal muscle stem cell state. Stem Cells. 30: 243–52. doi:10.1002/stem.775

Murphy M.P., 2009. How mitochondria produce reactive oxygen species. Biochem J. 417: 1–13. doi:10.1042/BJ20081386

Murphy M.M., Lawson J.A., Mathew S.J., et al., 2011. Satellite cells, connective tissue fibroblasts and their interactions are crucial for muscle regeneration. Development. 138: 3625–37. doi:10.1242/dev.064162

Nethery D., Callahan L.A., Stofan D., et al., 2000. PLA(2) dependence of diaphragm mitochondrial formation of reactive oxygen species. J Appl Physiol (1985). 89(1): 72–80. doi:10.1152/jappl.2000.89.1.72

Pallafacchina G., Francois S., Regnault B., et al., 2010. An adult tissue-specific stem cell in its niche: a gene profiling analysis of in vivo quiescent and activated muscle satellite cells. Stem Cell Res. 4: 77–91. doi:10.1016/j.scr.2009.10.003

Perales-Clemente E., Folmes C.F.D., Terzic A., 2014. Metabolic regulation of redox status in stem cells. Antioxid Redox Signal. 11: 1648–59. doi:10.1089/ars.2014.6000

Perdiguero E., Ruiz-Bonilla V., Gresh L., et al., 2007. Genetic analysis of p38 MAP kinases in myogenesis: fundamental role of p38alpha in abrogating myoblast proliferation. EMBO J. 26: 1245–56. doi:10.1038/sj.emboj.7601587

Piao Y.J., Seo Y.H., Hong F., et al., 2005. Nox 2 stimulates muscle differentiation via NF-kappaB/iNOS pathway. Free Radic Biol Med. 38: 989–1001. doi:10.1016/j.freeradbiomed.2004.11.011

Pickart C.M., 2001. Mechanisms underlying ubiquitination. Annu Rev Biochem. 70(1): 503–533. doi:10.1146/annurev.biochem.70.1.503

Relaix F., Zammit PS., 2012. Satellite cells are essential for skeletal muscle regeneration: the cell on the edge returns centre stage. Development. 139: 2845–56. doi:10.1242/dev.069088

Relaix F., Montarras D., Zaffran S., et al., 2006. Pax3 and Pax7 have distinct and overlapping functions in adult muscle progenitor cells. J Cell Biol. 172: 91–102. doi:10.1083/jcb.200508044

Remels A.H., Langen R.C., Schrauwen P., et al., 2010. Regulation of mitochondrial biogenesis during myogenesis. Mol Cell Endocrinol. 315: 113–20. doi:10.1016/j.mce.2009.09.029

Roos G., Messens J., 2011. Protein sulfenic acid formation: from cellular damage to redox regulation. Free Radic Biol Med. 51(2): 314–26. doi:10.1016/j.freeradbiomed.2011.04.031

Ryall J.G., Cliff T., Dalton S., et al., 2015. Metabolic reprogramming of stem cell epigenetics. Cell Stem Cell. 17: 651–662. doi:10.1016/j.stem.2015.11.012

Sabourin L.A., Girgis-Gabardo A., Seale P., et al., 1999. Reduced differentiation potential of primary MyoD-/- myogenic cells derived from adult skeletal muscle. J Cell Biol. 144: 631–43. doi:10.1083/jcb.144.4.631

Sahu A., Mamiya H., Shinde S.N., et al., 2018. Age-related declines in alpha-Klotho drive progenitor cell mitochondrial dysfunction and impaired muscle regeneration. Nat Commun. 9, 4859. doi:10.1038/s41467-018-07253-3

Sambasivan R., Yao R., Kissenpfennig A., et al., 2011. Pax7-expressing satellite cells are indispensable for adult skeletal muscle regeneration. Development. 138: 3647–56. doi:10.1242/dev.067587

Schieber M., Chandel N.S., 2014. ROS function in redox signaling and oxidative stress. Curr Biol. 24(10): R453–62. doi:10.1016/j.cub.2014.03.034

Schmidt M., Schüler S.C., Hüttner S.S., et al., 2019. Adult stem cells at work: regenerating skeletal muscle. Cell Mol Life Sci. 76: 2559–2570. doi:10.1007/s00018-019-03093-6

Schwörer S., Becker F., Feller C., et al., 2016. Epigenetic stress responses induce muscle stem-cell ageing by Hoxa9 developmental signals. Nature. 540: 428–432. doi:10.1038/nature20603

Seliskar M., Rozman D., 2007. Mammalian cytochromes P450 – importance of tissue specificity. Biochim Biophys Acta. 1770: 458–66. doi:10.1016/j.bbagen.2006.09.016

Shadel G.S., Horvath T.L., 2015. Mitochondrial ROS signaling in organismal homeostasis. Cell. 163(3): 560–9. doi:10.1016/j.cell.2015.10.001

Sies H., Jones D.P., 2020. Reactive oxygen species (ROS) as pleiotropic physiological signaling agents. Nat Rev Mol Cell Biol. 21: 363–383. doi:10.1038/s41580-020-0230-3

Singh C.K., Chhabra G., Ndiaye M.A., et al., 2018. The role of sirtuins in antioxidant and redox signaling. Antioxid Redox Signal. 28: 643–661. doi:10.1089/ars.2017.7290

Simsek T., Kocabas F., Zheng J., et al., 2010. The distinct metabolic profile of hematopoietic stem cells reflects their location in a hypoxic niche. Cell Stem Cell. 7: 380–90. doi:10.1016/j.stem.2010.07.011

Snijders T., Parise G., 2017. Role of muscle stem cells in sarcopenia. Curr Opin Clin Nutr Metab Care. 20: 186–190. doi:10.1097/MCO.0000000000000360

Sousa-Victor P., Gutarra S., García-Prat L., et al., 2014. Geriatric muscle stem cells switch reversible quiescence into senescence. Nature. 506: 316–21. doi:10.1038/nature13013

Takubo K., Goda N., Yamada W., et al., 2010. Regulation of the HIF-1alpha level is essential for hematopoietic stem cells. Cell Stem Cell. 7: 391–402. doi:10.1016/j.stem.2010.06.020

Tang A.H., Rando T.A., 2014. Induction of autophagy supports the bioenergetic demands of quiescent muscle stem cell activation. EMBO J. 33: 2782–97. doi:10.15252/embj.201488278

Tell G., Quadrifoglio F., Tiribelli C., et al., 2009. The many functions of APE1/Ref-1: not only a DNA repair enzyme. Antioxid Redox Signal. 11: 601–20. doi:10.1089/ars.2008.2194

Terrill J.R., Radley-Crabb H.G., Iwasaki T., et al., 2013. Oxidative stress and pathology in muscular dystrophies: focus on protein thiol oxidation and dysferlinopathies. FEBS J. 280: 4149–64. doi:10.1111/febs.12142

Togliatto G., Trombetta A., Dentelli P., et al., 2013. Unacylated ghrelin promotes skeletal muscle regeneration following hindlimb ischemia via SOD-2-mediated miR-221/222 expression. J Am Heart Assoc. 2(6): e000376. doi:10.1161/JAHA.113.000376

Trinei M., Berniakovich I., Beltrami E., et al., 2009. P66Shc signals to age. Aging (Albany NY) 1: 503–10. doi:10.18632/aging.100057

Urish K.L., Vella J.B., Okada M., et al., 2009. Antioxidant levels represent a major determinant in the regenerative capacity of muscle stem cells. Mol Biol Cell. 20: 509–20. doi:10.1091/mbc.e08-03-0274

Ustanina S., Carvajal J., Rigby P., et al., 2007. The myogenic factor Myf5 supports efficient skeletal muscle regeneration by enabling transient myoblast amplification. Stem Cells. 25: 2006–16. doi:10.1634/stemcells.2006-0736

Wang K., Zhang T., Dong Q., et al., 2013. Redox homeostasis: the linchpin in stem cell self-renewal and differentiation. Cell Death Dis. 4(3): e537. doi:10.1038/cddis.2013.50.

Willems P.H., Rossignol R., Dieteren C.E., et al., 2015. Redox homeostasis and mitochondrial dynamics. Cell Metab. 22: 207–18. doi:10.1016/j.cmet.2015.06.006

Won H., Lim S., Jang M., et al., 2012. Peroxiredoxin-2 upregulated by NF-kappaB attenuates oxidative stress during the differentiation of muscle-derived C2C12 cells. Antioxid Redox Signal. 16: 245–61. doi:10.1089/ars.2011.3952

Xiong G., Hindi S.M., Mann A.K., et al., 2017. The PERK arm of the unfolded protein response regulates satellite cell-mediated skeletal muscle regeneration. Elife. 6: e22871. doi:10.7554/eLife.22871

Yablonka-Reuveni Z., Rudnicki M.A., Rivera A.J., et al., 1999. The transition from proliferation to differentiation is delayed in satellite cells from mice lacking MyoD. Dev Biol. 210: 440–55. doi:10.1006/dbio.1999.9284

Yin Z., Machius M., Nestler E.J. et al., 2017. Activator protein-1: redox switch controlling structure and DNA-binding. Nucl Acids Res. 45(19): 11425–36. doi:10.1093/nar/gkx795

Yucel N., Wang Y.X., Mai T., et al., 2019. Glucose metabolism drives histone acetylation landscape transitions that dictate muscle stem cell function. Cell Rep. 27: 3939–3955. doi:10.1016/j.celrep.2019.05.092

Zhang H., Ryu D., Wu Y., et al., 2016. NAD(+) repletion improves mitochondrial and stem cell function and enhances life span in mice. Science. 352: 1436–1443. doi:10.1126/science.aaf2693

Zhang Y., Lee J.H., Paull T.T., et al., 2018. Mitochondrial redox sensing by the kinase ATM maintains cellular antioxidant capacity. Sci Signal. 11(538): eaaq0702. doi:10.1126/scisignal.aaq0702

Zhou X.L., Wei X.J., Li S.P., et al., 2019. Interactions between cytosolic phospholipase A2 activation and mitochondrial reactive oxygen species production in the development of ventilator-induced diaphragm dysfunction. Oxid Med Cell Longev. 2019: 2561929. doi:10.1155/2019/2561929

CHAPTER SIXTEEN

Protein Degradation in Cardiac Health and Disease

Xuejun Wang

CONTENTS

16.1 INTRODUCTION

Protein quality control (PQC) refers to a set of co-translational and post-translational mechanisms that act to sense and minimize the level and toxicity of misfolded proteins in the cell. PQC is an essential part of proteostasis. In general, PQC involves intricate collaborations between chaperones and the targeted removal of misfolded proteins. As illustrated in Figure 16.1, the degradation of terminally misfolded proteins is performed primarily by the ubiquitin–proteasome system (UPS) although the autophagic-lysosomal pathway (ALP) also plays a supplemental role (Wang and Robbins, 2014). A nascent polypeptide may achieve its native conformation through self-folding during translation but more often, the folding requires the help of a group of proteins known as molecular chaperones. The chaperones also play a vital role in the unfolding and refolding processes for the repair of misfolded proteins or during the cross-membrane translocation of proteins

such as the import of nuclear-encoded proteins into mitochondria. Once a misfolded protein is sensed by chaperones, its repair is attempted by them but when the repair fails, the misfolded protein is "terminally misfolded" and the least damaging outcome for the cell is its timely removal by the UPS or by chaperone-mediated autophagy (Kaushik and Cuervo, 2018). When misfolded proteins overwhelm chaperones and escape targeted degradation, they undergo aberrant aggregation via hydrophobic interactions, sequentially forming soluble oligomers (e.g. pre-amyloid oligomers [PAO]) and insoluble aggregates. The soluble intermediate species are believed to be highly toxic to the cell, more so than the insoluble aggregates. The aberrant aggregates can be removed through macroautophagy (Figure 16.1). If not removed onsite by macroautophagy, the aggregates from the periphery of the cell are moved via the microtubule-based transportation system to a location near the microtubule organization center and coalesce into a larger electron-dense

DOI: 10.1201/9781003048138-16

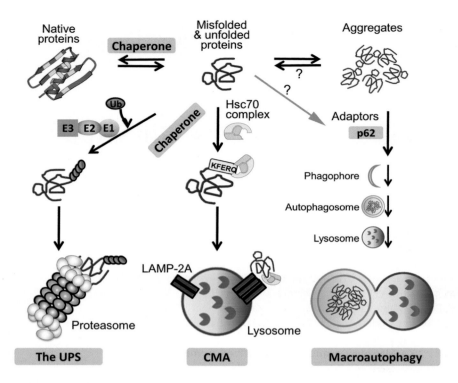

Figure 16.1 Illustration of protein quality control in the cell. Chaperones assist in the folding of nascent polypeptides, unfolding misfolded proteins and refolding them, and channeling terminally misfolded proteins for degradation by the ubiquitin–proteasome system (UPS) or chaperone-mediated autophagy (CMA). Misfolded proteins escaping from targeted degradation undergo aberrant aggregation. The protein aggregates may be selectively engulfed by autophagosomes and thereby targeted to, and degraded by, the lysosomes. *Abbreviation:* Hsc70, Heat-shock cognate 70; LAMP-2A, lysosome-associated membrane protein 2A. (Adapted from Wang et al., 2013.)

structure termed as aggresome (Wang et al., 2013). Aggresome formation probably reduces the toxicity of the aggregates and turns their removal by macroautophagy more efficient.

It is conceivable that both over production of misfolded proteins and reduced capacity to repair or remove misfolded proteins can cause PQC inadequacy, resulting in increased proteotoxic stress (IPTS). Both genetic and environmental factors can induce IPTS. Alpha B-crystallin (CRYAB) is the most abundant small heat shock protein expressed in cardiomyocytes and is believed to act as a chaperone protecting the integrity of the desmin filament and myofibrils (Iwaki et al., 1989; Vicart et al., 1998). The Arg120Gly missense mutation of CRYAB (CRYAB^{R120G}) is linked to desmin-related cardiomyopathy (DRC) and it was the first reported genetic mutation in a molecular chaperone that causes muscle disease in humans (Vicart et al., 1998). The CRYAB^{R120G}-associated DRC represents the best studied cardiac disease with IPTS (Wang

and Wang, 2020). The most studied cardiac PQC inadequacy caused by environmental factors is arguably myocardial ischemia–reperfusion (I–R) injury. Oxidative stress derived from I–R not only increases protein misfolding but also impairs the PQC machinery, including the proteasome (Divald et al., 2010). The prominent presence of PAO inside the cardiomyocytes of more than 80% of explanted failing human hearts (Sanbe et al., 2004) is compellingly indicative of PQC inadequacy in a large subset of heart diseases during their progression to heart failure, although the underlying cause remains largely unknown. Since PAO result from aberrant protein aggregation that is known to impair UPS function (Bence et al., 2001; Thibaudeau et al., 2018), the presence of PAO in cardiomyocytes *per se* can serve as an underlying cause of PQC inadequacy. Indeed, both clinical pathological studies and pre-clinical studies have provided convincing evidence that both UPS malfunction and impaired autophagy contribute

PROTEOSTASIS AND PROTEOLYSIS

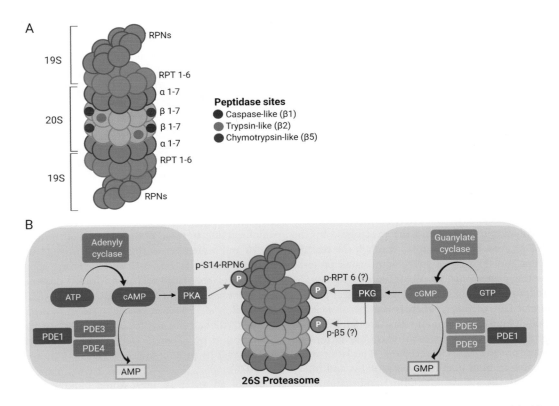

Figure 16.2 The composition of the 26S proteasome (A) and an illustration of a 26S proteasome nanodomain (B). (A) The 26S proteasome contains a cylinder-shaped 20S core particle capped by a 19S regulatory particle at one or both ends. The 19S consists of a lid and a base subcomplex; the lid is formed primarily by the non-ATPase subunits RPNs and the base is composed mainly by six ATPase subunits (RPT1–RPT6). (B) An illustration of a 26S proteasome nanodomain. Main regulators of the cAMP/PKA and the cGMP/PKG signaling modules are highlighted in the pink and the blue zone, respectively. (Adapted from Wang and Wang, 2020.)

to PQC inadequacy seen in diseased hearts (Wang and Robbins, 2014). This chapter focuses on the role of UPS in cardiac health and disease.

16.2 THE UPS IN PQC

UPS-mediated proteolysis occurs in two main steps. First, a chain of ubiquitin (Ub) is covalently attached with a substrate protein molecule via an ATP-consuming process known as ubiquitination and catalyzed sequentially by the Ub-activating enzyme (E1), Ub-conjugating enzyme (E2) and Ub ligases (E3; it confers substrate specificity). Second, the ubiquitinated protein is recognized and degraded by a multi-subunit protease complex, the 26S proteasome. Ubiquitination is reversible and the reverse process (deubiquitination) is catalyzed by deubiquitinases. In eukaryotes, the 26S proteasome is composed of the 20S

catalytic core particle and the 19S regulatory particle (RP) flanking the 20S at one or both ends (Figure 16.2A). The assembly of 20S proteasomes starts with the formation of half-proteasomes; each half-proteasome consists of an α-ring and a β-ring. Both the α- and β-rings are composed of seven unique protein subunits: α1–α7 and β1–β7, respectively. The three protease subunits are β1, β2 and β5. The axial stacking of αββα rings via a symmetric coupling between two half-proteasomes forms a barrel-shaped 20S proteasome and sequesters the proteolytic activities to the interior of the barrel chamber, with the α-rings gating substrate entry. The 19S RP sits atop of the α-ring and controls the opening of the gate. Compared with the 20S, the composition of 19S RP is much less defined due to reversible association with non-stoichiometric subunits; nonetheless, they all contain two essential parts: the base and the lid. The base is formed by 10 subunits: 4 non-ATPase subunits (RPN1, RPN2, RPN10 and RPN13) and 6 triple-A ATPase subunits (RTP1–RPT6) that directly interact with the α-ring of 20S proteasomes in an assembled 26S proteasome. The lid is formed mainly by nine non-ATPase subunits (RPN3, RPN5-9, RPN11, RPN12 and RPN15) (Figure 16.2A). The 20S proteasome may also form complexes with proteasome activators (PA) other than the 19S RP, including PA28αβ, PA28γ and PA200 under specific conditions, comprising alternative proteasome complexes with variable functions (Bohn et al., 2010; Tomko and Hochstrasser, 2013).

Polyubiquitination and the proteolysis by the 26S proteasome are coupled through the recruitment and docking of polyubiquitinated proteins to the proteasome by Ub receptors. Rpn10/S5a and Rpn13 of the 19S RP serve as intra-proteasomal Ub receptors, whereas the recruitment of polyubiquitinated proteins that are far from the proteasome seems to require shuttle factors or extra-proteasomal Ub receptors. The latter must be able to bind ubiquitinated proteins and interact with the 19S RP. Ubiquilin1 and mammalian homologs of Rad23 and Ddi1 are best suited to do so because their C-terminal Ub-associated (UBA) domain or domains can bind polyubiquitinated proteins while their N-terminal Ub-like (UBL) domain can interact with the 19S RP via Rpn10 or Rpn1 (Wang and Terpstra, 2013). Studies employing conditional gene targeting in mice have demonstrated that Ubiquilin1 is required for proper function of the UPS in the heart (Hu et al., 2018).

The role of UPS-mediated proteolysis in PQC is exemplified by the endoplasmic reticulum (ER)-associated degradation (ERAD), a key component of ER stress responses. Terminally misfolded proteins are retro-translocated from the ER lumen to the cytosol where the unfolded polypeptides are immediately ubiquitinated by E3 ligase complexes anchored to the ER membrane. HRD1 (also known as synoviolin), CHIP (C-terminus of Hsc70-interacting protein) and Parkin are among the identified E3s participating in ERAD in mammals (Wang and Terpstra, 2013). For cytosolic PQC, however, it remains an open question whether the Ub ligases mediating the ubiquitination of a misfolded protein differ from the ones responsible for ubiquitination of its native counterpart or whether there is a group of E3s specifically responsible for the ubiquitination of misfolded proteins. A few

Ub ligases were identified for their ability to target specific misfolded proteins for proteasomal degradation but it is still unknown whether they can do so for all misfolded proteins. Hul5 (HECT Ub ligase 5) and San1 were respectively identified as a cytosolic and a nuclear Ub ligase for degradation of misfolded proteins in yeast (Fang et al., 2011; Fredrickson et al., 2011); Ltn1 was identified as an E3 at the ribosome for co-translational degradation of non-stop polypeptides resulting from a mutation of, or missing, the stop codon (Bengtson and Joazeiro, 2010). In cardiomyocytes as well as in many other cell types, CHIP seems well positioned to be an E3 ligase for PQC because of its unique ability to serve as both a co-chaperone of Hsc70 and a member of the U-box family of Ub ligases, where the Hsc70-CHIP chaperone complex conceivably helps recruit misfolded proteins and makes them ready for ubiquitination by CHIP (Connell et al., 2001). A recent study shows that mouse CHIP half-life and its affinity to Hsc70 can be increased by phosphorylation at Ser20 (Ser19 in humans) by protein kinase G (PKG) (Ranek et al., 2020).

Recent advances in proteasome biology have revealed that the 26S proteasome activity is highly regulated and, as opposed to previous views, this regulation constitutes a rate-limiting step for UPS-mediated degradation of at least a subset of proteins including the misfolded ones (Collins and Goldberg, 2017). For example, cell culture studies have demonstrated that both the cyclic adenosine monophosphate (cAMP)-dependent protein kinase (PKA) and PKG can activate the 26S proteasome (Figure 16.2B) and thereby increase UPS-mediated degradation of human disease-linked mutant proteins (Lokireddy et al., 2015; Ranek et al., 2013; VerPlank et al., 2020). The in vivo requirement and physiological significance of proteasome phosphorylation remains largely undefined but emerging evidence suggests that proteasome phosphoregulation may be exploited to treat diseases with IPTS (see Section 16.7 of this chapter).

In addition to PQC, the other fundamental role of the UPS in the cell is the regulatory selective degradation of normal but no longer needed proteins (Wang and Wang, 2020). By estimate, the degradation of over 80% of cellular proteins is through the UPS. Through regulated protein degradation, the UPS plays essential roles in virtually all cellular processes. It has become increasingly clear that the UPS is pivotal to cardiac physiology and that UPS dysfunction, especially the proteasome functional insufficiency (PFI), contributes significantly to cardiac pathogenesis.

16.3 CLINICAL EVIDENCE OF UPS DYSFUNCTION IN CARDIAC PATHOGENESIS

Heart failure is the final common adversity for virtually all heart disease and represents a leading cause of death and disability in humans (Wang and Wang, 2020). Studies in human myocardial samples have indicated a major pathogenic role of UPS dysfunction in a range of heart diseases. Western blot analyses revealed significant increases in myocardial Ub conjugates in explanted human hearts with end-stage heart failure due to ischemic heart disease or dilated cardiomyopathy (DCM) (Weekes et al., 2003), indicative of increased ubiquitination, decreased deubiquitination or impaired proteasomal function in the diseased hearts.

At the ubiquitination step, genetic mutations in muscle-specific E3 ligases are linked to familial cardiomyopathies. MuRF1 (Muscle RING Finger 1) encoded by TRIM63 is known to target myofibrillar proteins for UPS-mediated degradation (Bodine et al., 2001). Two missense (p.A48V and p.I130M) variants and a deletion (p.Q247*) mutation in TRIM63 were identified exclusively in the probands of hypertrophic cardiomyopathy (S. N. Chen et al., 2012). A homozygous missense mutation (p.G243R) in the FBXO32 gene which encodes Atrogin-1 (or MAFbx), a muscle-specific F-box protein that serves as a major substrate receptor of the Skp1–Cullin1–F-box family of E3s for muscle proteins, was found to segregate with recessive familial DCM (Al-Hassnan et al., 2016). Potentially acquired dysregulation in ubiquitination was also implicated in the progression of more common forms of heart disease and heart failure.

In humans with end-stage heart failure resulting from DCM, besides the increase of myocardial Ub conjugates (Weekes et al., 2003), impairment at the proteasomal degradation step can also occur. This is evidenced by decreased myocardial proteasomal peptidase activities and impaired 26S proteasome assembly (Day et al., 2013; Predmore et al., 2010). Lastly, proteasome dysfunction is sufficient to cause cardiac malfunction in humans as indicated by evidence that cardiotoxicity, including arrhythmia and heart failure, occurs in a

significant portion of multiple myeloma patients receiving proteasome inhibitors as part of their chemotherapy, especially in those with pre-existing cardiac conditions (Chang et al., 2017; Cornell et al., 2019). Hence, there is a preponderance of clinical evidence that UPS impairment and especially PFI play a role in the progression of a large subset of heart disease to heart failure.

16.4 OCCURRENCE OF PFI IN ANIMAL MODELS OF HEART DISEASE

In animal models of nearly all forms of heart disease, such as myocardial infarction, I–R injury, pressure-overloaded cardiomyopathy, diabetic cardiomyopathy and DRC, increases in myocardial Ub conjugates are often observed, indicative of UPS dysfunction. Cardiac UPS malfunction in mouse models of myocardial I–R injury (Tian et al., 2012), diabetic cardiomyopathy (Li et al., 2017), pressure overload-induced heart failure (Ranek et al., 2015) and DRC (Q. Chen et al., 2005; Liu et al., 2006) has been supported further by accumulation of GFPdgn or GFPu. GFPdgn/GFPu is a UPS substrate protein engineered by carboxyl fusion of degron CL1 to the green fluorescence protein (GFP) (Kumarapeli et al., 2005). Notably, in some of these cases, the proteasome peptidase activities of the diseased hearts are comparable to or even greater than that of non-diseased controls, which is why PFI rather than proteasome impairment more accurately describes the situation. Due to increased production of misfolded proteins or impairment of other parts of PQC, normal or even compensatory increases of proteasome activities may not be able to meet the overwhelmingly increased demand on the proteasome, leading to PFI and accumulation of Ub conjugates.

Notably, PFI can be a manifestation of inadequate coupling between polyubiquitination and proteasomal degradation. Cardiomyocyte-restricted ablation of Ubiquilin1, which encodes a major extra-proteasomal Ub receptor, does not reduce cardiac proteasome peptidase activities but impairs UPS proteolytic function, thus accumulating polyubiquitinated proteins in the myocardium. Myocardial Ubiquilin1 was increased in mice with myocardial I–R; cardiac knockout of Ubiquilin1 exacerbated myocardial I–R injury and, conversely, overexpression of Ubiquilin1 facilitated proteasomal degradation of oxidized proteins and attenuated myocardial I–R injury (Hu et al., 2018).

These experimental findings demonstrate that an inadequacy in the coupling between ubiquitination and proteasomal degradation contributes to UPS malfunction and myocardial I–R injury.

When the UPS is impaired or overwhelmed, terminally misfolded proteins undergo aberrant aggregation, which in turn impairs the 26S proteasome (Bence et al., 2001; Thibaudeau et al., 2018). Hence, PFI and aberrant protein aggregation promote each other and form a vicious circle, inevitably impairing various functions of the affected cell and ultimately leading to cell death.

16.5 PFI PLAYS A MAJOR PATHOGENIC ROLE IN A RANGE OF HEART DISEASES

Pharmacological inhibition of the proteasome causes cardiac dysfunction even heart failure in both humans (Cole and Frishman, 2018) and animals (Herrmann et al., 2013; Tang et al., 2010). Genetically induced moderate proteasome inhibition in cardiomyocytes exacerbated myocardial I–R injury (Tian et al., 2012) and the progression of pressure overloaded hearts and led to heart failure in mice (Ranek et al., 2015). More recently, mice with cardiomyocyte-restricted ablation of the Psmc1 gene encoding Rpt2, an essential 19S proteasome subunit, were found to display embryonic and early postnatal lethality and DCM (Pan et al., 2020). These experimental studies have clearly demonstrated the sufficiency of PFI to injure the heart.

It was quite later that the necessity of PFI in pathogenesis was demonstrated as this would require a method to enhance or activate the proteasome, an invention that came decades after proteasome inhibitors were developed (Li et al., 2011a; Li et al., 2011b). The first reported method to enhance proteasome proteolytic function was the overexpression of PA28α. Overexpression of PA28α stabilizes PA28β and thereby increases the assembly of 11S proteasomes, leading to increased 11S-associated 20S proteasomes and thereby in hybrid proteasomes which are formed by association of 20S proteasomes with the 19S at one end and the 11S at the other. Proteasome enhancement by PA28α overexpression has been validated in various cell types (Lobanova et al., 2018). The net results of PA28α overexpression are the increased proteasome peptidase activities and enhanced proteasomal degradation of misfolded proteins in both cultured cardiomyocytes

(Li et al., 2011b) and transgenic mouse hearts (Li et al., 2011a). Using the cardiomyocyte-restricted PA28α overexpression mouse model, researchers have established a major pathogenic role for cardiac PFI in a large subset of heart disease including myocardial I–R injury and DRC (Li et al., 2011a), pressure overloaded cardiomyopathy (Rajagopalan et al., 2013) and diabetic cardiomyopathy (Li et al., 2017). In the CryAB[R120G]-based DRC mice, PA28α overexpression significantly reduced myocardial load of Ub conjugates and aberrant protein aggregates, improved cardiac function and delayed premature death (Li et al., 2011a). PA28α overexpression reduced infarct size and alleviated cardiac malfunction induced by acute myocardial I–R injury (Li et al., 2011a). In a pulmonary artery constriction-induced right ventricular hypertrophy and heart failure model, PA28α overexpression also delayed heart failure and reduced mortality (Rajagopalan et al., 2013). Similarly, PA28α overexpression significantly reduced Ub conjugate accumulation and cardiac function impairment in a streptozotocin-induced diabetic cardiomyopathy mouse model (Li et al., 2017).

16.6 MECHANISMS UNDERLYING CARDIAC PFI PATHOGENESIS

PFI inevitably slows down the degradation of both misfolded proteins and normal proteins. Accumulation of the former leads to increased proteotoxicity and of the latter to disruption of virtually all cellular functions. Both can conceivably cause cardiomyocyte dysfunction and, ultimately, cell death. For example, pharmacological proteasome inhibition was shown to activate the calcineurin–NFAT (nuclear factor of activated T cells) pathway in cultured cardiomyocytes and mouse hearts (Tang et al., 2010). This pathway is well known to mediate pathological cardiac hypertrophy and remodeling and, indeed, administration of the first-generation proteasome inhibitor bortezomib was sufficient to induce cardiac hypertrophy and malfunction. Moreover, activation of the calcineurin–NFAT pathway was observed in the heart of a transgenic mouse model of DRC known to have cardiac PFI (Tang et al., 2010).

Both cardiomyocyte apoptosis and necrosis were observed in diseased mouse hearts with PFI (Maloyan et al., 2005; Ranek et al., 2015). PFI can conceivably induce cell death via multiple mechanisms. A classic example of the proteasomal degradation-mediated regulation of signaling transduction is the NF-κB pathway where phosphorylation-triggered ubiquitination and proteasomal degradation of IκB (the inhibitor of NF-κB) allows NF-κB to translocate from the cytoplasm to the nucleus where it activates its target genes including caspase inhibitor cFLIP, thereby inhibiting apoptosis. This represents the cell survival arm of cell signaling downstream of the TNF-α (tumor necrosis factor α) receptor 1 (TNFR1) when activated by TNF-α. Due to endocrinal, paracrinal and autocrinal mechanisms, myocardial TNF-α is often increased during the progression of many forms of heart disease, which is also often associated with cardiac PFI. In cardiomyocytes with PFI, activation of the NF-κB pathway is blocked so that TNF-α stimulation will only activate the cell death arms of signaling, leading to increased cardiomyocyte death (Bergmann et al., 2001). Additionally, the degradation of caspases and many pro-apoptotic factors are UPS-dependent; thus, PFI can promote apoptosis through accumulation of caspases and proapoptotic factors (Sohns et al., 2010).

The COP9 signalosome dictates the proper function of Cullin-RING ligases, the largest family of Ub ligases, through Cullin de-NEDDylation, that is the removal of the covalently bound NEDD8 from Cullin. In mice with cardiomyocyte-specific ablation of the *Cops8* gene encoding an essential subunit of the COP9 signalosome, myocardial UPS function is impaired, and cardiomyocytes undergo massive necrosis (Su et al., 2011a, 2011b, 2013). Further investigation revealed that the necrosis is specifically necroptosis and is mediated by the RIPK1–RIPK3 pathway (Xiao et al., 2020), illustrating that UPS malfunction may also injure the heart via activation of the necroptotic pathway.

16.7 EXPLOITATION OF PROTEASOME PHOSPHOREGULATION TO TARGET PFI IN HEART DISEASE

The establishment of PFI as a major contributor to cardiac pathogenesis has prompted investigation into whether and how proteasome function is regulated while this is expected to reveal drug targets to enhance proteasome function and thereby to reverse PFI. For a long time, it was believed that ubiquitination is the rate-limiting step for UPS-mediated protein degradation, but recent advances in UPS biology have challenged this notion. It has been shown that the proteolytic function of 26S proteasomes is highly regulated. For at least a subset of UPS substrates including misfolded proteins, their half-life can be shortened by enhancing proteasome function. The best-studied regulation of the proteasome is phosphoregulation. To date, several kinases have been identified that can phosphorylate the proteasome at specific sites within specific subunits, and, with rare exceptions, the phosphorylation primes the proteasome or makes the proteasome working more efficiently (Collins and Goldberg, 2017).

16.7.1 PKA-Mediated Priming of Cardiac Proteasomes

PKA represents the first kinase implicated in proteasome phosphoregulation because it was the first kinase copurified with the proteasome and its activation was found to phosphorylate proteasome subunits and to increase proteasome activity (Pereira and Wilk, 1990). PKA can be copurified with cardiac proteasomes (Zong et al., 2006). It was recently demonstrated that PKA activates the 26S proteasome by selectively phosphorylating Ser14 of RPN6 (Lokireddy et al., 2015). Using a custom-made antibody against Ser14-phosphorylated RPN6 (p-Ser14-RPN6) in a more recent study, this group further showed that increase of cAMP by neurohormonal factors or intensive exercise can increase the abundance of p-Ser14-RPN6 and 26S proteasome activities in the target tissues/organs (VerPlank et al., 2019), suggesting that proteasome activation by the cAMP-PKA signaling is physiologically relevant. PKA activation led to a rapid increase in proteasome assembly and activity in canine hearts *in vivo* (Asai et al., 2009). In cultured cardiomyocytes, increased levels of cAMP were able to increase p-Ser14-RPN6 in a PKA-dependent manner and to increase UPS-mediated degradation of a surrogate misfolded protein (Zhang et al., 2019). cAMP and cyclic guanosine monophosphate (cGMP) are hydrolyzed by cyclic nucleotide phosphodiesterases (PDEs). Inhibition of PDE4, a predominant PDE for hydrolyzing cAMP in mouse hearts, was able to increase myocardial p-Ser14-RPN6 and 26S proteasome peptidase activities in mice with advanced DRC (Zhang et al., 2019), suggesting that PKA activation may be exploited to enhance proteasome-mediated degradation of misfolded proteins in the heart. It remains untested whether activation of the cAMP-PKA pathway alone benefits the treatment of heart disease with PFI but its benefits have been reported in mouse models of neurodegeneration (Myeku and Duff, 2018).

16.7.2 PKG-Mediated Priming of Cardiac Proteasomes

The first *in vivo* demonstration that the proteasome can be pharmacologically activated to treat disease came from a study showing that PKG positively regulates the function of the proteasome (Ranek et al., 2013). Both genetic and pharmacological activation of PKG increased proteasome peptidase activities and promoted the degradation of a surrogate UPS substrate and of a human disease-causing misfolded protein (CRYAB[R120G]) in cultured cardiomyocytes. Moreover, this study showed that increased cGMP and PKG activity via administration of a PDE5 inhibitor (sildenafil) was sufficient to decrease the accumulation of total ubiquitinated proteins and the abundance of aberrant protein aggregates and to slow down disease progression in the CRYAB[R120G]-based transgenic mouse model of DRC (Ranek et al., 2013). A subsequent study revealed that PKG mediates proteasome priming by muscarinic 2 receptor activation in cardiomyocytes and mouse hearts (Ranek et al., 2014), providing the first evidence that PKG-mediated proteasome regulation is physiologically relevant. Priming of the proteasome by PKG also occurs in neuronal cells (VerPlank et al., 2020). Although phosphorylation of RPT6 and PSMB5 by PKG was suggested (Ranek et al., 2013), the exact phosphosite(s) in the proteasome responsible for its activation by PKG remains to be defined. There is strong evidence that PKG does not phosphorylate the same site in the proteasome as PKA (VerPlank and Goldberg, 2018).

16.7.3 Priming the Proteasome by Duo-Activation of PKA and PKG to Target Cardiac PFI

Activation of the cAMP-PKA pathway through, for example, PDE3 inhibition is efficacious for treating acute heart failure but chronic use of PDE3 inhibitors (e.g. milrinone) causes fatal arrhythmia in patients with chronic heart failure (Amsallem et al., 2005). This raises a potential safety issue over the use of cAMP-PKA activation as a strategy to treat chronic heart disease with IPTS. Meanwhile, clinical trials on augmenting cGMP/PKG signaling via PDE5 inhibition (sildenafil) to treat heart failure with preserved ejection fraction (HFpEF) achieved only a neutral outcome (Redfield et al., 2013). HFpEF refers to the category of heart failure, in which cardiac muscle relaxation or diastolic function is impaired while cardiac contractile or systolic function remains relatively normal. A recent experimental study on PDE1 inhibition has unveiled a potential solution to these perceived hurdles (Zhang et al., 2019). PDE1 is a duo-substrate PDE, capable of hydrolyzing both cAMP and cGMP, and constitutes the great majority of cGMP hydrolytic and cAMP hydrolytic activities in the soluble fractions of human myocardium and the majority of cGMP hydrolytic activity in the microsomal fraction (Vandeput et al., 2007). Selective inhibition of PDE1 with a pharmacological inhibitor (IC86430) increased myocardial proteasome activity and UPS proteolytic function in mice. In cultured cardiomyocytes, PDE1 inhibition improved UPS function in both a PKG- and a PKA-dependent manner. Myocardial PDE1 expression in the CRYAB[R120G]-based mouse model of DRC was significantly increased. A 4-week continuous administration of IC86430 initiated at an advanced DRC stage was found to increase significantly myocardial p-Ser14-Rpn6, to attenuate myocardial accumulation of oligomeric CRYAB, to prevent or ameliorate cardiac diastolic malfunction, and to delay premature death (Zhang et al., 2019). The striking therapeutic benefits revealed by this pre-clinical study not only provide a compelling experimental demonstration of the efficacy of PDE1 inhibition in treating cardiac proteinopathy but also suggest that duo-activation of both PKA and PKG may be a more feasible strategy than solo-activation of PKA or PKG in targeting PFI to treat heart disease with IPTS. It is possible that the duo-activation of both PKA and PKG can improve proteasome function additively or even synergistically while mutually attenuating the undesirable effects from solo-activation of each. This is because activation of cAMP-PKA signaling is known to increase heart rate and cardiac hypertrophy, both detrimental to the heart; but on the other hand, activation of the cGMP-PKG signaling does the opposite, counteracting the effects of cAMP-PKA activation (Wang and Wang, 2020).

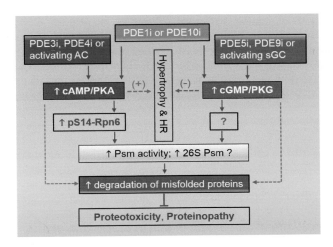

Figure 16.3 A schematic for the proposed duo-activation of PKA and PKG to treat cardiac disease with increased proteotoxic stress. PDE3 inhibition (PDE3i), PDE4i or activation of the adenylate cyclase (AC) induces cAMP/PKA signaling, increases Ser14-phosphorylated RPN6/PSMD11 (p-S14-RPN6) and thereby increases proteasome (Psm) activities and perhaps 26S Psm assembly. In parallel, PDE5i, PDE10i or use of an activator or stimulator of soluble guanylate cyclase (sGC) induces cGMP/PKG signaling, which increases proteasome activities by phosphorylating Psm subunit(s) that remain to be defined. Stimulation of cAMP/PKA promotes cardiac growth (hypertrophy) and increases heart rate (HR). Augmentation of cGMP/PKG signaling, on the other hand, suppresses hypertrophy and decreases HR. The duo-activation may be achievable by PDE1i or PDE10i. The dotted lines denote conceivable alternative pathways that currently have no clear support. (Adapted from Wang and Wang, 2020.)

It will be important to test whether an optimized equilibrium between PKA and PKG activation than that resulting from PDE1 inhibition can be achieved by a two-drug approach. As illustrated in Figure 16.3, cAMP-PKA and cGMP-PKG signaling can be induced in many ways.

16.8 CONCLUDING REMARKS

Heart disease and the resultant heart failure are the leading cause of morbidity and mortality globally, but the current therapies are far from satisfactory. IPTS is observed in a large subset of heart diseases, but no current treatment is aimed to target IPTS. Clinical evidence strongly implicates, and experimental studies have established that UPS malfunction, especially PFI and the resultant PQC inadequacy, contributes to IPTS and plays a major pathogenic role in the progression of a range of heart diseases to heart failure. Recent groundbreaking advances in proteasome biology have unveiled that the proteasome is the rate-limiting step in UPS-mediated degradation of many, if not all, misfolded proteins and that the proteasome can be primed or activated pharmacologically. Augmenting cAMP-PKA or cGMP-PKG signaling can improve UPS-mediated degradation

of misfolded proteins in cardiomyocytes but long-term in vivo activation of PKA or PKG exerts undesirable effects on the heart and the rest of the body. As illustrated by the efficacy of PDE1 inhibition in treating DRC in mice (Zhang et al., 2019), duo-activation of PKA and PKG may help improve the tolerance of patients with IPTS as some of the unfavorable off-proteasome effects (e.g. on cardiac growth and heart rate) of increased cAMP and cGMP signaling may cancel out each other (Figure 16.3). Given the emerging compartmentalized nature of cAMP and cGMP signaling in the cell, the ability to target the signaling events specifically or primarily at the proteasome nanodomain (Figure 16.2B) would also minimize the off-proteasome effects of PKA and PKG activation. Thus, it will be extremely important to map out the PKA and PKG regulation specifically in the proteasome nanodomains of cardiomyocytes.

As a growing field in cardiac (patho)physiology, the involvement of protein degradation, especially the proteasome-mediated one, in cardiac health and disease has countless important questions remaining to be answered. For example, what are the physiological and pathophysiological roles of proteasome activation by cAMP/PKA? How does cGMP/PKG prime the proteasome?

Does proteasome phosphoregulation by other kinases have a relevance to cardiac (patho)physiology? From a therapeutics development point of view, discovering and inventing small molecules that can selectively prime the proteasome will be extremely exciting.

ACKNOWLEDGMENTS

The research of Dr. Wang's laboratory at the University of South Dakota Sanford School of Medicine (Vermillion, SD, USA) is in part supported by grants HL072166, HL085629, HL131667 and HL153614 from the National Institutes of Health and an American Heart Association grant (20TPA35490091).

REFERENCES

Al-Hassnan, Z. N., Shinwari, Z. M., Wakil, S. M., et al. 2016. A substitution mutation in cardiac ubiquitin ligase, FBXO32, is associated with an autosomal recessive form of dilated cardiomyopathy. BMC Med Genet, 17, 3. doi:10.1186/s12881-016-0267-5

Amsallem, E., Kasparian, C., Haddour, G., et al. 2005. Phosphodiesterase III inhibitors for heart failure. Cochrane Database Syst Rev, (1), CD002230. doi:10.1002/14651858.CD002230.pub2

Asai, M., Tsukamoto, O., Minamino, T., et al. 2009. PKA rapidly enhances proteasome assembly and activity in in vivo canine hearts. J Mol Cell Cardiol, 46(4), 452–462. doi:10.1016/j.yjmcc.2008.11.001

Bence, N. F., Sampat, R. M., & Kopito, R. R. 2001. Impairment of the ubiquitin-proteasome system by protein aggregation. Science, 292(5521), 1552–1555. doi:10.1126/science.292.5521.1552

Bengtson, M. H., & Joazeiro, C. A. 2010. Role of a ribosome-associated E3 ubiquitin ligase in protein quality control. Nature, 467(7314), 470–473. doi:10.1038/nature09371

Bergmann, M. W., Loser, P., Dietz, R., et al. 2001. Effect of NF-kappa B inhibition on TNF-alpha-induced apoptosis and downstream pathways in cardiomyocytes. J Mol Cell Cardiol, 33(6), 1223–1232. doi:10.1006/jmcc.2001.1385

Bodine, S. C., Latres, E., Baumhueter, S., et al. 2001. Identification of ubiquitin ligases required for skeletal muscle atrophy. Science, 294(5547), 1704–1708. doi:10.1126/science.1065874

Bohn, S., Beck, F., Sakata, E., et al. 2010. Structure of the 26S proteasome from Schizosaccharomyces pombe at subnanometer resolution. Proc Natl Acad Sci U S A, 107(49), 20992–20997. doi:10.1073/pnas.1015530107

Chang, H. M., Moudgil, R., Scarabelli, T., et al. 2017. Cardiovascular complications of cancer therapy: best practices in diagnosis, prevention, and management: part 1. J Am Coll Cardiol, 70(20), 2536–2551. doi:10.1016/j.jacc.2017.09.1096

Chen, Q., Liu, J. B., Horak, K. M., et al. 2005. Intrasarcoplasmic amyloidosis impairs proteolytic function of proteasomes in cardiomyocytes by compromising substrate uptake. Circ Res, 97(10), 1018–1026. doi:10.1161/01.RES.0000189262.92896.0b

Chen, S. N., Czernuszewicz, G., Tan, Y., et al. 2012. Human molecular genetic and functional studies identify TRIM63, encoding muscle RING finger protein 1, as a novel gene for human hypertrophic cardiomyopathy. Circ Res, 111(7), 907–919. doi:10.1161/CIRCRESAHA.112.270207

Cole, D. C., & Frishman, W. H. 2018. Cardiovascular complications of proteasome inhibitors used in multiple myeloma. Cardiol Rev, 26(3), 122–129. doi:10.1097/CRD.0000000000000183

Collins, G. A., & Goldberg, A. L. 2017. The logic of the 26S proteasome. Cell, 169(5), 792–806. doi:10.1016/j.cell.2017.04.023

Connell, P., Ballinger, C. A., Jiang, J., et al. 2001. The co-chaperone CHIP regulates protein triage decisions mediated by heat-shock proteins. Nat Cell Biol, 3(1), 93–96. doi:10.1038/35050618

Cornell, R. F., Ky, B., Weiss, B. M., et al. 2019. Prospective study of cardiac events during proteasome inhibitor therapy for relapsed multiple myeloma. J Clin Oncol, 37(22), 1946–1955. doi:10.1200/JCO.19.00231

Day, S. M., Divald, A., Wang, P., et al. 2013. Impaired assembly and post-translational regulation of 26S proteasome in human end-stage heart failure. Circ Heart Fail, 6(3), 544–549. doi:10.1161/CIRCHEARTFAILURE.112.000119

Divald, A., Kivity, S., Wang, P., et al. 2010. Myocardial ischemic preconditioning preserves postischemic function of the 26S proteasome through diminished oxidative damage to 19S regulatory particle subunits. Circ Res, 106(12), 1829–1838. doi:10.1161/CIRCRESAHA.110.219485

Fang, N. N., Ng, A. H., Measday, V., et al. 2011. Hul5 HECT ubiquitin ligase plays a major role in the ubiquitylation and turnover of cytosolic misfolded proteins. Nat Cell Biol, 13(11), 1344–1352. doi:10.1038/ncb2343

Fredrickson, E. K., Rosenbaum, J. C., Locke, M. N., et al. 2011. Exposed hydrophobicity is a key determinant of nuclear quality control degradation. Mol Biol Cell, 22(13), 2384–2395. doi:10.1091/mbc.E11-03-0256

Herrmann, J., Wohlert, C., Saguner, A. M., et al. 2013. Primary proteasome inhibition results in cardiac dysfunction. *Eur J Heart Fail*, 15(6), 614–623. doi:10.1093/eurjhf/hft034

Hu, C., Tian, Y., Xu, H., et al. 2018. Inadequate ubiquitination-proteasome coupling contributes to myocardial ischemia-reperfusion injury. *J Clin Invest*, 128(12), 5294–5306. doi:10.1172/JCI98287

Iwaki, T., Kume-Iwaki, A., Liem, R. K., et al. 1989. Alpha B-crystallin is expressed in non-lenticular tissues and accumulates in Alexander's disease brain. *Cell*, 57(1), 71–78. doi:10.1016/0092-8674(89)90173-6

Kaushik, S., & Cuervo, A. M. 2018. The coming of age of chaperone-mediated autophagy. *Nat Rev Mol Cell Biol*, 19(6), 365–381. doi:10.1038/s41580-018-0001-6

Kumarapeli, A. R., Horak, K. M., Glasford, J. W., et al. 2005. A novel transgenic mouse model reveals deregulation of the ubiquitin-proteasome system in the heart by doxorubicin. *FASEB J*, 19(14), 2051–2053. doi:05-3973fje [pii] 10.1096/fj.05-3973fje

Li, J., Horak, K. M., Su, H., et al. 2011a. Enhancement of proteasomal function protects against cardiac proteinopathy and ischemia/reperfusion injury in mice. *J Clin Invest*, 121(9), 3689–3700. doi:10.1172/JCI45709

Li, J., Powell, S. R., & Wang, X. 2011b. Enhancement of proteasome function by PA28alpha overexpression protects against oxidative stress. *FASEB J*, 25(3), 883–893. doi:10.1096/fj.10-160895

Li, J., Ma, W., Yue, G., et al. 2017. Cardiac proteasome functional insufficiency plays a pathogenic role in diabetic cardiomyopathy. *J Mol Cell Cardiol*, 102, 53–60. doi:10.1016/j.yjmcc.2016.11.013

Liu, J., Chen, Q., Huang, W., et al. 2006. Impairment of the ubiquitin-proteasome system in desminopathy mouse hearts. *FASEB J*, 20(2), 362–364. doi: 10.1096/fj.05-4869fje

Lobanova, E. S., Finkelstein, S., Li, J., et al. 2018. Increased proteasomal activity supports photoreceptor survival in inherited retinal degeneration. *Nat Commun*, 9(1), 1738. doi:10.1038/s41467-018-04117-8

Lokireddy, S., Kukushkin, N. V., & Goldberg, A. L. 2015. cAMP-induced phosphorylation of 26S proteasomes on Rpn6/PSMD11 enhances their activity and the degradation of misfolded proteins. *Proc Natl Acad Sci U S A*, 112(52), E7176–E7185. doi:10.1073/pnas.1522332112

Maloyan, A., Sanbe, A., Osinska, H., et al. 2005. Mitochondrial dysfunction and apoptosis underlie the pathogenic process in alpha-B-crystallin desmin-related cardiomyopathy. *Circulation*, 112(22), 3451–3461. doi:10.1161/CIRCULATIONAHA.105.572552

Myeku, N., & Duff, K. E. 2018. Targeting the 26S proteasome to protect against proteotoxic diseases. *Trends Mol Med*, 24(1), 18–29. doi:10.1016/j.molmed.2017.11.006

Pan, B., Li, J., Parajuli, N., et al. 2020. The calcineurin-TFEB-p62 pathway mediates the activation of cardiac macroautophagy by proteasomal malfunction. *Circ Res*, 127(4), 502–518. doi:10.1161/CIRCRESAHA.119.316007

Pereira, M. E., & Wilk, S. 1990. Phosphorylation of the multicatalytic proteinase complex from bovine pituitaries by a copurifying cAMP-dependent protein kinase. *Arch Biochem Biophys*, 283(1), 68–74. doi:10.1016/0003-9861(90)90613-4

Predmore, J. M., Wang, P., Davis, F., et al. 2010. Ubiquitin proteasome dysfunction in human hypertrophic and dilated cardiomyopathies. *Circulation*, 121(8), 997–1004. doi:10.1161/CIRCULATIONAHA.109.904557

Rajagopalan, V., Zhao, M., Reddy, S., et al. 2013. Altered ubiquitin-proteasome signaling in right ventricular hypertrophy and failure. *Am J Physiol Heart Circ Physiol*, 305(4), H551–H562. doi:10.1152/ajpheart.00771.2012

Ranek, M. J., Terpstra, E. J., Li, J., et al. 2013. Protein kinase G positively regulates proteasome-mediated degradation of misfolded proteins. *Circulation*, 128(4), 365–376. doi:10.1161/CIRCULATIONAHA.113.001971

Ranek, M. J., Kost, C. K., Jr., Hu, C., et al. 2014. Muscarinic 2 receptors modulate cardiac proteasome function in a protein kinase G-dependent manner. *J Mol Cell Cardiol*, 69C, 43–51. doi:10.1016/j.yjmcc.2014.01.017

Ranek, M. J., Zheng, H., Huang, W., et al. 2015. Genetically induced moderate inhibition of 20S proteasomes in cardiomyocytes facilitates heart failure in mice during systolic overload. *J Mol Cell Cardiol*, 85, 273–281. doi:10.1016/j.yjmcc.2015.06.014

Ranek, M. J., Oeing, C., Sanchez-Hodge, R., et al. 2020. CHIP phosphorylation by protein kinase G enhances protein quality control and attenuates cardiac ischemic injury. *Nat Commun*, 11(1), 5237. doi:10.1038/s41467-020-18980-x

Redfield, M. M., Chen, H. H., Borlaug, B. A., et al. 2013. Effect of phosphodiesterase-5 inhibition on exercise capacity and clinical status in heart failure with preserved ejection fraction: a randomized clinical trial. *JAMA*, 309(12), 1268–1277. doi:10.1001/jama.2013.2024

Sanbe, A., Osinska, H., Saffitz, J. E., et al. 2004. Desmin-related cardiomyopathy in transgenic mice: a cardiac amyloidosis. *Proc Natl Acad Sci U S A*, 101(27), 10132–10136. doi:10.1073/pnas.0401900101

Sohns, W., van Veen, T. A., & van der Heyden, M. A. 2010. Regulatory roles of the ubiquitin-proteasome system in cardiomyocyte apoptosis. *Curr Mol Med*, 10(1), 1–13. doi:10.2174/156652410791065426

Su, H., Li, J., Menon, S., et al. 2011a. Perturbation of cullin deNEDDylation via conditional Csn8 ablation impairs the ubiquitin-proteasome system and causes cardiomyocyte necrosis and dilated cardiomyopathy in mice. *Circ Res*, 108(1), 40–50. doi:10.1161/CIRCRESAHA.110.230607

Su, H., Li, F., Ranek, M. J., et al. 2011b. COP9 signalosome regulates autophagosome maturation. *Circulation*, 124(19), 2117–2128. doi:10.1161/CIRCULATIONAHA.111.048934

Su, H., Li, J., Osinska, H., et al. 2013. The COP9 signalosome is required for autophagy, proteasome-mediated proteolysis, and cardiomyocyte survival in adult mice. *Circ Heart Fail*, 6(5), 1049–1057. doi:10.1161/CIRCHEARTFAILURE.113.000338

Tang, M., Li, J., Huang, W., et al. 2010. Proteasome functional insufficiency activates the calcineurin-NFAT pathway in cardiomyocytes and promotes maladaptive remodelling of stressed mouse hearts. *Cardiovasc Res*, 88(3), 424–433. doi:10.1093/cvr/cvq217

Thibaudeau, T. A., Anderson, R. T., & Smith, D. M. 2018. A common mechanism of proteasome impairment by neurodegenerative disease-associated oligomers. *Nat Commun*, 9(1), 1097. doi:10.1038/s41467-018-03509-0

Tian, Z., Zheng, H., Li, J., et al. 2012. Genetically induced moderate inhibition of the proteasome in cardiomyocytes exacerbates myocardial ischemia-reperfusion injury in mice. *Circ Res*, 111(5), 532–542. doi:10.1161/CIRCRESAHA.112.270983

Tomko, R. J., Jr., & Hochstrasser, M. 2013. Molecular architecture and assembly of the eukaryotic proteasome. *Annu Rev Biochem*, 82, 415–445. doi:10.1146/annurev-biochem-060410-150257

Vandeput, F., Wolda, S. L., Krall, J., et al. 2007. Cyclic nucleotide phosphodiesterase PDE1C1 in human cardiac myocytes. *J Biol Chem*, 282(45), 32749–32757. doi:10.1074/jbc.M703173200

VerPlank, J. J. S., & Goldberg, A. L. 2018. Exploring the regulation of proteasome function by subunit phosphorylation. *Methods Mol Biol*, 1844, 309–319. doi:10.1007/978-1-4939-8706-1_20

VerPlank, J. J. S., Lokireddy, S., Zhao, J., et al. 2019. 26S Proteasomes are rapidly activated by diverse hormones and physiological states that raise cAMP and cause Rpn6 phosphorylation. *Proc Natl Acad Sci U S A*, 116(10), 4228–4237. doi:10.1073/pnas.1809254116

VerPlank, J. J. S., Tyrkalska, S. D., Fleming, A., et al. 2020. cGMP via PKG activates 26S proteasomes and enhances degradation of proteins, including ones that cause neurodegenerative diseases. *Proc Natl Acad Sci U S A*, 117(25), 14220–14230. doi:10.1073/pnas.2003277117

Vicart, P., Caron, A., Guicheney, P., et al. 1998. A missense mutation in the alphaB-crystallin chaperone gene causes a desmin-related myopathy. *Nat Genet*, 20(1), 92–95. doi:10.1038/1765

Wang, X., & Terpstra, E. J. 2013. Ubiquitin receptors and protein quality control. *J Mol Cell Cardiol*, 55, 73–84. doi:10.1016/j.yjmcc.2012.09.012

Wang, X., & Robbins, J. 2014. Proteasomal and lysosomal protein degradation and heart disease. *J Mol Cell Cardiol*, 71, 16–24. doi:10.1016/j.yjmcc.2013.11.006

Wang, X., & Wang, H. 2020. Priming the proteasome to protect against proteotoxicity. *Trends Mol Med*, 26(7), 639–648. doi:10.1016/j.molmed.2020.02.007

Wang, X., Pattison, J. S., & Su, H. 2013. Posttranslational modification and quality control. *Circ Res*, 112(2), 367–381. doi:10.1161/CIRCRESAHA.112.268706

Weekes, J., Morrison, K., Mullen, A., et al. 2003. Hyperubiquitination of proteins in dilated cardiomyopathy. *Proteomics*, 3(2), 208–216. doi:10.1002/pmic.200390029

Xiao, P., Wang, C., Li, J., et al. 2020. COP9 signalosome suppresses RIPK1-RIPK3-mediated cardiomyocyte necroptosis in mice. *Circ Heart Fail*, 13(8), e006996. doi:10.1161/CIRCHEARTFAILURE.120.006996

Zhang, H., Pan, B., Wu, P., et al. 2019. PDE1 inhibition facilitates proteasomal degradation of misfolded proteins and protects against cardiac proteinopathy. *Sci Adv*, 5(5), eaaw5870. doi:10.1126/sciadv.aaw5870

Zong, C., Gomes, A. V., Drews, O., et al. 2006. Regulation of murine cardiac 20S proteasomes: role of associating partners. *Circ Res*, 99(4), 372–380. doi:10.1161/01.RES.0000237389.40000.02

CHAPTER SEVENTEEN

Autophagy in Aging and Oxidative Stress

Dimitra Ranti, Anna Gioran and Niki Chondrogianni

CONTENTS

17.1 INTRODUCTION

Cellular homeostasis depends on the maintenance of proteome fidelity (proteostasis). Autophagy constitutes a proteostatic, catabolic process able to remove long-lived aberrant proteins and protein aggregates and to clear defective organelles such as mitochondria and peroxisomes (Glick et al., 2010). It also serves as a nutrient-recycling mechanism providing energy during stress, while it is implicated in development and cell differentiation (Mizushima and Levine, 2010; Palm and Thompson, 2017).

Age-associated disturbances in cellular homeostasis include a decrease in autophagic function (Rubinsztein et al., 2011a). This impairment contributes to a gradual accumulation of aggregated proteins, thus decreasing the cellular ability to overcome the hazardous effects of stressful conditions. Among these stresses, oxidative stress is defined as a perturbation in the balance between oxidants and antioxidants in favor of the first affecting the structure of various cellular biomolecules (Lefaki et al., 2017). These structural alterations in combination with low autophagic activity have detrimental effects on cell survival (Filomeni et al., 2015). Progressive loss of proteostasis and insufficient antioxidant responses are directly associated with various age-related neurodegenerative diseases such as Alzheimer's (AD) and Parkinson's (PD) diseases (Kurtishi et al., 2019). This chapter briefly describes the process of autophagy and focuses on its involvement in oxidative stress and aging.

17.2 AUTOPHAGY: A BULK DEGRADATION SYSTEM

17.2.1 The Three Types of Autophagy

The process of autophagy is highly conserved in all eukaryotes and it is defined as the lysosome-dependent degradation of cytoplasmic

DOI: 10.1201/9781003048138-17

components; hence, it is commonly referred to as the autophagy-lysosome pathway (ALP). Three main types of autophagy are described in mammalian cells: macroautophagy, microautophagy and chaperone-mediated autophagy (CMA) (Mizushima and Komatsu, 2011).

Macroautophagy, usually referred to as autophagy, is the best-studied type of autophagy as it is highly conserved in all eukaryotes. It has been described as a "nutrient recycler", but its function is not limited to this role. Autophagy is not only crucial for the bulky degradation of cytosolic content, but it can also selectively eliminate aggregated proteins, RNA, lipids, ferritin, ribosomes, inflammasome complexes as well as organelles such as peroxisomes (pexophagy), mitochondria (mitophagy), the endoplasmic reticulum (ER) and parts of the nucleus (Ornatowski et al., 2020; Zaffagnini and Martens, 2016). Numerous studies have established the involvement of autophagy in diverse biological processes such as stress response, development, aging, resistance to pathogens, antigen presentation and tumor suppression, among others (Levine and Klionsky, 2004; Münz, 2006; Shintani and Klionsky, 2004).

Microautophagy is defined as the direct non-selective wrapping of cytosolic content into the lumen of lysosomes through membrane invagination (Marzella et al., 1981). Microautophagy can degrade organelles or parts of them in a selective manner by employing molecular components that are also used by macroautophagy. More specifically, peroxisomes (micropexophagy), part of the nucleus (piecemeal microautophagy of the nucleus) as well as mitochondria (micromitophagy), can be degraded through microautophagy (Li et al., 2012).

CMA facilitates the turnover of selected proteins that contain a specific amino-acid motif in their sequence. Specifically, the cytosolic chaperone HSC70 recognizes and binds the penta-peptide KEFRQ motif in the target protein and forms a chaperone–substrate complex that is further decorated with other co-chaperones (Chiang et al., 1989). This complex binds the lysosomal-associated membrane protein 2A (LAMP-2A) that functions as a receptor for protein substrates. Finally, substrate proteins are unfolded and translocated into the lysosome, where they are degraded by lysosomal hydrolases (Cuervo and Dice, 1996).

This chapter will focus on the role of macroautophagy (referred to as autophagy hereafter)

in aging and oxidative stress; hence, the main autophagic steps will be described below.

17.2.2 The Steps and Main Molecular Players of Autophagy

Numerous genes regulate autophagy, each one participating in one of the three sequential steps that comprise autophagy: initiation of phagophore nucleation, autophagosome formation and fusion with the lysosome. All autophagy steps are summarized in Figure 17.1.

Phagophore nucleation: Two interacting complexes mainly orchestrate the initiation of phagophore nucleation: (i) the Unc-51 like autophagy-activating kinase (ULK1) complex and (ii) the class III phosphoinositide 3-kinase (PI3K) complex. The ULK1 complex consists of ULK1 itself, autophagy-related gene 13 (ATG13), ATG101 and focal adhesion kinase family interacting protein of 200 kDa (FIP200). The class III PI3K complex consists of the class III PI3K VPS34, BECLIN-1, VPS15 and ATG14L (Zachari and Ganley, 2017). Localization of the ULK1 complex on phagophore nucleation initiation sites enables the building of the phagophore (Karanasios et al., 2013) and the direct activation of the PI3K complex. The ULK1 complex phosphorylates PI3K complex components as part of the PI3K complex recruitment to the initiation site (Park et al., 2016; Russell et al., 2013). The phagophore nucleation initiation site is on the ER where cytoplasm-facing projections give rise to a specialized cell compartment consisting of lipid bilayer membranes and enriched with phosphatidylinositol-3-phosphate (PI(3)P), known as "the omegasome" (Axe et al., 2008). A branched actin network growing from within the omegasome drives membrane elongation and results in the generation of the phagophore (Holland and Simonsen, 2015). The source of membranes that make up the phagophore is not entirely determined, but mitochondria-associated membranes (MAMs), ER-plasma membrane contact sites, as well as the plasma membrane or membranes of the Golgi complex, and endosomes are most likely involved (Abada and Elazar 2014).

Autophagosome formation: Phagophores continue elongating until they form double-membrane vesicles that engulf cytoplasmic content. The critical players for phagophore enlargement are two complexes: the Atg5–Atg12 complex that is associated with Atg16 in a noncovalent manner

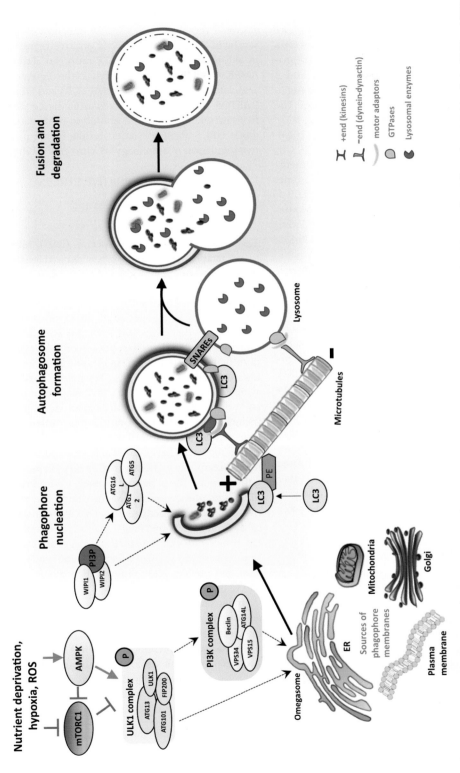

Figure 17.1 **Overview of the autophagic steps.** The nutrient and oxygen levels and the cellular redox status may initiate or inhibit autophagy. If autophagy is initiated, the ULK1 and PI3K complexes orchestrate the omegasome formation on the ER. Mitochondria, the Golgi complex or the plasma membrane may supply membranes for the phagophore formation (highlighted in light blue). PI3P, WIPI1, WIPI2 and ATG16L, ATG12, ATG5 contribute to the expansion and finally the formation of the autophagosome (highlighted in light purple). The lipidated form of LC3 participates in the formation of the autophagosome as well as its fusion with the lysosome. Elements of the cytoskeleton together with other proteins such as the SNAREs coordinate the fusion of the autophagosome together with the lysosome (highlighted in purple). In the lysosome, specialized enzymes digest the cargo of the autophagosome together with its inner membrane.

AUTOPHAGY IN AGING AND OXIDATIVE STRESS

227

and the Atg8 (or LC3 in mammals)-phosphatidyl-ethanolamine (PE) complex (Glick et al., 2010; Mizushima, 2007). The PI(3)P of the phagophore (present already in the omegasome) serves as a binding substrate that engages WD-repeat protein interacting with phosphoinositides 1 and 2 (WIPI1 and WIPI2) that in turn recruit the Atg5–Atg12–Atg16 complex to the phagophore. The Atg5–Atg12–Atg16 complex is essential for LC3/Atg8 lipidation (Proikas-Cezanne et al., 2015). The lipidated form of LC3, referred to as LC3-II, is thought to be involved in autophagosome membrane expansion and fusion with the lysosome. Although its exact function is still elusive, LC3-II is localized to phagophores and autophagosomes and is degraded by lysosomal enzymes, rendering it a widely used autophagosomal marker (Tanida et al., 2004).

Fusion with the lysosome: The cytoskeleton is an essential player in the fusion step. It is required since the newly formed autophagosome needs to be moved in a microtubule-dependent manner to the perinuclear region where the lysosomes are located. Specifically, the minus-end-directed dynein–dynactin motor complex facilitates the autophagosome's centripetal movement (Gross et al., 2007; Kimura et al., 2008). Amongst many, three main molecular components appear to drive the fusion event: Rab GTPases, soluble N-ethylmaleimide-sensitive factor attachment protein receptor (SNARE) proteins and membrane-tethering complexes (Nakamura and Yoshimori, 2017). The coordinated function of these proteins, along with several other regulators, results in the formation of the autolysosome in which the contents of the autophagosome and its inner membrane are digested.

17.2.3 Signaling in Autophagy

Autophagy signaling is regulated and affected by several molecules and intertwined with many signaling and metabolic pathways (Galluzzi et al., 2014). In this chapter, we will review only the two best-studied pathways that are the most relevant for the interplay of autophagy with oxidative stress and aging. The mammalian target of rapamycin complex 1 (mTORC1) phosphorylates ULK1 and ATG13 inhibiting the assembly of the ULK1 complex and consequently blocking autophagy activation (Fujioka et al., 2014). However, under certain stimuli like amino-acid deficiency, mTORC1 dissociates from the ULK1 complex, allowing autophagy (Hosokawa et al., 2009).

The fine-tuning between mTORC1 and the ULK1 complex depends on adenosine monophosphate-activated protein kinase (AMPK), an intracellular sensor of AMP:ATP ratio and a positive regulator of autophagy. A high AMP:ATP ratio can activate AMPK which in turn phosphorylates the regulatory associated protein of mTOR (RAPTOR) and ULK1. RAPTOR phosphorylation leads to its dissociation from mTOR and therefore inhibition of the latter, while ULK1 phosphorylation results in the positive regulation of autophagy (Gwinn et al., 2008).

17.3 AUTOPHAGY IN OXIDATIVE STRESS

17.3.1 Reactive Oxygen Species and Oxidative Stress

The term reactive oxygen species (ROS) refers to an array of molecular oxygen derivatives that occur as a normal by-product of aerobic life. Many organelles such as the peroxisomes, the ER, the lysosomes, the endosomes and the nucleus generate ROS (Navarro-Yepes et al., 2014). More than 40 enzymes including NADPH oxidases and the mitochondrial electron transport chain (ETC) produce hydrogen peroxide (H_2O_2) and the superoxide anion radical ($O_2^{•-}$) that are key redox-signaling agents (Sies and Jones, 2020). ROS play an important role in various intracellular signaling pathways including responses to growth factors, inflammation, proliferation and cell survival (Finkel, 2011).

Although cells have remarkably efficient antioxidant enzymes (e.g. the superoxide dismutase, glutathione peroxidase, peroxiredoxins, thioredoxins [Trxs] and glutaredoxins) that neutralize the reactive oxygen intermediates, partial accumulation of these anions is inevitable, leading to damaged macromolecules and organelles (Veal et al., 2018). This constitutes oxidative stress and results in damaged nucleic acids, lipids and proteins (Sies and Cadenas, 1985).

Understanding the impact and methods to counteract oxidative damage poses great translational interest for researchers. For example, increased oxidative damage is present in the aging human skin and vascular system while it has also been linked to age-related neurodegenerative diseases such as AD and PD, among others (Cheignon et al., 2018; Gu et al., 2020; Trist et al., 2019; Ungvari et al., 2019).

17.3.2 ROS Signaling and Autophagy Modulation

Increased cellular ROS levels are a consequence of various stimuli like starvation, hypoxia or

PROTEOSTASIS AND PROTEOLYSIS

mitochondrial deficiency. Elevated ROS levels such as H_2O_2 and superoxide anion induce autophagy (Filomeni et al., 2015; Scherz-Shouval et al., 2007). A vast amount of literature confirms the activation of autophagy after each of the above stimuli both in vitro and in vivo while the activation occurs via either transcriptional or post-translational regulatory events.

Elevated ROS levels activate transcription factors targeting genes involved in autophagy (Lee et al., 2012). For instance, moderate hypoxia can result in ROS production that activates the transcription factor HIF-1α, thus upregulating its target gene, namely BNIP3. BNIP3 prevents the binding of Beclin-1 on B-cell lymphoma 2 (Bcl-2), thereby allowing Beclin-1 to trigger autophagy (Bellot et al., 2009).

ROS can also modify proteins via a redox-based signal. Specifically, thiol (–SH) groups and thioethers (S^{2-}) found in reactive cysteine and methionine, respectively, are reversibly oxidized and reduced by ROS. Therefore, the existence of cysteine in the active sites of kinases, phosphatases and proteases in combination with the reversible oxidation/reduction reaction operates as a molecular "On-Off" switch, thus suggesting that ROS regulates several cellular signaling pathways (Marinho et al., 2014). Several autophagy-regulating proteins or proteins directly participating in specific autophagic steps have been found to respond to ROS, thereby bridging oxidative stress with the autophagic response. For instance, AMPK responds directly to H_2O_2 through oxidization of its active cysteine residues (Cys-299/Cys-304) on AMPK α-subunit (Cardaci et al., 2012). Studies in yeast revealed that a disulfide bond between Cys338 and Cys394 of Atg4, a protein with a pivotal role in autophagosome generation, is reduced by Trx (Perez-Perez et al., 2014). The reduced Atg4 (i.e. the activated form) enables both the cleavage of LC3/Atg8, thus exposing a glycine residue for conjugation with phospholipids (lipidation) and deconjugation (delipidation) of LC3/Atg8 that is involved in autophagosome biogenesis (Perez-Perez et al., 2014; Scherz-Shouval et al., 2007). In humans, Cys81 of ATG4A and Cys78 of ATG4B are regulated by ROS and redox-signaling dictates the "On/Off" state of these proteases (Marinho et al., 2014; Scherz-Shouval et al., 2007).

Although ROS can induce autophagy as described above, oxidation of specific core autophagic proteins may result in autophagy inhibition. More specifically, the oxidation of thiol groups of Atg3 and Atg7 prevents LC3 lipidation, that ultimately results in negative regulation of autophagy. This mechanism has been revealed in vitro but also in mouse aorta and it might explain how although ROS levels increase with age, they result in autophagy impairment instead of the expected activation (Frudd et al., 2018).

Another interesting mechanism links autophagy and ROS with apoptosis. Indeed, while autophagy is mainly considered a pro-survival mechanism, recent studies have revealed its essential role in cell death (Jung et al., 2020). Autophagy-dependent cell death is a form of regulated cell death that depends on the autophagic machinery (Galluzzi et al., 2018). One possible cause of the abnormal ROS accumulation that precedes autophagic cell death has been speculated to be the selective autophagic degradation of catalase, the major cellular ROS scavenger (Yu et al., 2006).

In conclusion, ROS act as signal transducers that moderate autophagy by targeting transcription factors and specific proteins. Autophagy serves mainly, although not only, as a cytoprotective process since it maintains cellular homeostasis by diminishing oxidative damage.

17.3.3 Autophagy-Dependent Regulation of ROS Production and Antioxidant Mechanisms

Autophagy adopts a pro-survival role during oxidative stress, since it eliminates damaged cellular components and organelles, thus limiting the source of oxidative stress and protecting the cell from further damage (Filomeni et al., 2015). Under stress conditions, autophagy may regulate ROS metabolism and control ROS levels in cells through various processes.

The removal of damaged mitochondria is facilitated by a type of selective degradation, known as mitophagy that can be induced by the accumulation of mitochondria-derived ROS (Pickles et al., 2018; Schofield and Schafer, 2020). Loss of mitochondrial membrane potential in parallel with ROS generation may cause accumulation of PTEN-induced kinase 1 (PINK1) on the outer membrane of damaged mitochondria (Roca-Agujetas et al., 2019). This leads to the recruitment of the ubiquitin E3 ligase, Parkin that ubiquitinates proteins of the outer mitochondrial membrane and induces their proteasomal degradation while at the same time it marks mitochondria for mitophagy. The marked mitochondria can be enwrapped by autophagosomes via the autophagy substrate receptor, p62 (Gao et al., 2020; Roca-Agujetas et al., 2019). Removal of the defective mitochondria that produce excessive ROS can restore the oxidized state of the cell while inability to do so may have detrimental consequences. For instance, in human T-cells defective for autophagy, mitochondria are not removed efficiently resulting in ROS accumulation. It has been shown that T-cell receptor stimulation in these cells induces apoptosis specifically due to defective mitophagy (Watanabe et al., 2014). Conclusively, through the selective clearance of dysfunctional mitochondria, cells avoid the consequences of toxic by-products.

Peroxisomes are organelles found in the cytoplasm of most eukaryotic cells hosting metabolic reactions like fatty acid β-oxidation. While peroxisomes play a central role in ROS neutralization, fatty acid β-oxidation can generate a significant amount of ROS. Therefore, the cell needs to control the number of peroxisomes to prevent excessive ROS accumulation (Islinger et al., 2018). The term "pexophagy" describes the specific elimination of peroxisomes through autophagy (Honsho et al., 2016). High levels of ROS activate pexophagy through the ATM serine/threonine kinase that phosphorylates PEX5, a peroxisome import receptor. ATM binds to the peroxisome, resulting in PEX5 mono-ubiquitination and its recognition

by p62. p62 functions as an autophagy adaptor in mammals consisting of a UBA domain that binds to the ubiquitinated substrate and an LIR domain that binds to LC3, comprising a substrate of autophagy (Ichimura et al., 2008). Subsequently, ATM activates ULK1 and inhibits mTORC1, driving the autophagosome to the peroxisome (Zhang et al., 2015).

Mounting evidence suggests an interplay between antioxidant responses following oxidative stress and enhanced autophagy levels (Filomeni et al., 2015; Pajares et al., 2018). For example, autophagy regulates the antioxidant response in mammals through the activation of transcription factors, such as the leucine zipper protein Nrf2, a major protector against oxidative stress. More specifically, Keap1, the inhibitory protein that is normally associated with Nrf2 and dictates its degradation, dissociates from it in a p62-dependent manner and subsequently Nrf2 is stabilized and translocated to the nucleus where it transcribes its target genes (Tonelli et al., 2018). Additionally, sestrins that are antioxidant-like proteins are induced by oxidative stress. They mediate an antioxidant response through the activation of transcription factors, such as p53 and Nrf2 and serve as positive regulators of autophagy upon stress conditions (Hou et al., 2015; Maiuri et al., 2009). The interplay of autophagy and oxidative stress is summarized in Figure 17.2.

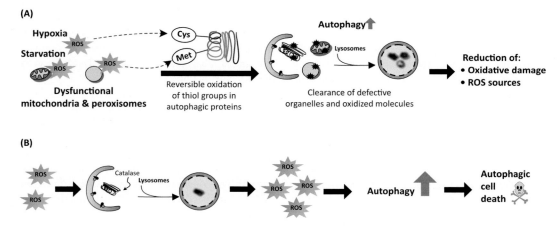

Figure 17.2 **ROS and autophagy interplay.** (A) Elevated ROS levels due to starvation or impaired mitochondria and/or peroxisomes or hypoxia may oxidize proteins carrying cysteine or methionine residues and alter their function. These modified proteins may trigger autophagy that eliminates damaged molecules or defective organelles. This eventually reduces oxidative damage and ROS sources. (B) Excessive ROS levels, possibly resulting from selective degradation of the ROS scavenger catalase, leads to a regulated cell death mediated by autophagic components.

PROTEOSTASIS AND PROTEOLYSIS

17.4 AUTOPHAGY AND AGING

17.4.1 Introduction in Aging

Aging can be defined as the progressive time-dependent deterioration of cell function and integrity (López-Otín et al., 2013). Many theories have attempted to explain why organisms age and although they may differ, there is significant consensus regarding the alterations an aging organism undergoes. The widely accepted aging hallmarks include genomic instability, telomere attrition, epigenetic alterations, deregulated nutrient sensing, mitochondrial dysfunction, cellular senescence, stem-cell exhaustion, altered intercellular communication and loss of proteostasis (López-Otín et al., 2013). Autophagy deterioration contributes to the loss of proteostasis, thereby establishing its involvement in aging.

17.4.2 Autophagy in Aging

Several studies have demonstrated that the autophagic capacity declines with aging in various animal models leading to insufficient protein degradation and subsequently to the accumulation of protein aggregates (Rubinsztein et al., 2011). Importantly, numerous long-lived strains in rats, flies and nematodes have been shown to require functional or even enhanced autophagy (Hansen, 2017; Hansen et al., 2018).

Autophagic deficiency can be the result of either diminished ATG expression or lysosomal dysfunction. For example in *Drosophila melanogaster*, the transcriptional levels of core autophagic genes like *Atg8a* (LC3/GABARAP in mammals), *Atg1* (Ulk1), *Atg5*, *Atg6* (BECN1) and *Atg7* are reduced in the aged brain and muscle tissue (Bai et al., 2013; Simonsen et al., 2008). In accordance, LC3 and ATG7 protein levels are decreased in aged mouse hypothalamus and muscle as well as in aged human muscle (Carnio et al., 2014; Kaushik et al., 2012). A spatiotemporal analysis of autophagy genes in nematodes showed an age-related decline in the expression of ATGs in the intestine, body-wall muscle, pharynx and neurons (Chang et al., 2017), while lysosomal protease activities were also found decreased with age (Sarkis et al., 1988). Additionally, an accumulation of autophagic vacuoles is detected in aged nematodes, probably as a result of insufficient autophagy (Chang et al., 2017; Hansen et al., 2018). In support of this, electron microscopy images revealed lower rates of autophagic vesicle formation in mouse hepatocytes from old mice compared to young ones (Terman, 1995).

Impaired autophagy is also central in age-related and neurodegenerative diseases. For example, the impairment of autophagy in neurons is highly linked to the accumulation of toxic protein aggregates. Neuron-specific knockout of *Atg-5*, *Atg-7* and FIP-200 in mice promotes accumulation of polyubiquitinylated and abnormal intracellular proteins (Hara et al., 2006; Komatsu et al., 2006; Liang et al., 2010). Importantly, in affected brain regions of AD patients, impaired autophagosome maturation and formation may contribute to accumulation of extracellular β-amyloid (Aβ) plaques and hyperphosphorylated tau since both constitute substrates of autophagy (Menzies et al., 2015; Nishiyama et al., 2007; Pickford et al., 2008). Furthermore, BECN1 mRNA levels are diminished in the brain of AD patients and dysregulated Becn1 in mice causes disruption of lysosomes and neurodegeneration. Moreover, low Becn1 levels increase intraneuronal Aβ accumulation and extracellular Aβ deposition in a PD mouse model (Pickford et al., 2008). Finally, Beclin-1 acetylation compromises the autophagic flux via lysosomal confinement, contributing to AD-related neurodegeneration in cytoplasmic hybrid (cybrid) cell lines (ρ0 cells – mtDNA-depleted recipient cells – that have received exogenous mtDNA) (Esteves et al., 2019).

Many neurodegenerative diseases are characterized by dysfunctional mitochondria with abnormal morphology, defective respiration and exaggerated ROS production that accumulate with aging (G. Chen et al., 2020). Impaired mitophagy leads to progressive accumulation of damaged mitochondria driving neuronal cell death and finally neurodegeneration (Palikaras et al., 2017). Abnormal mitophagy is observed in several PD models (Chinta et al., 2010), while many PD-associated gene products like PINK1 and Parkin are involved in mitophagy (Pickrell and Youle, 2015). Mutations in either PINK1 or Parkin are associated with the characteristic for PD dopaminergic neurodegeneration (Sliter et al., 2018).

Hampered mitophagy is also associated with cardiovascular aging and with a wide spectrum of age-related pathologies, such as sarcopenia and perturbed adipose tissue (Leidal et al., 2018). Reduced levels of Pink1 enhance ROS levels and mitochondrial dysfunction within cardiac myocytes that predispose to heart failure (Billia et al., 2011). In accordance, Atg5 knockout mice manifest cardiomyopathy (Taneike et al., 2010). On the contrary, knock-in of a Beclin-1 allele, whose interaction with its negative regulator Bcl-2 is disrupted, leads to hyperactivate autophagy and it is sufficient to maintain mitochondrial integrity and likely to alleviate age-related cardiomyocyte genotoxic stress and cardiovascular-related phenotypes (Fernández et al., 2018).

17.4.3 Autophagy in Longevity

Multiple experiments in model organisms like nematodes, flies and mice show that autophagy enhancement can prolong the organismal lifespan and healthspan or delay protein aggregation-mediated neurodegeneration. Genetic manipulation of ATGs as well as autophagy-enhancing compounds have been extensively studied in this context.

Overexpression of core autophagic genes can confer lifespan extension in animal models. For instance, Atg5 overexpression in mice extends their lifespan and healthspan as shown by maintenance of phenotypes like leanness, enhanced insulin sensitivity and good motor function (Pyo et al., 2013). Moreover, neuron-specific Atg8a overexpression results in lifespan extension and reduced accumulation of insoluble ubiquitinated proteins in the central nervous system of adult D. melanogaster (Simonsen et al., 2008). Additionally, expression of a Becn1 gain-of-function allele in mice activates autophagy, reduces amyloid accumulation in the brain, averts cognitive decline and extends survival in an AD mouse model (Rocchi et al., 2017). These examples indicate that restoration/maintenance of autophagy can decelerate the progression of age-related phenotypes.

Given that aging and several age-related and neurodegenerative diseases are characterized by limited autophagic capacity, many autophagy-targeting therapeutic strategies have been proposed (Leidal et al., 2018). These frequently involve caloric restriction mimetics or compounds that trigger autophagy (Levine et al., 2015). Examples of the positive effects of enhanced autophagy in longevity are the downregulation of the nutrient-sensing pathways mTOR and IGF1 in multiple models ranging from nematodes to mice and the activation of AMPK in invertebrate models, both of which are crucial for lifespan and healthspan extension (Burkewitz et al., 2016; Kennedy and Lamming, 2016; Piper et al., 2008). Two well-studied compounds that are widely used to stimulate autophagy are rapamycin, a natural mTOR inhibitor, and the natural disaccharide trehalose that induces autophagy through AMPK activation (Khandia et al., 2019). Rapamycin has been found

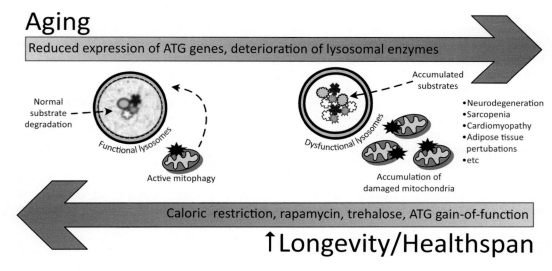

Figure 17.3 **Autophagy in aging.** During aging progression, autophagic activity is diminished either through the downregulation of ATGs or the reduction of lysosomal enzyme activity. As a result, undigested autophagic cargo and damaged mitochondria accumulate, events associated with neurodegeneration or cardiomyopathies, among others. In contrast, anti-aging interventions (e.g. caloric restriction) sustain autophagy while boost of autophagy (e.g. rapamycin treatment) promotes lifespan and healthspan extension.

PROTEOSTASIS AND PROTEOLYSIS

to prevent the formation of tau neurofibrillary tangles protecting from AD and other tauopathies (Frederick et al., 2015). Trehalose enhances LC3I–LC3II conversion in an mTOR-independent manner; it has been shown to delay the progression of amyotrophic lateral sclerosis, protect dopaminergic neurons, ameliorate tau pathologies, accelerate the clearance of mutant α-synuclein and huntingtin, avert neural tube defects and prevent cellular prion infection (X. Chen et al., 2016). Conclusively, restoration of the autophagic flux through either genetic or pharmacological means may boost organismal lifespan and healthspan and be protective against neurodegenerative diseases. The role of autophagy in aging and longevity is summarized in Figure 17.3.

17.5 CONCLUDING REMARKS AND FUTURE PERSPECTIVES

The multilayered role of autophagy in oxidative stress and aging is broadly acknowledged. Redox signaling can directly affect autophagy and vice versa. Several ATGs are redox-sensitive and can alter their function through post-translational modifications, while autophagy can buffer the harmful effects of excessive ROS production and preserve cell homeostasis or drive autophagy-dependent cell death. Understanding the interplay between autophagy and oxidative stress may lead to the development of interventions that protect from the consequences of oxidative stress in the context of aging or other diseases that are characterized by increased oxidative damage.

Several lines of evidence indicate that a decline of the autophagic flux comprises a hallmark of aging while ATGs are necessary for elongated lifespan in species ranging from yeast to mice. It would be interesting to understand the temporal intricacies of autophagy enhancement to optimize its benefits. Such findings could have serious implications in treatment planning, especially for age-related diseases that have a late onset.

REFERENCES

Abada, A., Elazar, Z., 2014. Getting ready for building: signaling and autophagosome biogenesis. EMBO Rep. 15, 839–852. doi:10.15252/embr.201439076

Axe, E.L., Walker, S.A., Manifava, M., et al., 2008. Autophagosome formation from membrane compartments enriched in phosphatidylinositol 3-phosphate and dynamically connected to the endoplasmic reticulum. J. Cell Biol. doi:10.1083/jcb.200803137

Bai, H., Kang, P., Hernandez, A.M., et al., 2013. Activin signaling targeted by insulin/dFOXO regulates aging and muscle proteostasis in Drosophila. PLoS Genet. 9, e1003941. doi:10.1371/journal.pgen.1003941

Bellot, G., Garcia-Medina, R., Gounon, P., et al., 2009. Hypoxia-induced autophagy is mediated through hypoxia-inducible factor induction of BNIP3 and BNIP3L via their BH3 domains. Mol. Cell. Biol. doi:10.1128/mcb.00166-09

Billia, F., Hauck, L., Konecny, F., et al., 2011. PTEN-inducible kinase 1 (PINK1)/Park6 is indispensable for normal heart function. Proc. Natl. Acad. Sci. U. S. A. 108, 9572–9577. doi:10.1073/pnas.1106291108

Burkewitz, K., Weir, H.J.M., Mair, W.B., 2016. AMPK as a pro-longevity target. EXS 107, 227–256. doi:10.1007/978-3-319-43589-3_10

Cardaci, S., Filomeni, G., Ciriolo, M.R., 2012. Redox implications of AMPK-mediated signal transduction beyond energetic clues. J. Cell Sci. 125, 2115–2125. doi:10.1242/jcs.095216

Carnio, S., LoVerso, F., Baraibar, M.A., et al., 2014. Autophagy impairment in muscle induces neuromuscular junction degeneration and precocious aging. Cell Rep. 8, 1509–1521. doi:10.1016/j.celrep.2014.07.061

Chang, J.T., Kumsta, C., Hellman, A.B., et al., 2017. Spatiotemporal regulation of autophagy during Caenorhabditis elegans aging. Elife 6. doi:10.7554/eLife.18459

Cheignon, C., Tomas, M., Bonnefont-Rousselot, D., et al., 2018. Oxidative stress and the amyloid beta peptide in Alzheimer's disease. Redox Biol. 14, 450–464. doi:10.1016/j.redox.2017.10.014

Chen, X., Li, M., Li, L., et al., 2016. Trehalose, sucrose and raffinose are novel activators of autophagy in human keratinocytes through an mTOR-independent pathway. Sci. Rep. 6, 28423. doi:10.1038/srep28423

Chen, G., Kroemer, G., Kepp, O., 2020. Mitophagy: an emerging role in aging and age-associated diseases. Front. Cell Dev. Biol. 8, 200. doi:10.3389/fcell.2020.00200

Chiang, H.L., Terlecky, S.R., Plant, C.P., et al., 1989. A role for a 70-kilo-Dalton heat shock protein in lysosomal degradation of intracellular proteins. Science (80–) 246, 382–385. doi:10.1126/science.2799391

Chinta, S.J., Mallajosyula, J.K., Rane, A., et al., 2010. Mitochondrial alpha-synuclein accumulation impairs complex I function in dopaminergic neurons and results in increased mitophagy in vivo. Neurosci. Lett. doi:10.1016/j.neulet.2010.09.061

Cuervo, A.M., Dice, J.F., 1996. A receptor for the selective uptake and degradation of proteins by lysosomes. Science (80–). doi:10.1126/science.273.5274.501

Esteves, A.R., Filipe, F., Magalhães, J.D., et al., 2019. The role of beclin-1 acetylation on autophagic flux in Alzheimer's disease. Mol. Neurobiol. 56, 5654–5670. doi:10.1007/s12035-019-1483-8

Fernández, Á.F., Sebti, S., Wei, Y., et al., 2018. Disruption of the beclin 1-BCL2 autophagy regulatory complex promotes longevity in mice. Nature 558, 136–140. doi:10.1038/s41586-018-0162-7

Filomeni, G., De Zio, D., Cecconi, F., 2015. Oxidative stress and autophagy: the clash between damage and metabolic needs. Cell Death Differ. 22, 377–388. doi:10.1038/cdd.2014.150

Finkel, T., 2011. Signal transduction by reactive oxygen species. J. Cell Biol. 194, 7–15. doi:10.1083/jcb.201102095Frederick, C., Ando, K., Leroy, K., et al., 2015. Rapamycin ester analog CCI-779/temsirolimus alleviates tau pathology and improves motor deficit in mutant tau transgenic mice. J. Alzheimer's Dis. 44, 1145–1156. doi:10.3233/JAD-142097

Frudd, K., Burgoyne, T., Burgoyne, J.R., 2018. Oxidation of Atg3 and Atg7 mediates inhibition of autophagy. Nat. Commun. 9. doi:10.1038/s41467-017-02352-z

Fujioka, Y., Suzuki, S.W., Yamamoto, H., et al., 2014. Structural basis of starvation-induced assembly of the autophagy initiation complex. Nat. Struct. Mol. Biol. doi:10.1038/nsmb.2822

Galluzzi, L., Pietrocola, F., Levine, B., et al., 2014. Metabolic control of autophagy. Cell 159, 1263–1276. doi:10.1016/j.cell.2014.11.006

Galluzzi, L., Vitale, I., Aaronson, S., et al., 2018. Molecular mechanisms of cell death: recommendations of the Nomenclature Committee on Cell Death 2018. Cell Death Differ. 25, 486–541. doi:10.1038/s41418-017-0012-4

Gao, F., Zhang, Y., Hou, X., et al., 2020. Dependence of PINK1 accumulation on mitochondrial redox system. Aging Cell 19, e13211. doi:10.1111/acel.13211

Glick, D., Barth, S., Macleod, K.F., 2010. Autophagy: cellular and molecular mechanisms. J. Pathol. 221, 3–12. doi:10.1002/path.2697

Gross, S.P., Vershinin, M., Shubeita, G.T.T., 2007. Cargo transport: two motors are sometimes better than one. Curr. Biol. 17, R478–R486. doi:10.1016/j.cub.2007.04.025

Gu, Y., Han, J., Jiang, C., et al., 2020. Biomarkers, oxidative stress and autophagy in skin aging. Ageing Res. Rev. 59, 101036. doi:10.1016/j.arr.2020.101036

Gwinn, D.M., Shackelford, D.B., Egan, D.F., et al., 2008. AMPK phosphorylation of raptor mediates a metabolic checkpoint. Mol. Cell 30, 214–226. doi:10.1016/j.molcel.2008.03.003

Hansen, M., 2017. Autophagy and ageing, in: Olsen, A., Gill, M.S. (Eds.), Ageing: Lessons from C. elegans. Springer, Cham, pp. 331–354.

Hansen, M., Rubinsztein, D.C., Walker, D.W., 2018. Autophagy as a promoter of longevity: insights from model organisms. Nat. Rev. Mol. Cell Biol. 19, 579–593. doi:10.1038/s41580-018-0033-y

Hara, T., Nakamura, K., Matsui, M., et al., 2006. Suppression of basal autophagy in neural cells causes neurodegenerative disease in mice. Nature 441, 885–889. doi:10.1038/nature04724

Holland, P., Simonsen, A., 2015. Actin shapes the autophagosome. Nat. Cell Biol. 17, 1094–1096. doi:10.1038/ncb3224

Honsho, M., Yamashita, S. ichi, Fujiki, Y., 2016. Peroxisome homeostasis: mechanisms of division and selective degradation of peroxisomes in mammals. Biochim. Biophys. Acta – Mol. Cell Res. 1863, 984–991. doi:10.1016/j.bbamcr.2015.09.032

Hosokawa, N., Hara, T., Kaizuka, T., et al., 2009. Nutrient-dependent mTORCl association with the ULK1-Atg13-FIP200 complex required for autophagy. Mol. Biol. Cell 20, 1981–1991. doi:10.1091/mbc.E08-12-1248

Hou, Y.-S., Guan, J.-J., Xu, H.-D., et al., 2015. Sestrin2 protects dopaminergic cells against rotenone toxicity through AMPK-dependent autophagy activation. Mol. Cell. Biol. doi:10.1128/mcb.00285-15

Ichimura, Y., Kumanomidou, T., Sou, Y.S., et al., 2008. Structural basis for sorting mechanism of p62 in selective autophagy. J. Biol. Chem. 283, 22847–22857. doi:10.1074/jbc.M802182200

Islinger, M., Voelkl, A., Fahimi, H.D., et al., 2018. The peroxisome: an update on mysteries 2.0. Histochem. Cell Biol. 150, 443–471. doi:10.1007/s00418-018-1722-5

Jung, S., Jeong, H., Yu, S.W., 2020. Autophagy as a decisive process for cell death. Exp. Mol. Med. doi:10.1038/s12276-020-0455-4

Karanasios, E., Stapleton, E., Manifava, M., et al., 2013. Dynamic association of the ULK1 complex with omegasomes during autophagy induction. J. Cell Sci. doi:10.1242/jcs.132415

Kaushik, S., Arias, E., Kwon, H., et al., 2012. Loss of autophagy in hypothalamic POMC neurons impairs lipolysis. EMBO Rep. 13, 258–265. doi:10.1038/embor.2011.260

Kennedy, B.K., Lamming, D.W., 2016. The mechanistic target of rapamycin: the Grand ConducTOR of metabolism and aging. Cell Metab. 23, 990–1003. doi:10.1016/j.cmet.2016.05.009

Khandia, R., Dadar, M., Munjal, A., et al., 2019. A comprehensive review of autophagy and its various roles in infectious, non-infectious, and lifestyle diseases: current knowledge and prospects for disease prevention, Novel Drug Des. Ther. Cells 8, 674. doi:10.3390/cells8070674

Kimura, S., Noda, T., Yoshimori, T., 2008. Dynein-dependent movement of autophagosomes mediates efficient encounters with lysosomes. Cell Struct. Funct. 33, 109–122. doi:10.1247/csf.08005

Komatsu, M., Waguri, S., Chiba, T., et al., 2006. Loss of autophagy in the central nervous system causes neurodegeneration in mice. Nature 441, 880–884. doi:10.1038/nature04723

Kurtishi, A., Rosen, B., Patil, K.S., et al., 2019. Cellular proteostasis in neurodegeneration. Mol. Neurobiol. 56, 3676–3689. doi:10.1007/s12035-018-1334-z

Lee, J., Giordano, S., Zhang, J., 2012. Autophagy, mitochondria and oxidative stress: cross-talk and redox signalling. Biochem. J. 441, 523–540. doi:10.1042/BJ20111451

Lefaki, M., Papaevgeniou, N., Chondrogianni, N., 2017. Redox regulation of proteasome function. Redox Biol. 13, 452–458. doi:10.1016/j.redox.2017.07.005

Leidal, A.M., Levine, B., Debnath, J., 2018. Autophagy and the cell biology of age-related disease. Nat. Cell Biol. 20, 1338–1348. doi:10.1038/s41556-018-0235-8

Levine, B., Klionsky, D.J., 2004. Development by self-digestion: molecular mechanisms and biological functions of autophagy. Dev. Cell 6, 463–477. doi:10.1016/S1534-5807(04)00099-1

Levine, B., Packer, M., Codogno, P., 2015. Development of autophagy inducers in clinical medicine. J. Clin. Invest. 125, 14–24. doi:10.1172/JCI73938

Li, W.W., Li, J., Bao, J.K., 2012. Microautophagy: lesser-known self-eating. Cell. Mol. Life Sci. 69, 1125–1136. doi:10.1007/s00018-011-0865-5

Liang, C.C., Wang, C., Peng, X., et al., 2010. Neural-specific deletion of FIP200 leads to cerebellar degeneration caused by increased neuronal death and axon degeneration. J. Biol. Chem. 285, 3499–3509. doi:10.1074/jbc.M109.072389

López-Otín, C., Blasco, M.A., Partridge, L., et al., 2013. The hallmarks of aging. Cell 153, 1194–1217. doi:10.1016/j.cell.2013.05.039

Maiuri, M.C., Malik, S.A., Morselli, E., et al., 2009. Stimulation of autophagy by the p53 target gene Sestrin2. Cell Cycle. doi:10.4161/cc.8.10.8498

Marinho, H.S., Real, C., Cyrne, L., et al., 2014. Hydrogen peroxide sensing, signaling and regulation of transcription factors. Redox Biol. 2, 535–562. doi:10.1016/j.redox.2014.02.006

Marzella, L., Ahlberg, J., Glaumann, H., 1981. Autophagy, heterophagy, microautophagy and crinophagy as the means for intracellular degradation. Virchows Arch. B Cell Pathol. Incl. Mol. Pathol. 36, 219–234. doi:10.1007/BF02912068

Menzies, F.M., Fleming, A., Rubinsztein, D.C., 2015. Compromised autophagy and neurodegenerative diseases. Nat. Rev. Neurosci. 16, 345–357. doi:10.1038/nrn3961

Mizushima, N., 2007. Autophagy: process and function. Genes Dev. 21, 2861–2873. doi:10.1101/gad.1599207

Mizushima, N., Levine, B., 2010. Autophagy in mammalian development and differentiation. Nat. Cell Biol. 12, 823–830. doi:10.1038/ncb0910-823

Mizushima, N., Komatsu, M., 2011. Autophagy: renovation of cells and tissues. Cell 147, 728–741. doi:10.1016/j.cell.2011.10.026

Münz, C., 2006. Autophagy and antigen presentation. Cell. Microbiol. 8, 891–898. doi:10.1111/j.1462-5822.2006.00714.x

Nakamura, S., Yoshimori, T., 2017. New insights into autophagosome-lysosome fusion. J. Cell Sci. 130, 1209–1216. doi:10.1242/jcs.196352

Navarro-Yepes, J., Burns, M., Anandhan, A., et al., 2014. Oxidative stress, redox signaling, and autophagy: cell death versus survival. Antioxid. Redox Signal. 21, 66–85. doi:10.1089/ars.2014.5837

Nishiyama, J., Miura, E., Mizushima, N., et al., 2007. Aberrant membranes and double-membrane structures accumulate in the axons of Atg5-null Purkinje cells before neuronal death. Autophagy 3, 591–596. doi:10.4161/auto.4964

Ornatowski, W., Lu, Q., Yegambaram, M., et al., 2020. Complex interplay between autophagy and oxidative stress in the development of pulmonary disease. Redox Biol. 36, 101679. doi:10.1016/j.redox.2020.101679

Pajares, M., Cuadrado, A., Engedal, N., et al., 2018. The role of free radicals in autophagy regulation: implications for ageing. Oxid. Med. Cell. Longev. doi:10.1155/2018/2450748

Palikaras, K., Daskalaki, I., Markaki, M., et al., 2017. Mitophagy and age-related pathologies: development of new therapeutics by targeting mitochondrial turnover. Pharmacol. Ther. doi:10.1016/j.pharmthera.2017.04.005

Palm, W., Thompson, C.B., 2017. Nutrient acquisition strategies of mammalian cells. Nature 546, 234–242. doi:10.1038/nature22379

Park, J.-M., Jung, C.H., Seo, M., et al., 2016. The ULK1 complex mediates MTORC1 signaling to the autophagy initiation machinery via binding and phosphorylating ATG14. Autophagy 12, 547–564. doi:10.1080/15548627.2016.1140293

Perez-Perez, M.E., Zaffagnini, M., Marchand, C.H., et al., 2014. The yeast autophagy protease Atg4 is regulated by thioredoxin. Autophagy 10, 1953–1964. doi:10.4161/auto.34396

Pickford, F., Masliah, E., Britschgi, M., et al., 2008. The autophagy-related protein beclin 1 shows reduced expression in early Alzheimer disease and regulates amyloid β accumulation in mice. J. Clin. Invest. 118, 2190–2199. doi:10.1172/JCI33585

Pickles, S., Vigié, P., Youle, R.J., 2018. Mitophagy and quality control mechanisms in mitochondrial maintenance. Curr. Biol. 28, R170–R185. doi:10.1016/j.cub.2018.01.004

Pickrell, A.M., Youle, R.J., 2015. The roles of PINK1, Parkin, and mitochondrial fidelity in Parkinson's disease. Neuron. doi:10.1016/j.neuron.2014.12.007

Piper, M.D.W., Selman, C., McElwee, J.J., et al., 2008. Separating cause from effect: how does insulin/IGF signalling control lifespan in worms, flies and mice? J. Intern. Med. 263, 179–191. doi:10.1111/j.1365-2796.2007.01906.x

Proikas-Cezanne, T., Takacs, Z., Dönnes, P., et al., 2015. WIPI proteins: essential PtdIns3P effectors at the nascent autophagosome. J. Cell Sci. 128, 207–217. doi:10.1242/jcs.146258

Pyo, J.O., Yoo, S.M., Ahn, H.H., et al., 2013. Overexpression of Atg5 in mice activates autophagy and extends lifespan. Nat. Commun. 4, 2300. doi:10.1038/ncomms3300

Roca-Agujetas, V., De Dios, C., Lestón, L., et al., 2019. Recent insights into the mitochondrial role in autophagy and its regulation by oxidative stress. Oxid. Med. Cell. Longev. 2019, 3809308. doi:10.1155/2019/3809308

Rocchi, A., Yamamoto, S., Ting, T., et al., 2017. A Becn1 mutation mediates hyperactive autophagic sequestration of amyloid oligomers and improved cognition in Alzheimer's disease. PLoS Genet. 13, e1006962. doi:10.1371/journal.pgen.1006962

Rubinsztein, D.C., Mariño, G., Kroemer, G., 2011b. Autophagy and aging. Cell 146, 682–695. doi:10.1016/j.cell.2011.07.030

Russell, R.C., Tian, Y., Yuan, H., et al., 2013. ULK1 induces autophagy by phosphorylating Beclin-1 and activating VPS34 lipid kinase. Nat. Cell Biol. doi:10.1038/ncb2757

Sarkis, G.J., Ashcom, J.D., Hawdon, J.M., et al., 1988. Decline in protease activities with age in the nematode Caenorhabditis elegans. Mech. Ageing Dev. 45, 191–201. doi:10.1016/0047-6374(88)90001-2

Scherz-Shouval, R., Shvets, E., Fass, E., et al., 2007. Reactive oxygen species are essential for autophagy and specifically regulate the activity of Atg4. EMBO J. 26, 1749–1760. doi:10.1038/sj.emboj.7601623

Schofield, J.H., Schafer, Z.T., 2020. Mitochondrial reactive oxygen species and mitophagy: a complex and nuanced relationship. Antioxid. Redox Signal. doi:10.1089/ars.2020.8058

Shintani, T., Klionsky, D.J., 2004. Autophagy in health and disease: a double-edged sword. Science (80-) 306, 990–995. doi:10.1126/science.1099993

Sies, H., Cadenas, E., 1985. Oxidative stress: damage to intact cells and organs. Philos. Trans. R. Soc. London. Ser. B, Biol. Sci. 311, 617–631. doi:10.1098/rstb.1985.0168

Sies, H., Jones, D.P., 2020. Reactive oxygen species (ROS) as pleiotropic physiological signalling agents. Nat. Rev. Mol. Cell Biol. 21, 363–383. doi:10.1038/s41580-020-0230-3

Simonsen, A., Cumming, R.C., Brech, A., et al., 2008. Promoting basal levels of autophagy in the nervous system enhances longevity and oxidant resistance in adult Drosophila. Autophagy 4, 176–184. doi:10.4161/auto.5269

Sliter, D.A., Martinez, J., Hao, L., et al., 2018. Parkin and PINK1 mitigate STING-induced inflammation. Nature 561, 258–262. doi:10.1038/s41586-018-0448-9

Taneike, M., Yamaguchi, O., Nakai, A., et al., 2010. Inhibition of autophagy in the heart induces age-related cardiomyopathy. Autophagy 6, 600–606. doi:10.4161/auto.6.5.11947

Tanida, I., Ueno, T., Kominami, E., 2004. LC3 conjugation system in mammalian autophagy. Int. J. Biochem. Cell Biol. 36, 2503–2518. doi:10.1016/j.biocel.2004.05.009

Terman, A., 1995. The effect of age on formation and elimination of autophagic vacuoles in mouse hepatocytes. Gerontology 41, 319–325. doi:10.1159/000213753

Tonelli, C., Chio, I.I.C., Tuveson, D.A., 2018. Transcriptional regulation by Nrf2. Antioxidants Redox Signal. doi:10.1089/ars.2017.7342

Trist, B.G., Hare, D.J., Double, K.L., 2019. Oxidative stress in the aging substantia nigra and the etiology of Parkinson's disease. Aging Cell 18, e13031. doi:10.1111/acel.13031

Ungvari, Z., Tarantini, S., Nyúl-Tóth, Á., et al., 2019. Nrf2 dysfunction and impaired cellular resilience to oxidative stressors in the aged vasculature: from increased cellular senescence to the pathogenesis of age-related vascular diseases. GeroScience 41, 727–738. doi:10.1007/s11357-019-00107-w

Veal, E., Jackson, T., Latimer, H., 2018. Role/s of 'antioxidant' enzymes in ageing. Subcell. Biochem. 90, 425–450. doi:10.1007/978-981-13-2835-0_14

Watanabe, R., Fujii, H., Shirai, T., et al., 2014. Autophagy plays a protective role as an anti-oxidant system in human T cells and represents a novel strategy for induction of T-cell apoptosis. Eur. J. Immunol. 44, 2508–2520. doi:10.1002/eji.201344248

Yu, L., Wan, F., Dutta, S., et al., 2006. Autophagic programmed cell death by selective catalase degradation. Proc. Natl. Acad. Sci. U. S. A. 103, 4952–4957. doi:10.1073/pnas.0511288103

Zachari, M., Ganley, I.G., 2017. The mammalian ULK1 complex and autophagy initiation. Essays Biochem. doi:10.1042/EBC20170021

Zaffagnini, G., Martens, S., 2016. Mechanisms of Selective Autophagy. J. Mol. Biol. 428, 1714–1724. doi:10.1016/j.jmb.2016.02.004

Zhang, J., Tripathi, D.N., Jing, J., et al., 2015. ATM functions at the peroxisome to induce pexophagy in response to ROS. Nat. Cell Biol. 17, 1259–1269. doi:10.1038/ncb3230

CHAPTER EIGHTEEN

Autophagy in Aging and Longevity Exemplified by the Aging Heart

Steffen Häseli and Christiane Ott

CONTENTS

18.1 INTRODUCTION

Macroautophagy, herein referred to as autophagy, is an either non-selective or cargo-specific mechanism of intracellular protein and organelle quality control. Due to its nutrient-dependent regulation, role in the utilization of cellular energy stores and maintenance of mitochondrial homeostasis, autophagy also contributes to cellular energy balance (Delbridge et al., 2015; Singh and Cuervo, 2011). There is growing evidence across diverse species, including mammals, that autophagy plays a decisive role in the regulation of lifespan and aging, although the underlying mechanisms are not fully understood (Bareja et al., 2019; Hansen et al., 2018).

In their review on the discoveries of aging research, López-Otín et al. described nine hallmarks of aging. In this, deregulated nutrient sensing, loss of proteostasis, mitochondrial dysfunction, altered intercellular communication, cellular senescence, stem-cell exhaustion, epigenetic alterations, genomic instability and telomere attrition are summarized as crucial elements of cellular aging (López-Otín et al., 2013). Remarkably, physiological or moderately elevated levels of autophagy may

counteract almost all hallmarks of aging (Barbosa et al., 2018; Bareja et al., 2019). Yet, there is insufficient evidence to determine whether selective activation or correction of autophagy in humans is geroprotective. Nevertheless, an increasing number of studies suggest that autophagy stimulation in selected tissues may be beneficial in the extension of health- and lifespan.

18.2 CARDIAC MANIFESTATION OF THE AGING PROCESS

Cardiovascular complications are the leading cause of death from noncommunicable diseases worldwide (World Health Organization, 2020). With a rapid onset of cardiac mortality over the age of 60, aging is considered the leading risk factor of cardiovascular death (McAloon et al., 2016; World Health Organization, 2016). High blood pressure, obesity and prevalence of metabolic abnormalities increase with age, facilitate the development of heart disease and are causally linked with premature cardiac aging (Sun et al., 2019). Still, physiological aging of the heart can be

DOI: 10.1201/9781003048138-18

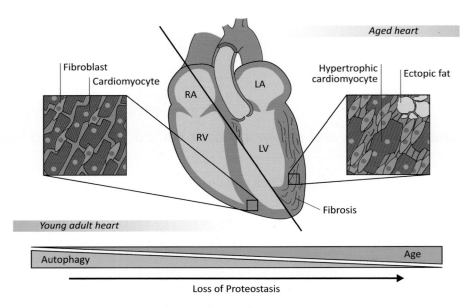

Figure 18.1 **Cardiac aging is accompanied by autophagic functional decline.** In the heart, the aging process primarily manifests as left-ventricular hypertrophy and fibrosis of the myocardium. On a cellular level, age-dependent structural changes of the heart relate to an increase in cardiomyocyte cell size with a loss of overall myocyte cell number, proliferation of cardiac fibroblasts, remodulation of extracellular matrix and assembly of ectopic fat. Autophagy is downregulated in the aging heart and contributes to an aging-dependent loss of proteostasis in the cardiac system. *Abbreviations:* RA, Right atrium; RV, right ventricle; LA, left atrium; LV, left ventricle.

seen as an intrinsic process on the basis of molecular and cellular changes that result in impaired organ function and prompt the development of cardiac diseases in the elderly (Figure 18.1).

On a functional level, the aging heart is characterized by a gradual decline in cardiac performance. Primary feature of age-related cardiac compromise is a restriction of left-ventricular diastolic function, which, under physical exercise, extends to a reduction in heart rate and systolic cardiac reserve (Nakou et al., 2016; Obas and Vasan, 2018). Structurally, the aging heart manifests an asymmetric left-ventricular hypertrophy accompanied by fibrosis of the myocardium (Strait and Lakatta, 2012). At cellular level, an aging-related proliferation of cardiac fibroblasts, a loss of cardiomyocytes in the working myocardium as well as a decline in the sinoatrial pacemaker cells lower the threshold for cardiac functional failure (Keller and Howlett, 2016; Steenman and Lande, 2017). Aging cardiomyocytes progressively lose their structural and functional properties (Bernhard and Laufer, 2008). Among other attributes, aged cardiomyocytes display increased cell size, accumulation of misfolded proteins, dysregulated calcium homeostasis, reorganization of the contractile

apparatus, decreased intercellular coupling, mitochondrial dysfunction, markers of cellular senescence, increased apoptosis and necrotic cell death (Bernhard and Laufer, 2008; Keller and Howlett, 2016; Shirakabe et al., 2016). Additionally, a growing pro-oxidative and pro-inflammatory milieu increases the susceptibility of the aging heart for cardiac pathophysiology (Obas and Vasan, 2018).

Recent discoveries in the field give interesting insides into possible causal links between molecular pathways of cardiac aging and autophagy. Being reliant on the comparison of autophagy-related proteins in presence and absence of autophagosomal or lysosomal inhibitors (Klionsky et al., 2016), the measurement of autophagic flux is an intricate challenge to be achieved in vivo or in dissected samples. Therefore, the majority of studies on autophagy in cardiac aging focus on animal-based data. Age-associated changes in mouse and rat cardiac physiology closely recapitulate the human aging heart (Dai and Rabinovitch, 2009; Keller and Howlett, 2016). Still, since rodents have a unique cardiovascular physiology, including differences in dynamics of the cardiac cycle, lack of atherosclerotic narrowing of the vasculature during aging, and distinct expression patterns of ion channels as well as isoforms of

PROTEOSTASIS AND PROTEOLYSIS

contractile proteins (Riehle and Bauersachs, 2019), caution should be taken when transferring recited results to human cardiology.

18.3 AUTOPHAGY IS DOWNREGULATED IN THE AGING HEART

Autophagy is typically referred to as a protective mechanism, induced in conditions of metabolic, oxidative, proteotoxic and genotoxic stress (Parzych and Klionsky, 2014). A time-dependent accumulation of cellular damage is detrimental to biological function and considered to have a crucial role in the aging process (López-Otín et al., 2013). Cardiomyocytes are generally classified as post-mitotic, long-lived cells (Terman and Brunk, 2005), even though, cardiomyocyte regeneration does take place in the adult heart (Bergmann et al., 2009, 2015). Still, turnover of cardiomyocytes declines with aging (Bergmann et al., 2015) and cardiomyocyte renewal seems insufficient to offset the impact of age-dependent accumulation of cytotoxic material (Terman and Brunk, 2005). Therefore, cardiac homeostasis is particularly reliant on cellular mechanisms of organelle and protein quality control (Yamaguchi, 2019). It is assumed that autophagy plays a fundamental role in the maintenance of heart homeostasis even under physiological conditions.

Gene-targeting of cardiac autophagy has revealed a protective role of autophagic quality control against premature cardiac aging. ATG5 is an essential protein in canonical phagophore expansion and autophagosome formation (Dikic and Elazar, 2018). In mice, cardiac-specific knockout of *Atg5* early in embryogenesis did not alter cardiac structure or function in 10-week-old animals (Nakai et al., 2007) and cardiac phenotype was inconspicuous until 3 months of age (Taneike et al., 2010). In older animals however, cardiac *Atg5* deficiency led to premature cardiac dysfunction and aging-related cardiomyopathy (Taneike et al., 2010). Hearts lacking *Atg5* showed dilated chambers, fibrosis, decreased systolic and diastolic dimensions, and impaired heart contractility as measured by a decline in diastolic to systolic diameter. Cardiomyocyte hypertrophy was accompanied by an increase in polyubiquitinated proteins, markers of oxidative stress, aggregation of abnormal mitochondria and reduced respiratory function (Taneike et al., 2010). Mice deficient in cardiac *Atg5* developed heart failure at 6 months of age and showed an overall decreased lifespan (Taneike et al., 2010). When cardiac *Atg5* was knocked out late in adulthood, mice developed a fast onset cardiac hypertrophy associated with functional decline and heart failure (Nakai et al., 2007).

Different interventions that led to an impairment of autophagy in the heart showed similar results even though the setting was more complex and mechanisms other than autophagy could be accountable for the observed effects. Transgenic mice with overexpressed mitochondrial aldehyde dehydrogenase (Aldh2) showed suppressed cardiac autophagy accompanied by cardiac hypertrophy, decline in heart function and overall accelerated cardiac aging (Zhang et al., 2017). Despite favorable effects on aging-related cardiac inflammation, knockout of macrophage migration inhibitory factor (MIF) accelerated murine cardiac aging with an exacerbated reduction in cardiac autophagy and premature adverse changes in cardiac morphology and function (Xu et al., 2016). Cardiac overexpression of miR-199a in transgenic mice inhibited cardiac autophagy, led to cardiac hypertrophy in 3-month-old animals, fibrosis, decline in left-ventricular contractile function and heart failure at 7 months of age (Li et al., 2017b).

On the other hand, stimulation of autophagy in the heart has been shown to be beneficial for cardiac aging. Knock-in of a single point mutated form of the BECN1 gene (BECN1[F121A]) was utilized to disrupt BECN1-BCL-2 autophagy regulatory complex in mice (Fernandez et al., 2018). BECN1[F121A] was associated with increased cardiac autophagic flux, decreased aging-related cardiomyocyte apoptosis, attenuated cardiac hypertrophy and less cardiac fibrosis, cumulatively delaying cardiac aging (Fernandez et al., 2018). Administration of spermidine, a natural polyamine and potent autophagy activator, provided cardioprotection in old mice (Eisenberg et al., 2016). In concomitance with induced cardiomyocyte autophagic flux, oral supplementation of spermidine reduced hypertrophy in aged hearts, improved diastolic function, preserved cardiomyocyte mechano-elastical properties and negated an aging-related decline in mitochondrial respiratory function (Eisenberg et al., 2016). Gene knockout of *Atg5* nullified spermidine effects, confirming the underlying role of autophagy (Eisenberg et al., 2016). It should be noted that both studies above reported results of systemic interventions with an overall extension of lifespan as primary outcome (Eisenberg et al.,

2016; Fernandez et al., 2018). Exclusive activation of autophagy in the cardiac system has been proven to be an experimental challenge and is subject of ongoing research. For spermidine, selective effects on cardiomyocyte function via polyamine importers in concert with electrochemical gradients are discussed (Nilsson and Persson, 2018). Interestingly, dietary spermidine uptake was correlated with attenuated blood pressure and reduced incidence of cardiovascular diseases, including heart failure, in humans (Eisenberg et al., 2016; Madeo et al., 2018). This indicates an attractive transferability of cardioprotection through autophagy from animal models to human physiology. In a prospective population-based study with 829 participants, the same research group showed that a diet rich in spermidine was linked to reduced cardiovascular and all-cause mortality in humans (Kiechl et al., 2018). Collectively, available literature indicates constitutive autophagy to be of central importance in the regulation of cardiomyocyte cell size, mitochondrial metabolic function and proteostasis under baseline conditions, thereby maintaining global cardiac function during aging.

During lifespan an age-dependent reduction of autophagic flux in the cardiac system is assumed (Ghosh et al., 2020), although this presumption mainly derives from studies in mice (Li et al., 2017a; Liang et al., 2020; Ren et al., 2017; Taneike et al., 2010; Zhang et al., 2017) and sufficient evidence for the human heart is still missing. In a human cohort for heart and aging research, a genome-wide association study found a single nucleotide polymorphism near *Atg4C* to be associated with major aging-related events, including heart failure and myocardial infarction (Walter et al., 2011). Recently, sequencing analysis of atrial appendages from patients of age 40–70 years identified over 40 modified miRNA–mRNA-mediated pathways of which autophagy showed the second most prevalent change during aging (Yao et al., 2019).

Studies on autophagy generally compare young versus old individuals and it is mostly unclear when or under which stimuli autophagy changes in the progression of cardiac aging. Molecular changes that contribute to a reduction of autophagy in the aging heart appear to be complex and not fully understood. An open hypothesis assumes that chronic activation of autophagy via a consistent or increasing stress-response leads to exhaustion of the autophagy-lysosomal pathway (ALP) (Shirakabe et al., 2016). This idea is based on the observation that reactive oxygen species (ROS) (Scherz-Shouval and Elazar, 2011) and protein aggregates (Lamark and Johansen, 2012) can induce autophagy in a compensatory manner, yet decline in autophagic flux of aged hearts is associated with dysfunctional mitochondria, high levels of oxidative stress and accumulation of misfolded proteins (Abdellatif et al., 2018). Negative regulation of autophagy under chronic distress could be the result of upstream signaling pathways that become altered during aging (Abdellatif et al., 2018). Age-related changes in autophagy regulatory hubs AMPK and mTORC1 have been reported to reduce overall cardiac stress resistance (Gonzalez et al., 2004; Yang and Ming, 2012).

18.3.1 Impairment of Autophagic Regulatory Signaling During Cardiac Aging

18.3.1.1 Impact of mTOR Signaling on Autophagy in the Aging Heart

The multiprotein complex mTORC1 is a major regulator of cellular energy status, responsible for nutrient balance and redox homeostasis. The signaling network surrounding mTORC1 has been implicated in the regulation of the aging process (Papadopoli et al., 2019). In mice, systemic inhibition of mTORC1 by rapamycin administration extended life span (Harrison et al., 2009); slowed diverse phenotypes of biological aging (Wilkinson et al., 2012); inhibited hepatocyte mitochondrial ROS production, lipoxidation, and lipofuscin accumulation; and was associated with an overall increase in Atg13 levels in the liver of old mice (Martinez-Cisuelo et al., 2016). Moreover, rapamycin treatment attenuated age-related cardiac dysfunction in association with decreased cardiac hypertrophy, inflammation, and improved mitochondrial respiratory efficiency in mice (Flynn et al., 2013; Quarles et al., 2020). In contrast, augmentation of Akt/mTORC1 pathway either by cardiac-specific overexpression of Akt itself (Hua et al., 2011) or miR-199a, which inhibited glycogen synthase kinase 3β (GSK3β) and thereby activated mTOR signaling (Li et al., 2017b), led to accentuated cardiac morphological changes and contractile dysfunction accompanied by mitigated autophagy in aged murine hearts (Hua et al., 2011; Li et al., 2017b).

Whether mTORC1 activity increases with age in the heart is still under debate. Some studies report an aging-related decrease in phosphorylated mTORC1 inhibitory constituent tuberous sclerosis complex (TSC) 2 together with increased phosphorylation of kinase mTOR in mouse heart (Hua et al., 2011; Zhang et al., 2017). Others could not detect significant changes in the phosphorylation status of mTOR or its targets p70S6k and Ulk1 during cardiac aging in mice (Liang et al., 2020). The reports on age-related changes in cardiac mTOR activity seem to depend significantly on the examined experimental setting (Baar et al., 2016). New insights into the molecular mechanisms surrounding mTORC1-related regulation of cardiac aging suggest that mTOR itself is finely tuned by input of upstream regulatory signaling cascades (see below).

Pro-inflammatory signaling by the NLR family pyrin domain containing 3 (NLRP3) inflammasome has been linked to various cardiovascular diseases, including myocardial infarction and congestive heart failure, although little is known about its role in physiological cardiac aging (Liu et al., 2018). In association with overall increased life span, NLRP3$^{-/-}$ knockout mice showed a reduced phosphorylation of cardiac mTOR that further declined during aging and was linked to an improved autophagic activity together with preserved cardiac integrity in old mice (Marin-Aguilar et al., 2020). In addition to NLRP3, the transforming growth factor beta (TGF-β) signaling pathway was implicated in inflammation-mediated

suppression of autophagy in aging hearts. Cardiac-specific impairment of the TGF-β-INHB/activin pathway improved mTOR complex 2 (mTORC2) signaling in D. melanogaster, activated autophagy and preserved cardiac function at advanced age (Chang et al., 2020). Rho-associated coiled-coil-containing protein kinases (ROCK) are regulators of cellular actin network and associated with adverse organization of cardiomyocyte non-sarcomeric cytoskeleton (Lai et al., 2017). Cardiomyocyte-specific ROCK1$^{-/-}$ and ROCK2$^{-/-}$ double knockout mice displayed reduced Akt/mTOR/Ulk1 signaling that persisted throughout aging, increased aging-related as well as starvation-induced autophagy and decreased cardiac fibrosis (Shi et al., 2019). Altogether, these studies suggest mTORC1 to be a regulatory switch in cardiomyocyte aging that incorporates changes in energy and nutrient supply, oxidative stress, inflammation and mechanical load, thereupon negatively regulating autophagy in the aging heart and ultimately contributing to cardiac hypertrophy and dysfunction.

18.3.1.2 Impact of AMPK Signaling on Autophagy in the Aging Heart

As another central sensor of cellular energy status, AMPK is activated at high AMP/ATP ratios and antagonizes mTOR activity (Figure 18.2). By

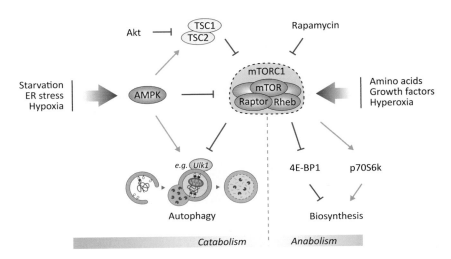

Figure 18.2 **AMPK- and mTOR-dependent regulation of autophagy.** Under starvation, ER and oxidative stress, as well as hypoxia, AMPK positively regulates catabolic processes, including autophagy. AMPK directly phosphorylates subunits of the multiprotein complex mTORC1 and activates the inhibitory TSC, collectively suppressing mTORC1 signaling. In situations of sufficient nutrient, growth factor and oxygen supply, mTORC1 upregulates anabolism, including protein biosynthesis and overall cell growth, simultaneously blocking autophagic catabolism. The immunosuppressant rapamycin blocks mTORC1 signaling through subunit raptor.

positively regulating responses to oxidative stress, inflammation, pro-apoptotic and -autophagic signaling, AMPK is thought to play a protective role in heart homeostasis (Li et al., 2020). Consensus was reached that AMPK activity is downregulated in the heart of aging mice (Turdi et al., 2010; Zhang et al., 2017). In elderly patients, cardioprotective AMPK activation was restricted in association with increased susceptibility to myocardial ischemia (Ma et al., 2010).

Genetic inactivation of AMPK enzymatic activity exacerbated cardiomyocyte contractile decline, loss of calcium homeostasis, hypertrophy, mitochondrial dysfunction and oxidative stress in aged murine hearts (Turdi et al., 2010). Additionally, short-term treatment of old mice with metformin, a well-established activator of AMPK, improved cardiomyocyte contractility, supporting the cardioprotective role of AMPK (Turdi et al., 2010). By suppressing a nuclear E3 ubiquitin ligase, AMPK signaling regulates the stability of coactivator-associated arginine methyltransferase 1 (CARM1), a crucial epigenetic activator of autophagy-related genes (Shin et al., 2016). In mice, a reduction of fasting-induced autophagy in aged hearts was directly linked to decreased CARM1 stability and pharmacological inhibition of AMPK in young mice reflected the limited autophagic response of old hearts. AMPK activation in old mice restored cardiac CARM1 levels and rejuvenated fasting-induced autophagic regulation (Li et al., 2017a). Double knockout *AMPK subunit* α *(AMPKα)2*$^{-/-}$ and *Akt2*$^{-/-}$ mice displayed premature aging-related changes in markers of cardiac autophagy, which predisposed hearts to accelerated cardiac hypertrophy, fibrosis and contractile decline during aging (S. Wang et al., 2019a). Evidently, AMPK signaling protects cardiac homeostasis and counteracting reduced AMPK activity in the aging heart is an attractive target to benefit cardiac autophagy and function in the elderly.

18.3.1.3 Impact of Oxidative Stress on Autophagy in the Aging Heart

In the heart, mitochondria are the primary source of ROS (Dey et al., 2018) and mitochondria-derived ROS accumulate during cardiac aging (Martin-Fernandez and Gredilla, 2016). Converging on diverse signaling cascades, including the mTOR and AMPK pathway, mild oxidative stress is thought to primarily activate selective forms of autophagy while higher oxidative stress triggers bulk, non-selective autophagy-lysosomal degradation (Scherz-Shouval et al., 2011; Sies and Jones, 2020). In the heart specifically, oxidative stress seems to regulate autophagy in a dose-, time- and context-dependent manner. For example, *in vivo* cardiac ischemia decreased ATP levels in mouse hearts, leading to severe energy stress, enhanced ROS production, increased AMPK signaling and autophagy induction (Matsui et al., 2007). Heightened autophagy activation persisted during subsequent reperfusion, a phase mainly characterized by an imbalance between excessive ROS production and antioxidative capacity (Bagheri et al., 2016), but was marked by a decline in AMPK activity and upregulation of BECN1, opposite to the ischemic phase (Matsui et al., 2007). This suggests ROS-mediated regulation of cardiac autophagy to act through AMPK-dependent and -independent mechanisms, functionally linking cardiac oxidative stress and metabolism.

In murine cardiomyocytes, both mTOR and AMPK activities were shown to be negatively regulated under oxidative stress (Oka et al., 2017; Shao et al., 2014). Hydrogen peroxide treatment of cultured primary cardiomyocytes induced mTOR disulfide bonds, inhibiting mTOR kinase activity as indicated by the reduced phosphorylation of mTOR substrates p70S6k and 4E-BP1 (Oka et al., 2017). Similarly, endogenous oxidation of redox-sensitive cysteine residues led to aggregation of cardiomyocytic AMPK through intermolecular disulfide bonds, thereby preventing the interaction of AMPK with upstream kinases and inhibiting its phosphorylation and subsequent activation (Shao et al., 2014). Low-dose, short-term treatments of cardiomyocytes with hydrogen peroxide tended to increase mTORC1 activity, while high-dose, long-term treatments significantly decreased it (Oka et al., 2017). *In vivo*, endogenous oxidoreductase thioredoxin-1 (Trx1) was shown to reduce oxidized cystine disulfide bonds for both mTOR and AMPK in the heart, maintaining their enzymatic function (Oka et al., 2017; Shao et al., 2014). Interestingly, others reported redox modulation of AMPK activity, not to be regulated through direct oxidation of AMPK itself but as a secondary response of redox effects on mitochondrial respiration and decreased energy generation, ultimately increasing AMPK signaling (Hinchy et al., 2018).

Normally, oxidative signaling promotes autophagy in a ROS-dependent manner in order to remove damaged cellular structures, mainly dysfunctional mitochondria, herein mitigating ROS production and protecting cells from pathological oxidant levels (Guo et al., 2020). In the aging heart, chronic pro-autophagic signaling caused by the accumulation of ROS may dampen the autophagic response as part of a negative feedback loop. In fact, *in vitro* and *in vivo* data from cultured rat aortic smooth muscle cells and mouse aorta, respectively, revealed Atg3 and Atg7 to be inhibited by ROS-mediated oxidation of catalytic thiol residues, effectively preventing LC3 lipidation (Frudd et al., 2018). It was shown that an oxidizing environment led to hyperoxidation of Atg3 and Atg7 in aged tissue, inhibiting starvation-induced LC3 lipidation resulting in suppressed autophagosome formation (Frudd et al., 2018). Moreover, mammalian Atg4 cleavage activity was also reported to be inhibited through oxidation of a catalytic cysteine site via redox signaling (Scherz-Shouval et al., 2007). By priming the LC3 precursor for lipidation at the growing autophagic phagophore as well as mediating the delipidation step, Atg4 fulfills a bilateral function in the autophagic process (Klionsky et al., 2016). Interestingly, oxidation of Atg4 caused by starvation-induced rise of hydrogen peroxide resulted in the accumulation of lipidated LC3, altogether increasing autophagic clearance capacity (Scherz-Shouval et al., 2007).

Collectively, these studies suggest cardiac autophagy to be finely balanced based on the cellular redox and energy status. On diverse levels, including signaling pathways upstream mTOR and AMPK as well as direct modulation of integral Atg proteins, ROS-mediated protein modification can impact autophagy either in a positive or negative manner, depending on the dose, duration and context of oxidative stress. Furthermore, loss of redox homeostasis, as observed in the aging heart, is assumed to impair this regulatory network in favor of autophagy suppression.

18.3.2 Mitophagy and the Aging Heart

The development of mitochondrial abnormalities is a prominent and well-described pathological feature of cardiac aging (Tocchi et al., 2015; Y. Wang et al., 2019b). Making up about one-third of cardiomyocyte cellular volume (Schaper et al., 1985), mitochondria are of central importance to cardiac physiology and the heart is particularly susceptible to mitochondrial dysfunction (Tocchi et al., 2015). By oxidizing mitochondrial DNA, proteins and lipids, excessive formation of ROS disrupts mitochondrial function, ultimately impairing respiration-related electron transport and further increasing ROS production (Martin-Fernandez and Gredilla, 2016). The mitochondrial free radical theory of aging postulates this vicious feed-forward loop of ROS-induced ROS production to be an integral part of biological aging (Harman, 1992).

In addition to specialized forms of unfolded protein response, proteolysis by mitochondrial proteases, and degradation via the ubiquitin-proteasomal pathway, the quality of mitochondrial proteins is controlled by a mitochondria-specific form of autophagy, termed mitophagy (Pickles et al., 2018). In the canonical mitophagic pathway, upon mitochondrial depolarization or accumulation of misfolded mitochondrial proteins, steady-state import of phosphatase and tensin homolog (PTEN)-induced kinase 1 (PINK1) into the inner mitochondrial membrane is compromised, which leads to its enrichment at the outer mitochondrial membrane (Kondapalli et al., 2012; Narendra et al., 2010). PINK1 phosphorylates preexisting ubiquitin residues on the mitochondrial surface and thereby recruits E3 ubiquitin-protein ligase Parkin for the polyubiquitination of diverse mitochondrial proteins, making them interaction sites for autophagic adaptors (Gladkova et al., 2018; Ordureau et al., 2015). Herein, cargo-specific machineries selectively label damaged subdomains of the mitochondrial network for autophagosome incorporation and subsequent degradation via the ALP (Figure 18.3). Mitochondrial quality control that is mediated by mitophagy is compromised in the aging heart (Hoshino et al., 2013; Ren et al., 2017; Sun et al., 2015; Terman et al., 2010).

Restriction of mitophagy observed in aged cardiac tissue is accompanied by loss of mitochondrial integrity (Hoshino et al., 2013), a phenotype which can be reproduced by targeted modulation of central control variables of the mitophagic pathway. Systemic knockout of PINK1$^{-/-}$ in mice showed an early onset of cardiac hypertrophy and left ventricular dysfunction, characterized by large, functionally impaired mitochondria as well as increased ROS production in cardiomyocytes (Billia et al., 2011; Siddall et al., 2013). On the other hand, *Parkin*$^{-/-}$ knockout mice developed a less pronounced cardiac phenotype, accompanied by smaller, disorganized,

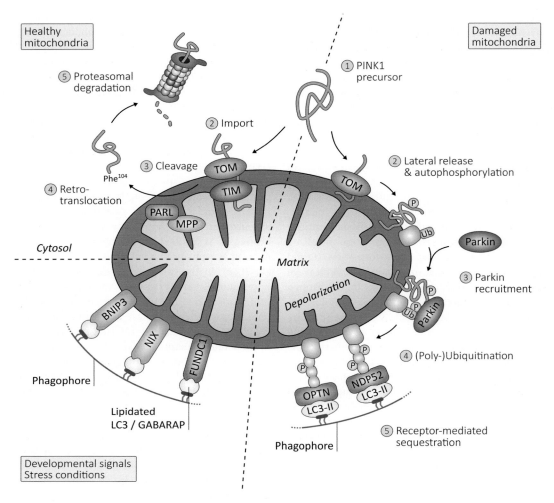

Figure 18.3 Principles of mitophagy, a mitochondria-specific form of autophagy. Under steady-state conditions, phosphatase and tensin homolog (PTEN)-induced kinase 1 (PINK1) is transferred into the inner mitochondrial membrane, where it is constitutively cleaved by mitochondrial proteases and transported back to the cytosol for N-end rule mediated degradation via the proteasome. Upon mitochondrial damage steady-state import and cleavage of PINK1 is compromised, leading to its enrichment on the outer mitochondrial membrane. By phosphorylating preexisting ubiquitin residues on the mitochondrial surface, PINK1 recruits E3 ubiquitin-protein ligase Parkin and promotes the polyubiquitination of diverse mitochondrial proteins. Polyubiquitinated mitochondria are recognized by autophagic adaptor proteins, e.g. nuclear dot protein 52 (NDP52) and optineurin (OPTN), linking damaged mitochondrial domains to lipidated LC3-II or GABARAP at the growing phagophore. BCL-2 interacting protein 3 (BNIP3), BNIP3-like (BNIP3L)/NIX and FUN14 domain containing 1 (FUNDC1) are mitochondria-anchored interaction partners for LC3 and GABARAP, directly targeting mitochondria for degradation. *Abbreviations:* MPP, Mitochondrial processing peptidase; PARL, presenilin associated rhomboid like protease; TIM, translocase of the inner membrane; TOM, translocase of the outer membrane.

yet functionally unaffected mitochondrial networks (Kubli et al., 2013), although these mice accumulated mitochondrial anomalies later in life and showed premature cardiac aging (Hoshino et al., 2013). As expected, transgenic mice with increased Parkin expression were resistant to cardiac aging (Hoshino et al., 2013).

Destruction of mitochondria by severe damage or following autophagy-lysosomal sequestration results in the release of mitochondrial content. Through mitophagy, mitochondrial peroxides and catalytic iron are assumed to promote crosslinking of long-lived, intra-lysosomal proteins, resulting in the formation of protein aggregates, such as the autofluorescent age-pigment lipofuscin, a non-degradable material (Terman et al., 2010). Accumulation of lipofuscin during cardiac aging impairs lysosomal function and hence inhibits autophagic turnover (Terman et al., 2010). Failure to remove damaged mitochondrial is likely to contribute to a progressive loss of proteostasis in the aging heart.

18.4 CONCLUSIONS

Autophagy is positively linked to the regulation of lifespan and aging. However, autophagic turnover is suppressed in the aging heart, which is associated with age-related changes in cardiac micromilieu, effecting among others mTOR and AMPK signaling, deregulated ROS-mediated autophagy modulation and accumulation of non-degradable protein aggregates, such as lipofuscin, collectively exhausting the ALP. Studies on autophagy in cardiac aging generally compare young and old individuals and it is unclear at which age autophagy becomes altered in the progression of human heart aging. Information for factors that initiate a decline in proteolytic activity is still missing. The rate of autophagy functional decline could be linked to age-related cardiac comorbidities, such as hypertension or diabetes, and delineation of such factors would simplify the use of autophagy in interventional approaches in the elderly. Moreover, since there are experimental difficulties to measure autophagic turnover in the human heart in vivo, our knowledge so far on the alteration of autophagy in the aging heart is based on preclinical models. Analyses in mice linked downregulation of autophagy in the hematopoietic system to premature aging of non-hematopoietic organs, including the heart (Fang et al., 2019). Therefore, early approaches aimed to measure autophagic flux in human blood cells are expected to provide exciting insights into the field, although such methods still lack validity and efforts are ongoing to improve these measurements.

REFERENCES

Abdellatif, M., Sedej, S., Carmona-Gutierrez, D., et al., 2018. Autophagy in cardiovascular aging. Circ Res 123 (7), 803–824. doi:10.1161/CIRCRESAHA.118.312208

Baar, E. L., Carbajal, K. A., Ong, I. M., et al., 2016. Sex- and tissue-specific changes in mTOR signaling with age in C57BL/6J mice. Aging Cell 15 (1), 155–166. doi:10.1111/acel.12425

Bagheri, F., Khori, V., Alizadeh, A. M., et al., 2016. Reactive oxygen species-mediated cardiac-reperfusion injury: mechanisms and therapies. Life Sci 165, 43–55. doi:10.1016/j.lfs.2016.09.013

Barbosa, M. C., Grosso, R. A., Fader, C. M., 2018. Hallmarks of aging: an autophagic perspective. Front Endocrinol (Lausanne) 9, 790. doi:10.3389/fendo.2018.00790

Bareja, A., Lee, D. E., White, J. P., 2019. Maximizing longevity and healthspan: multiple approaches all converging on autophagy. Front Cell Dev Biol 7, 183. doi:10.3389/fcell.2019.00183

Bergmann, O., Bhardwaj, R. D., Bernard, S., et al., 2009. Evidence for cardiomyocyte renewal in humans. Science 324 (5923), 98–102. doi:10.1126/science.1164680

Bergmann, O., Zdunek, S., Felker, A., et al., 2015. Dynamics of cell generation and turnover in the human heart. Cell 161 (7), 1566–1575. doi:10.1016/j.cell.2015.05.026

Bernhard, D., Laufer, G., 2008. The aging cardiomyocyte: a mini-review. Gerontology 54 (1), 24–31. doi:10.1159/000113503

Billia, F., Hauck, L., Konecny, F., et al., 2011. PTEN-inducible kinase 1 (PINK1)/Park6 is indispensable for normal heart function. Proc Natl Acad Sci U S A 108 (23), 9572–9577. doi:10.1073/pnas.1106291108

Chang, K., Kang, P., Liu, Y., et al., 2020. TGFB-INHB/activin signaling regulates age-dependent autophagy and cardiac health through inhibition of MTORC2. Autophagy 16 (10), 1807–1822. doi:10.1080/15548627.2019.1704117

Dai, D. F., Rabinovitch, P. S., 2009. Cardiac aging in mice and humans: the role of mitochondrial oxidative stress. Trends Cardiovasc Med 19 (7), 213–220. doi:10.1016/j.tcm.2009.12.004

Delbridge, L. M., Mellor, K. M., Taylor, D. J., et al., 2015. Myocardial autophagic energy stress responses – Macroautophagy, mitophagy, and glycophagy. Am J Physiol Heart Circ Physiol 308 (10), H1194–H1204. doi:10.1152/ajpheart.00002.2015

Dey, S., DeMazumder, D., Sidor, A., et al., 2018. Mitochondrial ROS drive sudden cardiac death and chronic proteome remodeling in heart failure. Circ Res 123 (3), 356–371. doi:10.1161/CIRCRESAHA.118.312708

Dikic, I., Elazar, Z., 2018. Mechanism and medical implications of mammalian autophagy. Nat Rev Mol Cell Biol 19 (6), 349–364. doi:10.1038/s41580-018-0003-4

Eisenberg, T., Abdellatif, M., Schroeder, S., et al., 2016. Cardioprotection and lifespan extension by the natural polyamine spermidine. Nat Med 22 (12), 1428–1438. doi:10.1038/nm.4222

Fang, Y., Zhu, L., An, N., et al., 2019. Blood autophagy defect causes accelerated non-hematopoietic organ aging. Aging (Albany, NY) 11 (14), 4910–4922. doi:10.18632/aging.102086

Fernandez, A. F., Sebti, S., Wei, Y., et al., 2018. Disruption of the beclin 1-BCL2 autophagy regulatory complex promotes longevity in mice. Nature 558 (7708), 136–140. doi:10.1038/s41586-018-0162-7

Flynn, J. M., O'Leary, M. N., Zambataro, C. A., et al., 2013. Late-life rapamycin treatment reverses age-related heart dysfunction. Aging Cell 12 (5), 851–862. doi:10.1111/acel.12109

Frudd, K., Burgoyne, T., Burgoyne, J. R., 2018. Oxidation of Atg3 and Atg7 mediates inhibition of autophagy. Nat Commun 9 (1), 95. doi:10.1038/s41467-017-02352-z

Ghosh, R., Vinod, V., Symons, J. D., et al., 2020. Protein and mitochondria quality control mechanisms and cardiac aging. Cells 9 (4). doi:10.3390/cells9040933

Gladkova, C., Maslen, S. L., Skehel, J. M., et al., 2018. Mechanism of Parkin activation by PINK1. Nature 559 (7714), 410–414. doi:10.1038/s41586-018-0224-x

Gonzalez, A. A., Kumar, R., Mulligan, J. D., et al., 2004. Effects of aging on cardiac and skeletal muscle AMPK activity: basal activity, allosteric activation, and response to in vivo hypoxemia in mice. Am J Physiol Regul Integr Comp Physiol 287 (5), R1270–R1275. doi:10.1152/ajpregu.00409.2004

Guo, Q. Q., Wang, S. S., Zhang, S. S., et al., 2020. ATM-CHK2-beclin 1 axis promotes autophagy to maintain ROS homeostasis under oxidative stress. EMBO J 39 (10), e103111. doi:10.15252/embj.2019103111

Hansen, M., Rubinsztein, D. C., Walker, D. W., 2018. Autophagy as a promoter of longevity: insights from model organisms. Nat Rev Mol Cell Biol 19 (9), 579–593. doi:10.1038/s41580-018-0033-y

Harman, D., 1992. Free-radical theory of aging. Mut Res 275 (3–6), 257–266. doi:10.1016/0921-8734(92)90030-S

Harrison, D. E., Strong, R., Sharp, Z. D., et al., 2009. Rapamycin fed late in life extends lifespan in genetically heterogeneous mice. Nature 460 (7253), 392–395. doi:10.1038/nature08221

Hinchy, E. C., Gruszczyk, A. V., Willows, R., et al., 2018. Mitochondria-derived ROS activate AMP-activated protein kinase (AMPK) indirectly. J Biol Chem 293 (44), 17208–17217. doi:10.1074/jbc.RA118.002579

Hoshino, A., Mita, Y., Okawa, Y., et al., 2013. Cytosolic p53 inhibits Parkin-mediated mitophagy and promotes mitochondrial dysfunction in the mouse heart. Nat Commun 4, 2308. doi:10.1038/ncomms3308

Hua, Y., Zhang, Y., Ceylan-Isik, A. F., et al., 2011. Chronic Akt activation accentuates aging-induced cardiac hypertrophy and myocardial contractile dysfunction: role of autophagy. Basic Res Cardiol 106 (6), 1173–1191. doi:10.1007/s00395-011-0222-8

Keller, K. M., Howlett, S. E., 2016. Sex differences in the biology and pathology of the aging heart. Can J Cardiol 32 (9), 1065–1073. doi:10.1016/j.cjca.2016.03.017

Kiechl, S., Pechlaner, R., Willeit, P., et al., 2018. Higher spermidine intake is linked to lower mortality: a prospective population-based study. Am J Clin Nutr 108 (2), 371–380. doi:10.1093/ajcn/nqy102

Klionsky, D. J., Abdelmohsen, K., Abe, A., et al., 2016. Guidelines for the use and interpretation of assays for monitoring autophagy (3rd edition). Autophagy 12 (1), 1–222. doi:10.1080/15548627.2015.1100356

Kondapalli, C., Kazlauskaite, A., Zhang, N., et al., 2012. PINK1 is activated by mitochondrial membrane potential depolarization and stimulates Parkin E3 ligase activity by phosphorylating Serine 65. Open Biol 2 (5), 120080. doi:10.1098/rsob.120080

Kubli, D. A., Zhang, X., Lee, Y., et al., 2013. Parkin protein deficiency exacerbates cardiac injury and reduces survival following myocardial infarction. J Biol Chem 288 (2), 915–926. doi:10.1074/jbc.M112.411363

Lai, D., Gao, J., Bi, X., et al., 2017. The Rho kinase inhibitor, fasudil, ameliorates diabetes-induced cardiac dysfunction by improving calcium clearance and actin remodeling. J Mol Med (Berl) 95 (2), 155–165. doi:10.1007/s00109-016-1469-1

Lamark, T., Johansen, T., 2012. Aggrephagy: selective disposal of protein aggregates by macroautophagy. Int J Cell Biol 2012, 736905. doi:10.1155/2012/736905

Li, C., Yu, L., Xue, H., et al., 2017a. Nuclear AMPK regulated CARM1 stabilization impacts autophagy in aged heart. Biochem Biophys Res Commun 486 (2), 398–405. doi:10.1016/j.bbrc.2017.03.053

Li, Z., Song, Y., Liu, L., et al., 2017b. miR-199a impairs autophagy and induces cardiac hypertrophy through mTOR activation. Cell Death Differ 24 (7), 1205–1213. doi:10.1038/cdd.2015.95

Li, T., Mu, N., Yin, Y., et al., 2020. Targeting AMP-activated protein kinase in aging-related cardiovascular diseases. Aging Dis 11 (4), 967–977. doi:10.14336/AD.2019.0901

Liang, W., Moyzis, A. G., Lampert, M. A., et al., 2020. Aging is associated with a decline in Atg9b-mediated autophagosome formation and appearance of enlarged mitochondria in the heart. Aging Cell 19 (8), e13187. doi:10.1111/acel.13187

Liu, D., Zeng, X., Li, X., et al., 2018. Role of NLRP3 inflammasome in the pathogenesis of cardiovascular diseases. Basic Res Cardiol 113 (1), 5. doi:10.1007/s00395-017-0663-9

López-Otín, C., Blasco, M. A., Partridge, L., et al., 2013. The hallmarks of aging. Cell 153 (6), 1194–1217. doi:10.1016/j.cell.2013.05.039

Ma, H., Wang, J., Thomas, D. P., et al., 2010. Impaired macrophage migration inhibitory factor-AMP-activated protein kinase activation and ischemic recovery in the senescent heart. Circulation 122 (3), 282–292. doi:10.1161/CIRCULATIONAHA.110.953208

Madeo, F., Eisenberg, T., Pietrocola, F., et al., 2018. Spermidine in health and disease. Science 359 (6374). doi:10.1126/science.aan2788

Marin-Aguilar, F., Lechuga-Vieco, A. V., Alcocer-Gomez, E., et al., 2020. NLRP3 inflammasome suppression improves longevity and prevents cardiac aging in male mice. Aging Cell 19 (1), e13050. doi:10.1111/acel.13050

Martin-Fernandez, B., Gredilla, R., 2016. Mitochondria and oxidative stress in heart aging. Age (Dordr) 38 (4), 225–238. doi:10.1007/s11357-016-9933-y

Martinez-Cisuelo, V., Gomez, J., Garcia-Junceda, I., et al., 2016. Rapamycin reverses age-related increases in mitochondrial ROS production at complex I, oxidative stress, accumulation of mtDNA fragments inside nuclear DNA, and lipofuscin level, and increases autophagy, in the liver of middle-aged mice. Exp Gerontol 83, 130–138. doi:10.1016/j.exger.2016.08.002

Matsui, Y., Takagi, H., Qu, X., et al., 2007. Distinct roles of autophagy in the heart during ischemia and reperfusion: roles of AMP-activated protein kinase and Beclin 1 in mediating autophagy. Circ Res 100 (6), 914–922. doi:10.1161/01.RES.0000261924.76669.36

McAloon, C. J., Osman, F., Glennon, P., et al., 2016. Global epidemiology and incidence of cardiovascular disease, in: Papageorgiou, N. (Ed.), Cardiovascular Disease: Genetic Susceptibility, Enviromental Factors and their Interaction. Academic Press, pp. 57–96. doi:10.1016/b978-0-12-803312-8.00004-5

Nakai, A., Yamaguchi, O., Takeda, T., et al., 2007. The role of autophagy in cardiomyocytes in the basal state and in response to hemodynamic stress. Nat Med 13 (5), 619–624. doi:10.1038/nm1574

Nakou, E. S., Parthenakis, F. I., Kallergis, E. M., et al., 2016. Healthy aging and myocardium: a complicated process with various effects in cardiac structure and physiology. Int J Cardiol 209, 167–175. doi:10.1016/j.ijcard.2016.02.039

Narendra, D. P., Jin, S. M., Tanaka, A., et al., 2010. PINK1 is selectively stabilized on impaired mitochondria to activate Parkin. PLoS Biol 8 (1), e1000298. doi:10.1371/journal.pbio.1000298

Nilsson, B. O., Persson, L., 2018. Beneficial effects of spermidine on cardiovascular health and longevity suggest a cell type-specific import of polyamines by cardiomyocytes. Biochem Soc Trans 47 (1), 265–272. doi:10.1042/BST20180622

Obas, V., Vasan, R. S., 2018. The aging heart. Clin Sci (Lond) 132 (13), 1367–1382. doi:10.1042/CS20171156

Oka, S. I., Hirata, T., Suzuki, W., et al., 2017. Thioredoxin-1 maintains mechanistic target of rapamycin (mTOR) function during oxidative stress in cardiomyocytes. J Biol Chem 292 (46), 18988–19000. doi:10.1074/jbc.M117.807735

Ordureau, A., Heo, J. M., Duda, D. M., et al., 2015. Defining roles of Parkin and ubiquitin phosphorylation by PINK1 in mitochondrial quality control using a ubiquitin replacement strategy. Proc Natl Acad Sci U S A 112 (21), 6637–6642. doi:10.1073/pnas.1506593112

Papadopoli, D., Boulay, K., Kazak, L., et al., 2019. mTOR as a central regulator of lifespan and aging. F1000Res 8. doi:10.12688/f1000research.17196.1

Parzych, K. R., Klionsky, D. J., 2014. An overview of autophagy: morphology, mechanism, and regulation. Antioxid Redox Signal 20 (3), 460–473. doi:10.1089/ars.2013.5371

Pickles, S., Vigie, P., Youle, R. J., 2018. Mitophagy and quality control mechanisms in mitochondrial maintenance. Curr Biol 28 (4), R170–R185. doi:10.1016/j.cub.2018.01.004

Quarles, E., Basisty, N., Chiao, Y. A., et al., 2020. Rapamycin persistently improves cardiac function in aged, male and female mice, even following cessation of treatment. Aging Cell 19 (2), e13086. doi:10.1111/acel.13086

Ren, J., Yang, L., Zhu, L., et al., 2017. Akt2 ablation prolongs life span and improves myocardial contractile function with adaptive cardiac remodeling: role of Sirt1-mediated autophagy regulation. Aging Cell 16 (5), 976–987. doi:10.1111/acel.12616

Riehle, C., Bauersachs, J., 2019. Small animal models of heart failure. Cardiovasc Res 115 (13), 1838–1849. doi:10.1093/cvr/cvz161

Schaper, J., Meiser, E., Stammler, G., 1985. Ultrastructural morphometric analysis of myocardium from dogs, rats, hamsters, mice, and from human hearts. Circ Res 56 (3), 377–391. doi:10.1161/01.res.56.3.377

Scherz-Shouval, R., Elazar, Z., 2011. Regulation of autophagy by ROS: physiology and pathology. Trends Biochem Sci 36 (1), 30–38. doi:10.1016/j.tibs.2010.07.007

Scherz-Shouval, R., Shvets, E., Fass, E., et al., 2007. Reactive oxygen species are essential for autophagy and specifically regulate the activity of Atg4. EMBO J 26 (7), 1749–1760. doi:10.1038/sj.emboj.7601623

Shao, D., Oka, S., Liu, T., et al., 2014. A redox-dependent mechanism for regulation of AMPK activation by Thioredoxin1 during energy starvation. Cell Metab 19 (2), 232–245. doi:10.1016/j.cmet.2013.12.013

Shi, J., Surma, M., Yang, Y., et al., 2019. Disruption of both ROCK1 and ROCK2 genes in cardiomyocytes promotes autophagy and reduces cardiac fibrosis during aging. FASEB J 33 (6), 7348–7362. doi:10.1096/fj.201802510R

Shin, H. J., Kim, H., Oh, S., et al., 2016. AMPK-SKP2-CARM1 signalling cascade in transcriptional regulation of autophagy. Nature 534 (7608), 553–557. doi:10.1038/nature18014

Shirakabe, A., Ikeda, Y., Sciarretta, S., et al., 2016. Aging and Autophagy in the Heart. Circulation Research 118 (10), 1563–1576. doi:10.1161/circresaha.116.307474

Siddall, H. K., Yellon, D. M., Ong, S. B., et al., 2013. Loss of PINK1 increases the heart's vulnerability to ischemia-reperfusion injury. PLoS One 8 (4), e62400. doi:10.1371/journal.pone.0062400

Sies, H., Jones, D. P., 2020. Reactive oxygen species (ROS) as pleiotropic physiological signalling agents. Nat Rev Mol Cell Biol 21 (7), 363–383. doi:10.1038/s41580-020-0230-3

Singh, R., Cuervo, A. M., 2011. Autophagy in the cellular energetic balance. Cell Metab 13 (5), 495–504. doi:10.1016/j.cmet.2011.04.004

Steenman, M., Lande, G., 2017. Cardiac aging and heart disease in humans. Biophys Rev 9 (2), 131–137. doi:10.1007/s12551-017-0255-9

Strait, J. B., Lakatta, E. G., 2012. Aging-associated cardiovascular changes and their relationship to heart failure. Heart Fail Clin 8 (1), 143–164. doi:10.1016/j.hfc.2011.08.011

Sun, N., Yun, J., Liu, J., et al., 2015. Measuring in vivo mitophagy. Mol Cell 60 (4), 685–696. doi:10.1016/j.molcel.2015.10.009

Sun, M., Tan, Y., Rexiati, M., et al., 2019. Obesity is a common soil for premature cardiac aging and heart diseases - Role of autophagy. Biochim Biophys Acta Mol Basis Dis 1865 (7), 1898–1904. doi:10.1016/j.bbadis.2018.09.004

Taneike, M., Yamaguchi, O., Nakai, A., et al., 2010. Inhibition of autophagy in the heart induces age-related cardiomyopathy. Autophagy 6 (5), 600–606. doi:10.4161/auto.6.5.11947

Terman, A., Brunk, U. T., 2005. Autophagy in cardiac myocyte homeostasis, aging, and pathology. Cardiovasc Res 68 (3), 355–365. doi:10.1016/j.cardiores.2005.08.014

Terman, A., Kurz, T., Navratil, M., et al., 2010. Mitochondrial turnover and aging of long-lived postmitotic cells: the mitochondrial-lysosomal axis theory of aging. Antioxid Redox Signal 12 (4), 503–535. doi:10.1089/ars.2009.2598

Tocchi, A., Quarles, E. K., Basisty, N., et al., 2015. Mitochondrial dysfunction in cardiac aging. Biochim Biophys Acta 1847 (11), 1424–1433. doi:10.1016/j.bbabio.2015.07.009

Turdi, S., Fan, X., Li, J., et al., 2010. AMP-activated protein kinase deficiency exacerbates aging-induced myocardial contractile dysfunction. Aging Cell 9 (4), 592–606. doi:10.1111/j.1474-9726.2010.00586.x

Walter, S., Atzmon, G., Demerath, E. W., et al., 2011. A genome-wide association study of aging. Neurobiol Aging 32 (11), 2109.e2115–2128. doi:10.1016/j.neurobiolaging.2011.05.026

Wang, S., Kandadi, M. R., Ren, J., 2019a. Double knockout of Akt2 and AMPK predisposes cardiac aging without affecting lifespan: role of autophagy and mitophagy. Biochim Biophys Acta Mol Basis Dis 1865 (7), 1865–1875. doi:10.1016/j.bbadis.2018.08.011

Wang, Y., Li, Y., He, C., et al., 2019b. Mitochondrial regulation of cardiac aging. Biochim Biophys Acta Mol Basis Dis 1865 (7), 1853–1864. doi:10.1016/j.bbadis.2018.12.008

Wilkinson, J. E., Burmeister, L., Brooks, S. V., et al., 2012. Rapamycin slows aging in mice. Aging Cell 11 (4), 675–682. doi:10.1111/j.1474-9726.2012.00832.x

World Health Organization, 2016. Global health estimates 2016: deaths by cause, age, sex, by country and by region, 2000–2016 [WWW document]. WHO. https://www.who.int/healthinfo/global_burden_disease/estimates/en/ (accessed 10.25.19).

World Health Organization, 2020. World Health Statistics 2020: Monitoring Health for the SDGs, Sustainable Development Goals. WHO, Geneva.

PROTEOSTASIS AND PROTEOLYSIS

Xu, X., Pang, J., Chen, Y., et al., 2016. Macrophage migration inhibitory factor (MIF) deficiency exacerbates aging-induced cardiac remodeling and dysfunction despite improved inflammation: role of autophagy regulation. Sci Rep 6, 22488. doi:10.1038/srep22488

Yamaguchi, O., 2019. Autophagy in the heart. Circ J 83 (4), 697–704. doi:10.1253/circj.CJ-18-1065

Yang, Z., Ming, X. F., 2012. mTOR signalling: the molecular interface connecting metabolic stress, aging and cardiovascular diseases. Obes Rev 13 (Suppl. 2), 58–68. doi:10.1111/j.1467-789X.2012.01038.x

Yao, Y., Jiang, C., Wang, F., et al., 2019. Integrative analysis of miRNA and mRNA expression profiles associated with human atrial aging. Front Physiol 10, 1226. doi:10.3389/fphys.2019.01226

Zhang, Y., Wang, C., Zhou, J., et al., 2017. Complex inhibition of autophagy by mitochondrial aldehyde dehydrogenase shortens lifespan and exacerbates cardiac aging. Biochim Biophys Acta Mol Basis Dis 1863 (8), 1919–1932. doi:10.1016/j.bbadis.2017.03.016

INDEX

Note: Locators in *italics* represent figures and **bold** indicate tables in the text.